Physics Research and Technology

Theoretical Physics and Nonlinear Optics

Theories and Models

PHYSICS RESEARCH AND TECHNOLOGY

Additional books in this series can be found on Nova's website under the Series tab.

Additional E-books in this series can be found on Nova's website under the E-book tab.

PHYSICS RESEARCH AND TECHNOLOGY

THEORETICAL PHYSICS AND NONLINEAR OPTICS

THEORIES AND MODELS

**THOMAS F. GEORGE,
RENAT R. LETFULLIN
AND
GUOPING ZHANG
EDITORS**

Nova Science Publishers, Inc.
New York

Copyright © 2012 by Nova Science Publishers, Inc.

All rights reserved. No part of this book may be reproduced, stored in a retrieval system or transmitted in any form or by any means: electronic, electrostatic, magnetic, tape, mechanical photocopying, recording or otherwise without the written permission of the Publisher.

For permission to use material from this book please contact us:
Telephone 631-231-7269; Fax 631-231-8175
Web Site: http://www.novapublishers.com

NOTICE TO THE READER

The Publisher has taken reasonable care in the preparation of this book, but makes no expressed or implied warranty of any kind and assumes no responsibility for any errors or omissions. No liability is assumed for incidental or consequential damages in connection with or arising out of information contained in this book. The Publisher shall not be liable for any special, consequential, or exemplary damages resulting, in whole or in part, from the readers' use of, or reliance upon, this material. Any parts of this book based on government reports are so indicated and copyright is claimed for those parts to the extent applicable to compilations of such works.

Independent verification should be sought for any data, advice or recommendations contained in this book. In addition, no responsibility is assumed by the publisher for any injury and/or damage to persons or property arising from any methods, products, instructions, ideas or otherwise contained in this publication.

This publication is designed to provide accurate and authoritative information with regard to the subject matter covered herein. It is sold with the clear understanding that the Publisher is not engaged in rendering legal or any other professional services. If legal or any other expert assistance is required, the services of a competent person should be sought. FROM A DECLARATION OF PARTICIPANTS JOINTLY ADOPTED BY A COMMITTEE OF THE AMERICAN BAR ASSOCIATION AND A COMMITTEE OF PUBLISHERS.

Additional color graphics may be available in the e-book version of this book.

LIBRARY OF CONGRESS CATALOGING-IN-PUBLICATION DATA

Theoretical physics and nonlinear optics : theories and models / editors,
Thomas F. George, Renat R. Letfullin, and Guoping Zhang.
    p. cm.
 Includes bibliographical references and index.
 ISBN 978-1-61122-939-4 (hardcover)
 1. Nuclear physics. 2. Nonlinear optics. I. George, Thomas F., 1947- II. Letfullin, Renat R. III. Zhang, Guoping, 1970-
 QC771.T48 2011
 530--dc22
                      2010047105

Published by Nova Science Publishers, Inc. † New York

Contents

Preface		vii
Chapter 1	Momentum Picture of Motion in Lagrangian Quantum Field Theory *Bozhidar Z. Iliev*	1
Chapter 2	The Dynamics of a IIA String Theory D0-brane Probing a Four Dimensional Black Hole *A. Chenaghlou*	21
Chapter 3	The Seiberg-Witten Map in Noncommutative Field Theory: An Alternative Interpretation *Subir Ghosh*	29
Chapter 4	Higher Dimensional Model Spaces and Defects *Dominic G.B. Edelen*	39
Chapter 5	One-Magnon Systems in an Anisotropic Non-Heisenberg Ferromagnetic Impurity Model *S.M. Tashpulatov*	55
Chapter 6	Ground State Mass Spectra of Low Lying Baryons in an Equally Mixed Scalar and Vector Potential Model *S.N. Jena, T.C. Tripathy and M.K. Muni*	69
Chapter 7	Bell's Theorem Refuted with a Kolmogorovian Counterexample *J.F. Geurdes*	85
Chapter 8	Ground States and Excitation Spectra of Light and Strange Baryons in a Relativistic Linear Potential Model *S.N. Jena, T.C. Tripathy and H.H. Muni*	99
Chapter 9	Neutrino Mass in a Six Dimensional E_6 Model *Snigdha Mishra and Sarita Mohanty*	115

Chapter 10	Determination of the Proton Bag Radius Based on MIT Bag Model *G.R. Boroun and M. Zandi*	**123**
Chapter 11	Searches for Higgs Bosons in Two-Higgs Doublet Models: The Fermiophobic Limit *L. Brücher and R. Santos*	**131**
Chapter 12	Quark-Pion Coupling Constant and Ground-State Baryon Masses in a Chiral Symmetric Potential Model *S.N. Jena, M.K. Muni and H.R. Pattnaik*	**165**
Chapter 13	Goldstone-Boson-Exchange and One-Gluon Exchange Contributions to Baryon Spectra in a Relativistic Quark Model *S.N. Jena, M.K. Muni and H.R. Pattnaik*	**189**
Chapter 14	Non-Linear Refractive Index Theory: (I) Generalised Optical Extinction Theorem and Dispersion Relation *S.S. Hassan, R.K. Bullough and R. Saunders*	**211**
Chapter 15	Non-Linear Refractive Index Theory: II -Numerical Studies *S.S. Hassan, R.K. Bullough, M.N. Ibrahim, R. Saunders, T. Jarad and N. Nayak*	**229**
Index		**239**

PREFACE

This book presents research in the field of theoretical physics and nonlinear optics. Topics discussed include quantum field theory; higher dimensional model spaces and defects; non-linear refractive index theory; quark-pion coupling constant and ground-state baryon masses; the search for higgs bosons in two-higgs doublet models and the neutrino mass in a six dimensional E6 model.

In Chapter 1, the basic aspects of the momentum picture of motion in Lagrangian quantum field theory are given. Under some assumptions, this picture is a 4-dimensional analogue of the Schrödinger picture: in it the field operators are constant, spacetime-independent, while the state vectors have a simple, exponential, spacetime-dependence. The role of these assumptions is analyzed. The Euler-Lagrange equations in the momentum picture are derived and attention is paid to the conserved operators in it.

In Chapter 2, the authors study the motion of the four dimensional black hole for the case of a IIA D0-brane probe. It is shown that the dynamics without angular momentum is reduced to the radial motion of the system. However in presence of the angular momentum, the dynamics has angular motion in addition to the radial motion. It is also noticed that instead of a spiral motion, the D-brane will have uniform motion in the asymptotic region.

In Chapter 3, an alternative interpretation of the Seiberg-Witten map in non-commutative field theory is provided. The authors show that the Seiberg-Witten map can be induced in a geometric way, by a field dependent co-ordinate transformation that connects noncommutative and ordinary space-times. Furthermore, in continuation of their earlier works, it has been demonstrated here that the above (field dependent co-ordinate) transformation can occur naturally in the Batalin-Tyutin extended space version of the relativistic spinning particle model, (in a particular gauge). They emphasize that the space-time non-commutativity emerges naturally from the particle spin degrees of freedom. Contrary to similarly motivated works, the non-commutativity is not imposed here in an *ad-hoc* manner.

In Chapter 4, higher dimensional model spaces are used to obtain general solutions of the kinematic equations of defects; that is, for dislocations and disclinations that result from local action of the Euclidean group $SO(3) \triangleleft T(3)$ as a gauge group. Appropriate choices of the Lie algebra valued generating matrices are used to obtain distortion 1- forms that satisfy the kinematic equations of defects and the equations of equilibrium of linear elasticity for both dislocations (local action of $T(3)$) and disclinations (local action of the full Euclidean group). These solutions lead to an improved understanding of the classical Volterra solutions, and to new classes of solutions that are self screening. It is shown that the undetermined functions in these solutions for screw dislocation problems can be determined by use of the field equations of the full gauge theory of defects, and explicit solutions are obtained.

In Chapter 5 the authors consider a one-magnon system in an anisotropic non-Heisenberg impurity model with an arbitrary spin s and investigate the spectrum and the localized impurity states of the system on the v-dimensional integer lattice Z^v ▷ The authors show that there are at most four types of localized impurity states (not counting the degeneracy multiplicities of their energy levels) in this system. They find the domains of these states and calculate the degeneracy multiplicities of their energy levels.

In Chapter 6, assuming the baryons as an assembly of independent quarks, confined in a first approximation by an equally mixed scalar and vector linear potential which presumably represents the non-perturbative multi-gluon interactions including gluon self-coupling, the ground state mass spectra of low-lying baryons are studied, taking into account perturbatively the contribution of the quark-gluon coupling due to one-gluon exchange and that of the Goldstone boson (π, η and k) exchange interaction arising from spontaneous breaking of chiral symmetry over and above that of the center-of-mass motion. The results obtained for the masses of the ground state baryons with a suitable choice of the strong coupling constant $\alpha_c = 0.16$ agree reasonably well with the corresponding experimental values.

In Chapter 7, the statistics behind Bell's inequality is demonstrated to allow a Kolmogorovian model of probabilities that recovers the quantum covariance. It shown that $P(A=+1)=P(A=-1)=1/2$ given $A=1$, or $A=-1$ is the result of measurement with device A. The occurrence of irregular integration prevents the use of Schwarz's inequality, hence, prevents the derivation of Bell's inequality. The obtained result implies that not all local hidden variables theories are ruled out by Aspect's (see Ref 4) experiment.

In Chapter 8, assuming the baryons as an assembly of individual quarks confined in a first approximation by an equally mixed scalar-vector potential in linear form which presumably represents the non-perturbative gluon interactions including gluon self-coupling, the mass spectra of the light and strange baryons are investigated. The contributions of the Goldston boson (π, η and K) exchange interactions between the constituent quarks arising from spontaneous breaking of chiral symmetry are taken into account over and above the center-of-mass correction to provide a unified description of the ground states and excitation spectra of baryons. The present model faces problems in explaining the correct level ordering of excited states; still the results obtained for the ground states of baryons are in reasonable agreement with the experimental values.

In Chapter 9, the authors consider a supersymmetric E_6 GUT in a six dimesional $M^4 \otimes T^2$ space. E_6 breaking is achieved by orbifolding the extra two dimensional torus T^2 via a $T^2/(Z_2 \times Z_2' \times Z_2'')$. In effective four dimension the authors obtain an extended Pati-Salam group with N = 1 SUSY, as a result of orbifold compactification. They then discuss the problem of neutrino mass with imposed parity assignment. A light Dirac neutrino is predicted with mass of the order of 10^{-2} eV.

As shown in Chapter 10, the bag radius of the proton can be determined by MIT bag model based on electric and magnetic form factors of the proton at $0.65 \leq Q^2 \leq 5 (\text{Gev}/c)^2$. The study of their results show that the bag radius decreases as Q^2 increases. Also, the electric and magnetic root-mean-squared radius is determined. In doing so, the static bag radius have been calculated and compared with other results. Comparison of the authors' results with those obtained by others suggests a suitable compatibility.

Over the last three decades physicists have tried to discover the famous minimal Higgs boson: the so-called Standard Model (SM) Higgs boson. So far, no scalar boson has been found. This inspired physicists to broaden their view and enlarge the Higgs sector of the SM. The simplest and most natural extension is to include another complex field. Such a model maintains most of the characteristics of the SM, and at the same time has a much richer phenomenology including charged scalars, pseudoscalars and the possibility of including CP-violation. These models are known as Two-Higgs Doublet Models (2HDM). Besides these new features, it comes as a bonus that some 2HDM mimic the Minimal Supersymmetric Model (MSSM) regarding its scalar sector.

Contrary to the SM and MSSM, 2HDM models have an extremely interesting facet: they allow a very weak coupling between scalars and fermions. These types of scalars are called fermiophobic Higgs bosons and the only way to detect them is via vector boson decays, in particular through the two-photon signature which was used by experimentalists at LEP to set a limit on its mass.

In Chapter 11 the phenomenology of the five Higgs particles of some 2HDM's will be shown and the current status of their searches will be reviewed. Special attention will be paid to the fermiophobic limit, where new searches at the next generation colliders (the Large Hadron Collider and the Linear Collider) are proposed.

Incorporating chiral symmetry into an independent quark model with logarithmic confining potential, the quark-pion coupling constant $G_{qq\pi}$ for quarks in a nucleon are estimated in Chapter 12. The value of $\dfrac{G_{qq\pi}^2}{4\pi}$ obtained with the approximation of a point pion is consistent with that extracted from the experimental vector meson decay width ratio $\Gamma(\rho \to \pi^+\pi^0) \triangleleft \Gamma(\rho \to \pi^- \nu)$ by Suzuki and Bhaduri. The ground state baryon masses are also calculated in this model taking into account the contributions of the residual quark-pion coupling arising out of the requirement of chiral symmetry and that of the quark-gluon coupling due to one gluon-exchange in a perturbative manner, over and above the necessary center-of-mass correction. The results obtained for the baryon masses agree reasonably well with the corresponding experimental data. The quarkgluon coupling constant $\alpha_c = 0 \triangleright 258$ required here to explain the QCD mass splittings is quite consistent with the idea of considering gluonic correction in lowest- order perturbation theory.

In Chapter 13, the energy contributions from the Goldstone-boson ($\pi \triangleleft \eta \triangleleft K$) exchange and that from one-gluon-exchange interactions between the constituent quarks to the baryon spectra are studied together with that from center-of-mass motion and the mass spectra of light-and strange-baryons are computed in a relativistic quark model. The baryons are assumed here as an assembly of independent quarks confined in a first approximation by an effective logarithmic potential which presumably represents the nonperturbative multigluon interactions including the gluon self-couplings. The results obtained for the ground states and excitation spectra of baryons agree reasonably well with the experimental values. The model yields correct level ordering of the excited states in $N \triangleleft \triangle$ and Λ spectra.

In Chapter 14, starting from the Bloch-Maxwell equations for two-level atoms forming an extended system, taken as a parallel-sided slab for a Fabry-Perot cavity configuration, the authors generalize the optical extinction theorem to the non-linear regime. Generalized form of the Lorentz-Lorenz dispersion relation for the refractive index (m) is derived. Within the context of optical multistability phenomenon, the input-output field relationship is derived in terms of (m).

In Chapter 15, the analytical formulae derived in the preceding paper [1] of this series, for the input-output field relation that is nonlinearly dependent on the refractive index is treated numerically through a self-consistent procedure scheme. The multi-stable behavior is exhibited for system data suitable for dense medium.

In: Theoretical Physics and Nonlinear Optics
Editors: Thomas F. George et al.
ISBN 978-1-61122-939-4
© 2012 Nova Science Publishers, Inc.

Chapter 1

MOMENTUM PICTURE OF MOTION IN LAGRANGIAN QUANTUM FIELD THEORY

Bozhidar Z. Iliev[*]
Laboratory of Mathematical Modeling in Physics,
Institute for Nuclear Research and Nuclear Energy,
Bulgarian Academy of Sciences,
Boul. Tzarigradsko chaussée 72, 1784 Sofia, Bulgaria

Abstract

The basic aspects of the momentum picture of motion in Lagrangian quantum field theory are given. Under some assumptions, this picture is a 4-dimensional analogue of the Schrödinger picture: in it the field operators are constant, spacetime-independent, while the state vectors have a simple, exponential, spacetime-dependence. The role of these assumptions is analyzed. The Euler-Lagrange equations in the momentum picture are derived and attention is paid to the conserved operators in it.

PACS 03.70.+k, 11.10.Ef, 11.90.+t, 12.90.+b.

Keywords: Quantum field theory, Pictures of motion, Pictures of motion in quantum field theory, Momentum picture, Equations of motion, Euler-Lagrange equations, Heisenberg equations/relations.

1. Introduction

The main item of the present work is a presentation of the basic aspects of the *momentum picture of motion* in Lagrangian quantum field theory, suggested in [1]. In a sense, under some assumptions, this picture is a 4-dimensional analogue of the Schrödinger picture: in it the field operators are constant, spacetime-independent, while the state vectors have a simple, exponential, spacetime-dependence. This state of affairs offers the known merits of the Schrödinger picture (with respect to Heisenberg one in quantum mechanics [2]) in the

---
[*]E-mail address: bozho@inrne.bas.bg; URL: http://theo.inrne.bas.bg/~bozho/

region of quantum field theory. Particular applications of the momentum picture to scalar, spinor and vector free quantum fields can be found in [3–5].

We should mention, in this paper it is considered only the Lagrangian (canonical) quantum field theory in which the quantum fields are represented as operators, called field operators, acting on some Hilbert space, which in general is unknown if interacting fields are studied. These operators are supposed to satisfy some equations of motion, from them are constructed conserved quantities satisfying conservation laws, etc. From the view-point of present-day quantum field theory, this approach is only a preliminary stage for more or less rigorous formulation of the theory in which the fields are represented via operator-valued distributions, a fact required even for description of free fields. Moreover, in non-perturbative directions, like constructive and conformal field theories, the main objects are the vacuum mean (expectation) values of the fields and from these are reconstructed the Hilbert space of states and the acting on it fields. Regardless of these facts, the Lagrangian (canonical) quantum field theory is an inherent component of the most of the ways of presentation of quantum field theory adopted explicitly or implicitly in books like [6–13]. Besides the Lagrangian approach is a source of many ideas for other directions of research, like the axiomatic quantum field theory [8, 12, 13].

In Sect. 2 are reviewed some basic moments of the Lagrangian formalism in quantum field theory. In Sect 3 are recalled part of the relations arising from the assumption that the conserved operators are generators of the corresponding invariance transformations of the action integral; in particular the Heisenberg relations between the field operators and momentum operator are written.

The momentum picture of motion is defined in Sect 4. Two basic restrictions on the considered quantum field theories is shown to play a crucial role for the convenience of that picture: the mutual commutativity between the components of the momentum operator and the Heisenberg commutation relation between them and the field operators. If these conditions hold, the field operators in momentum picture become spacetime-independent and the state vectors turn to have exponential spacetime-dependence. In Sect 5, the attention is called to the Euler-Lagrange equations and dynamical variables in momentum picture. In Sect. 6 is given an idea of the momentum representation in momentum picture and the similarity with that representation in Heisenberg picture is pointed. In Sect. 7 is made a comparison between the momentum picture in quantum field theory and the Schrödinger picture in quantum mechanics. Some closing remarks are given in Sect. 8. It is pointed that the above-mentioned restrictions are fundamental enough to be put in the basic postulates of quantum field theory, which may result in a new way of its (Lagrangian) construction.

The books [6–8] will be used as standard reference works on quantum field theory. Of course, this is more or less a random selection between the great number of (text)books and papers on the theme to which the reader is referred for more details or other points of view. For this end, e.g., [9, 14] or the literature cited in [6–9, 14] may be helpful.

Throughout this paper $\hbar$ denotes the Planck's constant (divided by 2π), c is the velocity of light in vacuum, and i stands for the imaginary unit. The superscript † means Hermitian conjugation (of operators or matrices), and the symbol ∘ denotes compositions of mappings/operators.

The Minkowski spacetime is denoted by M. The Greek indices run from 0 to $\dim M =$

4. All Greek indices will be raised and lowered by means of the standard 4-dimensional Lorentz metric tensor $\eta^{\mu\nu}$ and its inverse $\eta_{\mu\nu}$ with signature $(+---)$. The Einstein's summation convention over indices repeated on different levels is assumed over the whole range of their values.

2. Lagrangian Formalism

Let us consider a system of quantum fields, represented in Heisenberg picture of motion by field operators $\tilde{\varphi}_i(x)\colon \mathcal{F} \to \mathcal{F}$, with $i = 1,\ldots,n \in \mathbb{N}$, in system's Hilbert space $\mathcal{F}$ of states and depending on a point x in Minkowski spacetime M. Here and henceforth, all quantities in Heisenberg picture, in which the state vectors are spacetime-independent contrary to the field operators and observables, will be marked by a tilde (wave) "~" over their kernel symbols. Let

$$\tilde{\mathcal{L}} = \tilde{\mathcal{L}}(\tilde{\varphi}_i(x), \partial_\nu \tilde{\varphi}_j(x)) \tag{2.1}$$

be the system's Lagrangian, which is supposed to depend on the field operators and their first partial derivatives.[1] We expect that this dependence is polynomial or in a form of convergent power series, which can be treated term by term. The Euler-Lagrange equations for the Lagrangian (2.1), i.e.

$$\frac{\partial \tilde{\mathcal{L}}(\tilde{\varphi}_j(x), \partial_\nu \tilde{\varphi}_l(x))}{\partial \tilde{\varphi}_i(x)} - \frac{\partial}{\partial x^\mu} \frac{\partial \tilde{\mathcal{L}}(\tilde{\varphi}_j(x), \partial_\nu \tilde{\varphi}_l(x))}{\partial(\partial_\mu \tilde{\varphi}_i(x))} = 0, \tag{2.2}$$

are identified with the field equations (of motion) for the quantum fields $\tilde{\varphi}_i(x)$.[2]

For definiteness, above and below, we consider a quantum field theory *before* normal ordering and, possibly, without (anti)commutation relations (see Sect. 4). However, our investigation is, practically, independent of these procedures and can easily be modified to include them.

Following the standard procedure [6–8, 14] (see also [15]), from the Lagrangian (2.1) can be constructed the densities of the conserved quantities of the system, viz. the energy-momentum tensor $\tilde{T}_{\mu\nu}(x)$, charge current $\tilde{J}_\mu(x)$, the (total) angular momentum density operator

$$\tilde{\mathcal{M}}^\lambda_{\mu\nu} = \tilde{\mathcal{L}}^\lambda_{\mu\nu}(x) + \tilde{\mathcal{S}}^\lambda_{\mu\nu}(x), \tag{2.3}$$

where

$$\tilde{\mathcal{L}}^\lambda_{\mu\nu}(x) = x_\mu \tilde{T}^\lambda{}_\nu(x) - x_\nu \tilde{T}^\lambda{}_\mu(x) \tag{2.4}$$

---

[1] One can easily generalize the below presented material for Lagrangians depending on higher order derivatives.

[2] In (2.2) and similar expressions appearing further, the derivatives of functions of operators with respect to operator arguments are calculated in the same way as if the operators were ordinary (classical) fields/functions, only the order of the arguments should not be changed. This is a silently accepted practice in the literature [6–8]. In the most cases such a procedure is harmless, but it leads to the problem of non-unique definitions of the quantum analogues of the classical conserved quantities, like the energy-momentum and charge operators. For some details on this range of problems in quantum field theory, see [15]. In *loc. cit.* is demonstrated that these problems can be eliminated by changing the rules of differentiation with respect to *not*-commuting variables. The paper [15] contains an example of a Lagrangian (describing spin $\frac{1}{2}$ field) whose field equations are *not* the Euler-Lagrange equations (2.2) obtained as just described, but we shall not investigate such cases in the present work.

and $\tilde{\mathcal{S}}^{\lambda}_{\mu\nu}(x)$ are respectively the orbital and spin angular momentum density operators, and others, if such ones exist. The corresponding to these quantities integral ones, viz. the momentum, charge, (total) angular momentum, orbital and spin angular momentum operators, are respectively defined by:

$$\tilde{\mathcal{P}}_\mu := \frac{1}{c} \int_{x^0=\text{const}} \tilde{\mathcal{T}}_{0\mu}(x)\, \mathrm{d}^3 \boldsymbol{x}. \tag{2.5}$$

$$\tilde{\mathcal{Q}} := \frac{1}{c} \int_{x^0=\text{const}} \tilde{\mathcal{J}}_0(x)\, \mathrm{d}^3 \boldsymbol{x} \tag{2.6}$$

$$\tilde{\mathcal{M}}_{\mu\nu} = \tilde{\mathcal{L}}_{\mu\nu}(x) + \tilde{\mathcal{S}}_{\mu\nu}(x), \tag{2.7}$$

$$\tilde{\mathcal{L}}_{\mu\nu}(x) := \frac{1}{c} \int_{x^0=\text{const}} \{x_\mu \tilde{\mathcal{T}}^0{}_\nu(x) - x_\nu \tilde{\mathcal{T}}^0{}_\mu(x)\}\, \mathrm{d}^3 \boldsymbol{x} \tag{2.8a}$$

$$\tilde{\mathcal{S}}_{\mu\nu}(x) := \frac{1}{c} \int_{x^0=\text{const}} \tilde{\mathcal{S}}^0_{\mu\nu}(x)\, \mathrm{d}^3 \boldsymbol{x} \tag{2.8b}$$

and satisfy the conservation laws

$$\frac{\mathrm{d}\tilde{\mathcal{P}}_\mu}{\mathrm{d}x^0} = 0 \quad \frac{\mathrm{d}\tilde{\mathcal{Q}}}{\mathrm{d}x^0} = 0 \quad \frac{\mathrm{d}\tilde{\mathcal{M}}_{\mu\nu}}{\mathrm{d}x^0} = 0 \tag{2.9}$$

which, in view of (2.5)–(2.8), are equivalent to

$$\partial_\lambda \tilde{\mathcal{P}}_\mu = 0 \quad \partial_\lambda \tilde{\mathcal{Q}} = 0 \quad \partial_\lambda \tilde{\mathcal{M}}_{\mu\nu} = 0 \tag{2.10}$$

and also to

$$\partial^\lambda \tilde{\mathcal{T}}_{\lambda\mu} = 0 \quad \partial^\lambda \tilde{\mathcal{J}}_\lambda = 0 \quad \partial_\lambda \tilde{\mathcal{M}}^\lambda_{\mu\nu} = 0. \tag{2.11}$$

The Lagrangian, as well as the conserved quantities and there densities, are Hermitian operators; in particular, such is the momentum operator,

$$\tilde{\mathcal{P}}^\dagger_\mu = \tilde{\mathcal{P}}_\mu. \tag{2.12}$$

The reader can find further details on the Lagrangian formalism in, e.g., [6–8, 14, 15].

3. Heisenberg Relations

The conserved quantities (2.5)–(2.7) are often identified with the generators of the corresponding transformations, under which the action operator is invariant [6, 8, 14, 16]. This leads to a number of commutation relations between the conserved operators and between them and the field operators. The relations of the latter set are often referred as the Heisenberg relations or equations. Part of them are briefly reviewed below; for details, see *loc. cit.*

The consideration of $\tilde{\mathcal{P}}_\mu$, $\tilde{\mathcal{Q}}$ and $\tilde{\mathcal{M}}_{\mu\nu}$ as generators of translations, constant phase transformations and 4-rotations, respectively, leads to the following relations:

$$[\tilde{\varphi}_i(x), \tilde{\mathcal{P}}_\mu]_- = i\hbar \frac{\partial \tilde{\varphi}_i(x)}{\partial x^\mu} \tag{3.1}$$

$$[\tilde{\varphi}_i(x), \tilde{\mathcal{Q}}]_- = \varepsilon(\tilde{\varphi}_i) q_i \tilde{\varphi}_i(x) \tag{3.2}$$

$$[\tilde{\varphi}_i(x), \tilde{\mathcal{M}}_{\mu\nu}]_- = i\hbar \{ x_\mu \partial_\nu \tilde{\varphi}_i(x) - x_\nu \partial_\mu \tilde{\varphi}_i(x) + I^j_{i\mu\nu} \tilde{\varphi}_j(x) \}. \tag{3.3}$$

Here: $q_i = \text{const}$ is the charge of the ith field, $\varepsilon(\tilde{\varphi}_i) = 0$ if $\tilde{\varphi}_i^\dagger = \tilde{\varphi}_i$, $\varepsilon(\tilde{\varphi}_i) = \pm 1$ if $\tilde{\varphi}_i^\dagger \neq \tilde{\varphi}_i$ with $\varepsilon(\tilde{\varphi}_i) + \varepsilon(\tilde{\varphi}_i^\dagger) = 0$, and the constants $I^j_{i\mu\nu} = -I^j_{i\nu\mu}$ characterize the transformation properties of the field operators under 4-rotations. (It is a convention whether to put $\varepsilon(\tilde{\varphi}_i) = +1$ or $\varepsilon(\tilde{\varphi}_i) = -1$ for a fixed i.) Besides, the operators $\tilde{\mathcal{P}}_\mu$, $\tilde{\mathcal{Q}}$ and $\tilde{\mathcal{M}}_{\mu\nu}$ satisfy certain commutation relation between themselves, from which we shall write the following two:

$$[\tilde{\mathcal{P}}_\mu, \tilde{\mathcal{P}}_\nu]_- = 0 \tag{3.4}$$

$$[\tilde{\mathcal{Q}}, \tilde{\mathcal{P}}_\mu]_- = 0. \tag{3.5}$$

It should be clearly understood, the equations (3.1)–(3.5) are from pure geometrical origin and are completely external to the Lagrangian formalism. However, there are strong evidences that they should hold in a realistic Lagrangian quantum field theory (see [16, § 68] and [6, § 5.3 and § 9.4]). Moreover, (most of) the above relations happen to be valid for Lagrangians that are frequently used, e.g. for the ones describing free fields [16].

4. The Momentum Picture of Motion

Let $\tilde{\mathcal{P}}_\mu$ be the system's momentum operator, given by equation (2.5). Since $\tilde{\mathcal{P}}_\mu$ is Hermitian (see (2.12)), the operator

$$\mathcal{U}(x, x_0) = \exp\left(\frac{1}{i\hbar} \sum_\mu (x^\mu - x_0^\mu) \tilde{\mathcal{P}}_\mu \right), \tag{4.1}$$

where $x_0 \in M$ is arbitrarily fixed and $x \in M$, is unitary, i.e.

$$\mathcal{U}^\dagger(x_0, x) := (\mathcal{U}(x, x_0))^\dagger = (\mathcal{U}(x, x_0))^{-1} =: \mathcal{U}^{-1}(x, x_0). \tag{4.2}$$

Let $\tilde{\mathcal{X}} \in \mathcal{F}$ be a state vector in the system's Hilbert space $\mathcal{F}$ and $\tilde{\mathcal{A}}(x): \mathcal{F} \to \mathcal{F}$ be an operator on it. The transformations

$$\tilde{\mathcal{X}} \mapsto \mathcal{X}(x) = \mathcal{U}(x, x_0)(\tilde{\mathcal{X}}) \tag{4.3}$$

$$\tilde{\mathcal{A}}(x) \mapsto \mathcal{A}(x) = \mathcal{U}(x, x_0) \circ (\tilde{\mathcal{A}}(x)) \circ \mathcal{U}^{-1}(x, x_0), \tag{4.4}$$

evidently, preserve the Hermitian scalar product $\langle \cdot | \cdot \rangle : \mathcal{F} \times \mathcal{F} \to \mathbb{C}$ of $\mathcal{F}$ and the mean values of the operators, i.e.

$$\langle \tilde{\mathcal{X}} | \tilde{\mathcal{A}}(x)(\tilde{\mathcal{Y}}) \rangle = \langle \mathcal{X}(x) | \mathcal{A}(x)(\mathcal{Y}(x)) \rangle \tag{4.5}$$

for any $\tilde{\mathcal{X}}, \tilde{\mathcal{Y}} \in \mathcal{F}$ and $\tilde{\mathcal{A}}(x) \colon \mathcal{F} \to \mathcal{F}$. Since the physically predictable/measurable results of the theory are expressible via scalar products in $\mathcal{F}$ [6, 7, 14], the last equality implies that the theory's description via vectors and operators like $\tilde{\mathcal{X}}$ and $\tilde{\mathcal{A}}(x)$ above is completely equivalent to the one via the vectors $\mathcal{X}(x)$ and operators $\mathcal{A}(x)$, respectively. The description of quantum field theory via $\mathcal{X}$ and $\mathcal{A}(x)$ will be called the *momentum picture (of motion (of quantum field theory))* [1].

However, without further assumptions, this picture turns to be rather complicated. The mathematical cause for this is that derivatives of different operators are often met in the theory and, as a consequence of (4.4), they transform as

$$\partial_\mu \tilde{\mathcal{A}}(x) \mapsto \mathcal{U}(x, x_0) \circ (\partial_\mu \tilde{\mathcal{A}}(x)) \circ \mathcal{U}^{-1}(x, x_0) = \partial_\mu \mathcal{A}(x) + [\mathcal{A}(x), \mathcal{H}_\mu(x, x_0)]_{-} \quad (4.6)$$

$$\mathcal{H}_\mu(x, x_0) := (\partial_\mu \mathcal{U}(x, x_0)) \circ \mathcal{U}^{-1}(x, x_0) \quad (4.7)$$

from Heisenberg to momentum picture. Here $[\mathcal{A}, \mathcal{B}]_{-} := \mathcal{A} \circ \mathcal{B} - \mathcal{B} \circ \mathcal{A}$ is the commutator of $\mathcal{A}, \mathcal{B} \colon \mathcal{F} \to \mathcal{F}$. The entering in (4.6), via (4.7), derivatives of the operator (4.1) can be represented as the convergent power series

$$\partial_\mu \mathcal{U}(x, x_0) = \frac{1}{i\hbar} \tilde{\mathcal{P}}_\mu + \frac{1}{i\hbar} \sum_{n=1}^{\infty} \frac{1}{(i\hbar)^n} \frac{1}{(n+1)!} \sum_{m=0}^{n} \left((x^\lambda - x_0^\lambda) \tilde{\mathcal{P}}_\lambda\right)^m \circ \tilde{\mathcal{P}}_\mu \circ \left((x^\lambda - x_0^\lambda) \tilde{\mathcal{P}}_\lambda\right)^{n-m},$$

where $(\cdots)^n := (\cdots) \circ \cdots \circ (\cdots)$ (n-times) and $(\cdots)^0 := \mathrm{id}_\mathcal{F}$ is the identity mapping of $\mathcal{F}$, which cannot be written in a closed form unless the commutator $[\tilde{\mathcal{P}}_\mu, \tilde{\mathcal{P}}_\nu]_{-}$ has 'sufficiently simple' form. In particular, the relation (3.4) entails $\partial_\mu \mathcal{U}(x, x_0) = \frac{1}{i\hbar} \tilde{\mathcal{P}}_\mu \circ \mathcal{U}(x, x_0)$, so that (4.7) and (4.6) take respectively the form

$$\mathcal{H}_\mu(x, x_0) = \frac{1}{i\hbar} \tilde{\mathcal{P}}_\mu \quad (4.8)$$

$$\partial_\mu \tilde{\mathcal{A}}(x) \mapsto \partial_\mu \mathcal{A}(x) + \frac{1}{i\hbar} [\mathcal{A}(x), \tilde{\mathcal{P}}_\mu]_{-}. \quad (4.9)$$

Notice, the equality (4.8) is possible if and only if $\mathcal{U}_\mu(x, x_0)$ is a solution of the initial-value problem (see (4.7))

$$i\hbar \frac{\partial \mathcal{U}(x, x_0)}{\partial x^\mu} = \tilde{\mathcal{P}}_\mu \circ \mathcal{U}(x, x_0) \quad (4.10\mathrm{a})$$

$$\mathcal{U}(x_0, x_0) = \mathrm{id}_\mathcal{F}, \quad (4.10\mathrm{b})$$

the integrability conditions for which are exactly (3.4).[3] Since (3.4) and (4.1) imply

$$[\mathcal{U}(x, x_0), \tilde{\mathcal{P}}_\mu]_{-} = 0, \quad (4.11)$$

by virtue of (4.4), we have

$$\mathcal{P}_\mu = \tilde{\mathcal{P}}_\mu, \quad (4.12)$$

---

[3] For a system with a non-conserved momentum operator $\tilde{\mathcal{P}}_\mu(x)$ the operator $\mathcal{U}(x, x_0)$ should be defined as the solution of (4.10), with $\tilde{\mathcal{P}}_\mu(x)$ for $\tilde{\mathcal{P}}_\mu$, instead of by (4.1); in this case, equation (3.4) should be replace with

$$[\tilde{\mathcal{P}}_\mu(x), \tilde{\mathcal{P}}_\nu(x)]_{-} + \partial_\nu \tilde{\mathcal{P}}_\mu(x) - \partial_\mu \tilde{\mathcal{P}}_\nu(x) = 0.$$

Most of the material in the present section remains valid in that more general situation.

i.e. the momentum operators in Heisenberg and momentum pictures coincide, provided (3.1) holds.

It is worth to be mentioned, equation (4.12) is a special case of

$$\mathcal{A}(x) = \tilde{\mathcal{A}}(x) + [\mathcal{U}(x, x_0), \tilde{\mathcal{A}}(x)]_- \circ \mathcal{U}^{-1}(x, x_0), \tag{4.13}$$

which is a consequence of (4.4) and is quite useful if one knows explicitly the commutator $[\mathcal{U}(x, x_0), \tilde{\mathcal{A}}(x)]_-$. In particular, if

$$[[\tilde{\mathcal{A}}(x), \tilde{\mathcal{P}}_\mu]_-, \tilde{\mathcal{P}}_\nu]_- = 0 \tag{4.14}$$

and (3.4) holds, then, by expanding (4.1) into a power series, one can prove that

$$[\tilde{\mathcal{A}}(x), \mathcal{U}(x, x_0)]_- = \frac{1}{i\hbar}(x^\lambda - x_0^\lambda)[\tilde{\mathcal{A}}(x), \tilde{\mathcal{P}}_\lambda]_- \circ \mathcal{U}(x, x_0). \tag{4.15}$$

So, in this case, (4.13) reduces to

$$\mathcal{A}(x) = \tilde{\mathcal{A}}(x) - \frac{1}{i\hbar}(x^\lambda - x_0^\lambda)[\tilde{\mathcal{A}}(x), \tilde{\mathcal{P}}_\lambda]_-. \tag{4.16}$$

This formula allows to be found an operator in momentum picture if its commutator(s) with (the components of) the momentum operator is (are) explicitly known, provided (3.4) and (4.14) hold. The choice $\tilde{\mathcal{A}}(x) = \tilde{\mathcal{P}}_\mu$ reduces (4.16) to (4.12).

Of course, a transition from one picture of motion to other one is justified if there are some merits from this step; for instance, if some (mathematical) simplification, new physical interpretation etc. occur in the new picture. A classical example of this kind is the transition between Schrödinger and Heisenberg pictures in quantum mechanics [2] or, in a smaller extend, in quantum field theory [6]. Until now we have not present evidences that the momentum picture can bring some merits with respect to, e.g., Heisenberg picture. On the opposite, there was an argument that, without further restrictions, mathematical complications may arise in it.

In this connection, let us consider, as a second possible restriction, a theory in which the Heisenberg equation (3.1) is valid. In momentum picture, it reads

$$[\varphi_i(x), \tilde{\mathcal{P}}_\mu - i\hbar\mathcal{H}_\mu]_- = i\hbar\partial_\mu\varphi_i(x), \tag{4.17}$$

where

$$\tilde{\varphi}_i(x) \mapsto \varphi_i(x) = \mathcal{U}(x, x_0) \circ \tilde{\varphi}_i(x) \circ \mathcal{U}^{-1}(x, x_0). \tag{4.18}$$

are the field operators in momentum picture and the relations (4.4) and (4.6) were applied. The equation (4.17) shows that, if (4.8) holds, which is equivalent to the validity of (3.4), then

$$\partial_\mu\varphi_i(x) = 0, \tag{4.19}$$

i.e. in this case the field operators in momentum picture turn to be constant,

$$\varphi_i(x) = \mathcal{U}(x, x_0) \circ \tilde{\varphi}_i(x) \circ \mathcal{U}^{-1}(x, x_0) = \varphi_i(x_0) = \tilde{\varphi}_i(x_0) =: \varphi_i. \tag{4.20}$$

As a result of the last fact, all functions of the field operators and their derivatives, polynomial or convergent power series in them, become constant operators in momentum picture, which are algebraic functions of the field operators in momentum picture. This is an essentially new moment in the theory that reminds to a similar situation in the Schrödinger picture in quantum mechanics (see [2] and Sect. 7 below).

If $\tilde{\mathcal{P}}_\mu$ is considered, as usual [6,7], as a generator of 4-translations, then the constancy of the field operators in momentum picture is quite natural. In fact, in this case, the transition $\tilde{\varphi}_i(x) \mapsto \varphi_i(x)$, given by (4.18), means that the argument of $\tilde{\varphi}_i(x)$ is shifted by $(x_0 - x)$, i.e. that $\tilde{\varphi}_i(x) \mapsto \varphi_i(x) = \tilde{\varphi}_i(x + (x_0 - x)) = \tilde{\varphi}_i(x_0)$.

Let us turn our attention now to system's state vectors. By definition [6,7], such a vector $\tilde{\mathcal{X}}$ is a spacetime-constant one in Heisenberg picture,

$$\partial_\mu \tilde{\mathcal{X}} = 0. \tag{4.21}$$

In momentum picture, the situation is opposite, as, by virtue of (4.3), the operator (4.1) plays a role of spacetime 'evolution' operator, i.e.

$$\mathcal{X}(x) = \mathcal{U}(x, x_0)(\mathcal{X}(x_0)) = e^{\frac{1}{i\hbar}(x^\mu - x_0^\mu)\tilde{\mathcal{P}}_\mu}(\mathcal{X}(x_0)), \tag{4.22}$$

with

$$\mathcal{X}(x_0) = \mathcal{X}(x)|_{x=x_0} = \tilde{\mathcal{X}} \tag{4.23}$$

being considered as initial value of $\mathcal{X}(x)$ at $x = x_0$. Thus, if $\mathcal{X}(x_0) = \tilde{\mathcal{X}}$ is an eigenvector of the momentum operators $\tilde{\mathcal{P}}_\mu = \mathcal{P}_\mu(x)|_{x=x_0}\ (= \{\mathcal{U}(x, x_0) \circ \tilde{\mathcal{P}}_\mu \circ \mathcal{U}^{-1}(x, x_0)\}|_{x=x_0})$ with eigenvalues p_μ, i.e.

$$\tilde{\mathcal{P}}(\tilde{\mathcal{X}}) = p_\mu \tilde{\mathcal{X}} \quad \left(= p_\mu \mathcal{X}(x_0) = \mathcal{P}_\mu(x_0)(\tilde{\mathcal{X}}(x_0))\right), \tag{4.24}$$

we have the following *explicit* form of a state vector $\mathcal{X}$:

$$\mathcal{X}(x) = e^{\frac{1}{i\hbar}(x^\mu - x_0^\mu)p_\mu}(\mathcal{X}(x_0)). \tag{4.25}$$

It should be understood, this is the *general form of all state vectors in momentum picture*, as they are eigenvectors of all (commuting) observables [8, p. 59], in particular, of the momentum operator.

So, in momentum picture, the state vectors have a relatively simple *global* description. However, their differential (local) behavior is described via a differential equation that may turn to be rather complicated unless some additional conditions are imposed. Indeed, form (4.22), we get

$$\partial_\mu \mathcal{X}(x) = \mathcal{H}_\mu(x, x_0)(\mathcal{X}(x)) \tag{4.26}$$

in which equation the operator $\mathcal{H}_\mu(x, x_0)$ is given by (4.7) and may have a complicated explicit form (*vide supra*). The equality (4.26) has a form similar to the one of the Schrödinger equation, but in '4-dimensions', with '4-dimensional Hamiltonian' $i\hbar \mathcal{H}_\mu(x, x_0)$. It is intuitively clear, in this context, the operators $i\hbar \mathcal{H}_\mu(x, x_0)$ should be identified with the components of the momentum operator $\mathcal{P}_\mu$, i.e. the equality (4.8) is a natural one on this background.

Thus, if we accept (4.8), or equivalently (3.4), a state vector $\mathcal{X}(x)$ in momentum picture will be a solution of the initial-value problem

$$i\hbar \frac{\partial \mathcal{X}(x)}{\partial x^\mu} = \mathcal{P}_\mu(\mathcal{X}(x)) \qquad \mathcal{X}(x)|_{x=x_0} = \mathcal{X}(x_0) = \tilde{\mathcal{X}} \qquad (4.27)$$

and, respectively, the evolution operator $\mathcal{U}(x, x_0)$ of the state vectors will be a solution of (4.10). Consequently, the equation (3.4) entails not only a simplified description of the operators in momentum picture, but also a natural one of the state vectors in it.

The above discussion reveals that the momentum picture is worth to be employed in quantum field theories in which the conditions

$$[\tilde{\mathcal{P}}_\mu, \tilde{\mathcal{P}}_\nu]_- = 0 \qquad (4.28a)$$

$$[\tilde{\varphi}_i(x), \tilde{\mathcal{P}}_\mu]_- = i\hbar \partial_\mu \tilde{\varphi}_i(x) \qquad (4.28b)$$

are valid. In that case, the momentum picture can be considered as a 4-dimensional analogue of the Schrödinger picture [1]: the field operators are spacetime-constant and the state vectors are spacetime-dependent and evolve according to the '4-dimensional Schrödinger equation' (4.27) with evolution operator (4.1). More details on that item will be given in Sect. 7 below.

In connection with the conditions (4.28), it should be said that their validity is more a rule than an exception. For instance, in the axiomatic quantum field theory, they hold identically as in this approach, by definition, the momentum operator is identified with the generator of translations [12,13]. In the Lagrangian formalism, to which (4.28) are external restrictions, the conditions (4.28) seem to hold at least for the investigated free fields and most (all?) interacting ones [16]. For example, the commutativity between the components of the momentum operator, expressed via (4.28a), is a consequence of the (anti)commutation relations and, possibly, the field equations. Besides, it expresses the simultaneous measurability of the components of system's momentum. The Heisenberg relation (4.28b) is verified in [16] for a number of Lagrangians. Moreover, in *loc. cit.* it is regarded as one of the conditions for relativistic covariance in a translation-invariant Lagrangian quantum field theory. All these facts point that the conditions (4.28) are fundamental enough to be incorporated in the basic postulates of quantum field theory, as it is done (more implicitly than explicitly), e.g., in [6, 8, 14]. Some comments on that problem will be presented in Sect. 8 (see also [17, chapter 1]).

5. General Aspects of Lagrangian Formalism in Momentum Picture

In this section, some basic moments of the Lagrangian formalism in momentum picture will be considered, provided the equations (4.28) hold.

To begin with, let us recall, in the momentum picture, under the conditions (4.28), the field operators φ_i are constant, i.e. spacetime-independent (which is equivalent to (4.28a)), and the state vectors are spacetime-dependent, their dependence being of exponential type (see (4.22) and (4.25)). As a result of this, one can expect a simplification of the formalism, as it happens to be the case.

Combining (4.6), with $\mathcal{A} = \varphi_i$, (4.8), (4.19) and (4.12), we see that the first partial derivatives of the field operators transform from Heisenberg to momentum picture according to the rule

$$\partial_\mu \tilde{\varphi}_i(x) \mapsto y_{i\mu} := \frac{1}{i\hbar}[\varphi_i, \mathcal{P}_\mu]_-. \tag{5.1}$$

Therefore the operator ∂_μ, when applied to field operators, transforms into $\frac{1}{i\hbar}[\cdot, \tilde{\mathcal{P}}_\mu]_- = \frac{1}{i\hbar}[\cdot, \mathcal{P}_\mu]_-$, which is a differentiation of the operator space over $\mathcal{F}$. An important corollary of (5.1) is that any (finite order) differential expression of $\tilde{\varphi}_i(x)$ transforms in momentum picture into an *algebraic* one of φ_i. In particular, this concerns the Lagrangian (which is supposed to be polynomial or convergent power series in the field operators and their partial derivatives):

$$\begin{aligned}
\tilde{\mathcal{L}} \mapsto \mathcal{L} :&= \mathcal{L}(\varphi_i(x)) := \mathcal{U}(x, x_0) \circ \tilde{\mathcal{L}}(\tilde{\varphi}_i(x), \partial_\nu \tilde{\varphi}_j(x)) \circ \mathcal{U}^{-1}(x, x_0) \\
&= \tilde{\mathcal{L}}\big(\mathcal{U}(x, x_0) \circ \tilde{\varphi}_i(x) \circ \mathcal{U}^{-1}(x, x_0), \mathcal{U}(x, x_0) \circ \partial_\nu \tilde{\varphi}_j(x) \circ \mathcal{U}^{-1}(x, x_0)\big) \\
&= \tilde{\mathcal{L}}\big(\varphi_i, \frac{1}{i\hbar}[\varphi_j, \mathcal{P}_\nu]_-\big).
\end{aligned}$$

Thus, the Lagrangian (2.1) in momentum picture reads

$$\mathcal{L} = \mathcal{L}(\phi_i) = \tilde{\mathcal{L}}(\varphi_i, y_{j\nu}) \qquad y_{j\nu} = \frac{1}{i\hbar}[\varphi_j, \mathcal{P}_\nu]_-, \tag{5.2}$$

i.e. one has to make simply the replacements $\tilde{\varphi}_i(x) \mapsto \varphi_i$ and $\partial_\nu \tilde{\varphi}_i(x) \mapsto y_{i\nu}$ in (2.1).

Applying the general rule (4.4) to the Euler-Lagrange equations (2.2) and using (4.20) and (5.1), we find, after some simple calculations,[4] the *Euler-Lagrange equations in momentum picture* as

$$\left\{\frac{\partial \tilde{\mathcal{L}}(\varphi_j, y_{l\nu})}{\partial \varphi_i} - \frac{1}{i\hbar}\left[\frac{\partial \tilde{\mathcal{L}}(\varphi_j, y_{l\nu})}{y_{i\mu}}, \mathcal{P}_\mu\right]_-\right\}\bigg|_{y_{j\nu} = \frac{1}{i\hbar}[\varphi_j, \mathcal{P}_\nu]_-} = 0. \tag{5.3}$$

A feature of these equations is that they are *algebraic*, not differential, ones with respect to the field operators φ_i (in momentum picture), provided $\mathcal{P}_\mu$ is regarded as a given known operator. This is a natural fact in view of (4.19).

We shall illustrate the above general considerations on the almost trivial example of a free Hermitian scalar field $\tilde{\varphi}$, described in Heisenberg picture by the Lagrangian $\tilde{\mathcal{L}} = -\frac{1}{2}m^2c^4 \tilde{\varphi} \circ \tilde{\varphi} + c^2\hbar^2(\partial_\mu \tilde{\varphi}) \circ (\partial^\mu \tilde{\varphi}) = \tilde{\mathcal{L}}(\tilde{\varphi}, \tilde{y}_\nu)$, with $m = \text{const}$ and $\tilde{y}_\nu = \partial_\nu \tilde{\varphi}$, and satisfying the Klein-Gordon equation $(\tilde{\Box} + \frac{m^2c^2}{\hbar^2}\,\text{id}_\mathcal{F})\tilde{\varphi} = 0$, $\tilde{\Box} := \partial_\mu \partial^\mu$. In momentum picture $\tilde{\varphi}$ transforms into the constant operator

$$\varphi(x) = \mathcal{U}(x, x_0) \circ \tilde{\varphi} \circ \mathcal{U}^{-1}(x, x_0) = \varphi(x_0) = \tilde{\varphi}(x_0) =: \varphi \tag{5.4}$$

which, in view of (5.3), $\frac{\partial \tilde{\mathcal{L}}}{\partial \varphi} = -m^2c^4\varphi$, and $\frac{\partial \tilde{\mathcal{L}}}{\partial y_\nu} = c^2\hbar^2 y_\mu \eta^{\mu\nu}$ is a solution of

$$m^2c^2\varphi - [[\varphi, \mathcal{P}_\mu]_-, \mathcal{P}^\mu]_- = 0. \tag{5.5}$$

---

[4] For details, see [1].

This is the *Klein-Gordon equation in momentum picture*, which is considered in detail in [3] As a consequence of (4.12), this equation is valid in Heisenberg picture too, when it is also a corollary of the Klein-Gordon equation and the Heisenberg relation (4.28b).

The Euler-Lagrange equations (5.3) are not enough for determination of the field operators φ_i. This is due to the simple reason that in them enter also the components $\mathcal{P}_\mu$ of the (canonical) momentum operator (2.5), which are functions (functionals) of the field operators. Hence, a complete system of equations for the field operators should consists of (5.3) and an explicit connection between them and the momentum operator. This situation is considered on particular examples in [3–5].

Since the densities of the conserved operators of a system are polynomial functions of the field operators and their partial derivatives in Heisenberg picture (for a polynomial Lagrangian of type (2.1)), in momentum picture they became polynomial functions of φ_i and $y_{j\nu} = \frac{1}{i\hbar}[\varphi_j, \mathcal{P}_\nu]_-$. When working in momentum picture, in view of (4.4), the following representations turn to be useful:

$$\mathcal{P}_\mu = \tilde{\mathcal{P}}_\mu = \frac{1}{c} \int_{x^0=\text{const}} \mathcal{U}^{-1}(x, x_0) \circ \mathcal{T}_{0\mu} \circ \mathcal{U}(x, x_0) \, d^3x \tag{5.6}$$

$$\tilde{\mathcal{Q}} = \frac{1}{c} \int_{x^0=\text{const}} \mathcal{U}^{-1}(x, x_0) \circ \mathcal{J}_0 \circ \mathcal{U}(x, x_0) \, d^3x \tag{5.7}$$

$$\tilde{\mathcal{L}}_{\mu\nu}(x) = \frac{1}{c} \int_{x^0=\text{const}} \mathcal{U}^{-1}(x, x_0) \circ \{x_\mu \mathcal{T}^0{}_\nu - x_\nu \mathcal{T}^0{}_\mu\} \circ \mathcal{U}(x, x_0) \, d^3x \tag{5.8}$$

$$\tilde{\mathcal{S}}_{\mu\nu}(x) = \frac{1}{c} \int_{x^0=\text{const}} \mathcal{U}^{-1}(x, x_0) \circ \mathcal{S}^0_{\mu\nu} \circ \mathcal{U}(x, x_0) \, d^3x. \tag{5.9}$$

In particular, the combination of (5.6) and (5.3) (together with an explicit expression for the energy-momentum tensor $\mathcal{T}_{\mu\nu}$) provide a closed algebraic-functional system of equations for determination of the field operators φ_i in momentum picture. In fact, this is the *system of field equations in momentum picture*. Concrete types of such systems of field equations and their links with the (anti)commutation (and paracommutation) relations will be investigated elsewhere.

In principle, from (5.6)–(5.9) and the field equations (i.e. (5.6) and (5.3)) can be found the commutation relations between the conserved quantities and the momentum operator, i.e. $[\tilde{\mathcal{D}}, \tilde{\mathcal{P}}_\lambda]_-$ with $\tilde{\mathcal{D}} = \tilde{\mathcal{P}}_\mu, \tilde{\mathcal{Q}}, \tilde{\mathcal{M}}_{\mu\nu}$. If one succeeds in computing $[\tilde{\mathcal{D}}, \tilde{\mathcal{P}}_\lambda]_-$, one can calculate $[\tilde{\mathcal{D}}, \mathcal{U}(x, x_0)]_-$ and, via (4.13), the operator $\mathcal{D} = \mathcal{P}_\mu, \mathcal{Q}, \mathcal{M}_{\mu\nu}$ in momentum picture. If it happens that (4.14) holds for $\tilde{\mathcal{A}} = \tilde{\mathcal{D}}$, then one can use simply the formula (4.16). In particular, this is the case if the commutators $[\tilde{\mathcal{D}}, \tilde{\mathcal{P}}_\lambda]_-$ coincide with relations like (3.4) and (3.5) (see also [8, 16].[5] For instance, if (3.5) holds, then (4.16) yields $\mathcal{Q} = \tilde{\mathcal{Q}}$, i.e. the charge operator remains one and the same in momentum and in Heisenberg pictures. Obviously, the last result holds for any operator commuting with the momentum operator.

---

[5] In future work(s), it will be proved that, in fact, the so-calculated commutators $[\tilde{\mathcal{D}}, \tilde{\mathcal{P}}_\lambda]_-$ reproduce similar relations, obtained from pure geometrical reasons in Heisenberg picture, at least for the most widely used Lagrangians. However, for the above purpose, one cannot use directly the last relations, except (3.4) in this case, because they are external to the Lagrangian formalism, so that they represent additional restriction to its consequences.

A constant operator $\tilde{\mathcal{C}}$ in Heisenberg picture,

$$\partial_\mu \tilde{\mathcal{C}} = 0, \tag{5.10}$$

transforms in momentum picture into an operator $\mathcal{C}(x)$ such that

$$\partial_\mu \mathcal{C}(x) + \frac{1}{i\hbar}[\mathcal{C}(x), \mathcal{P}_\mu]_- = 0, \tag{5.11}$$

due to (4.9) and (4.12). In particular, the conserved quantities (e.g., the momentum, charge and angular momentum operators) are solutions of equation (5.11), i.e. a conserve operator need not to be a constant one in momentum picture, but it necessarily satisfies (5.11). Obviously, a constant operator $\tilde{\mathcal{C}}$ in Heisenberg picture is such in momentum picture if and only if it commutes with the momentum operator,

$$\partial_\mu \mathcal{C}(x) = 0 \iff [\mathcal{C}(x), \mathcal{P}_\mu]_- = 0. \tag{5.12}$$

Such an operator, by virtue of (4.14) and (4.16), is one and the same in Heisenberg and momentum pictures,

$$\mathcal{C}(x) = \tilde{\mathcal{C}}. \tag{5.13}$$

In particular, the dynamical variables which are simultaneously measurable with the momentum, i.e. commuting with $\tilde{\mathcal{P}}_\mu$, remain constant in momentum picture and, hence, coincide with their values in Heisenberg one. Of course, such an operator is $\tilde{\mathcal{P}}_\mu = \mathcal{P}_\mu$, as we suppose the validity of (4.28a), and the charge operator $\tilde{\mathcal{Q}} = \mathcal{Q}$, if (3.5) holds.

Evidently, equation (5.11) is a 4-dimensional analogue of $i\hbar \frac{\partial \mathcal{A}(t)}{\partial t} + [\mathcal{A}(t), \mathcal{H}(t)]_- = 0$, which is a necessary and sufficient condition (in Schrödinger picture) for an observable $\mathcal{A}(t)$ to be an integral of motion of a quantum system with Hamiltonian $\mathcal{H}(t)$ in non-relativistic quantum mechanics [2, 18].

At the end, let us consider the Heisenberg relations (3.1)–(3.3) in momentum picture. As we said above, the first of them reduces to (4.19) in momentum picture and simply expresses the constantcy of the field operators φ_i. Since (3.2) has polynomial structure with respect to $\tilde{\varphi}_i(x)$, the transition to momentum picture preserves it, i.e. we have (see (4.4))

$$[\varphi_i, \mathcal{Q}]_- = \varepsilon(\varphi_i) q_i \varphi_i \qquad \varepsilon(\varphi_i) = \varepsilon(\tilde{\varphi}_i). \tag{5.14}$$

At last, applying (4.4) to the both sides of (3.3) and taking into account (5.1), we obtain

$$[\varphi_i, \mathcal{M}_{\mu\nu}(x, x_0)]_- = x_\mu [\varphi_i, \mathcal{P}_\nu]_- - x_\nu [\varphi_i, \mathcal{P}_\mu]_- + i\hbar I^j_{i\mu\nu} \varphi_j. \tag{5.15}$$

However, in a pure Lagrangian approach, to which (5.14) and (5.15) are external restrictions, one is not allowed to apply (5.14) and (5.15) unless these equations are explicitly proved for the operators $\mathcal{M}_{\mu\nu}$ and $\mathcal{P}_\mu$ given via (2.5)–(2.8) and (4.4).

6. On the Momentum Representation and Particle Interpretation

An important role in quantum field theory plays the so-called *momentum representation* (in Heisenberg picture) [6, 14, 16]. Its essence is in the replacement of the field operators $\tilde{\varphi}_i(x)$

with their Fourier images $\tilde{\phi}_i(k)$, both connected by the Fourier transform ($kx := k_\mu x^\mu$)[6]

$$\tilde{\varphi}_i(x) = \int e^{-\frac{1}{i\hbar}kx}\, \tilde{\phi}_i(k)\, d^4k, \qquad (6.1)$$

and then the representation of the field equations, dynamical variables, etc. in terms of $\tilde{\phi}_i(k)$.

Applying the general rule (4.4) to (6.1), we see that the analogue of $\tilde{\phi}_i(k)$ in momentum picture is the operator

$$\phi_i(k) := e^{-\frac{1}{i\hbar}kx}\, \mathcal{U}(x, x_0) \circ \tilde{\phi}_i(k) \circ \mathcal{U}^{-1}(x, x_0), \qquad (6.2)$$

which is independent of x, depends generally on x_0 and is such that

$$\varphi_i = \int \phi_i(k)\, d^4k. \qquad (6.3)$$

A field theory in terms of the operators $\phi_i(k)$ will be said to be in the *momentum representation* in momentum picture.

The Heisenberg relation (4.28b) in momentum representation, evidently, reads

$$[\tilde{\phi}_i(k), \tilde{\mathcal{P}}_\mu]_- = -k_\mu\, \tilde{\phi}_i(k) \qquad [\phi_i(k), \tilde{\mathcal{P}}_\mu]_- = -k_m\, \phi_i(k) \qquad (6.4)$$

in Heisenberg and momentum picture, respectively.[7] Consider a state vector $\tilde{\mathcal{X}}_p$ with fixed 4-momentum $p = (p_0, \ldots, p_3)$, i.e. for which

$$\tilde{\mathcal{P}}_\mu(\tilde{\mathcal{X}}_p) = p_\mu\, \tilde{\mathcal{X}}_p \qquad \mathcal{P}_\mu(\mathcal{X}_p) = p_\mu\, \mathcal{X}_p. \qquad (6.5)$$

Combining these equations with (6.4), we get

$$\tilde{\mathcal{P}}_\mu\big(\tilde{\phi}_i(k)(\tilde{\mathcal{X}}_p)\big) = (p_\mu + k_\mu)\, \tilde{\phi}_i(k)(\tilde{\mathcal{X}}_p) \qquad \mathcal{P}_\mu\big(\phi_i(k)(\mathcal{X}_p)\big) = (p_\mu + k_\mu)\, \phi_i(k)(\mathcal{X}_p). \qquad (6.6)$$

So, the operators $\tilde{\phi}_i(k)$ and $\phi_i(k)$ increase the state's 4-momentum p_μ by k_μ. If it happens that $k_0 \geq 0$, we can say that these operators create a particle with 4-momentum $(\sqrt{k^2 - \boldsymbol{k}^2}, \boldsymbol{k})$. (Notice $k^2 = k_0^2 + \boldsymbol{k}^2$, $\boldsymbol{k} := (k^1, k^2, k^3)$, need not to be a constant in the general case, so the mass $m := \frac{1}{c}\sqrt{k^2}$ is, generally, momentum-dependent.) One can introduce the creation/annihilation operators by

$$\tilde{\phi}_i^\pm(k) := \begin{cases} \tilde{\phi}_i(\pm k) & \text{for } k_0 \geq 0 \\ \frac{1}{2}\tilde{\phi}_i(\pm k) & \text{for } k_0 = 0 \\ 0 & \text{for } k_0 < 0 \end{cases} \qquad \phi_i^\pm(k) := \begin{cases} \phi_i(\pm k) & \text{for } k_0 \geq 0 \\ \frac{1}{2}\phi_i(\pm k) & \text{for } k_0 = 0 \\ 0 & \text{for } k_0 < 0 \end{cases}. \qquad (6.7)$$

---

[6] For brevity, we omit the inessential for us factor, equal to a power of 2π, in the r.h.s. of (6.1).

[7] The equations (6.4) are a particular realization of a general rule, according to which any linear combination, possibly with operator coefficients, of $\tilde{\varphi}_i(x)$ and their partial derivatives (up to a finite order) transforms into a polynomial in k_μ, the coefficients of which are proportional to $\tilde{\phi}_i(k)$. By virtue of (6.2), the same result holds in terms of $\phi_i(k)$ instead of $\tilde{\phi}_i(k)$, i.e. in momentum picture.

In terms of them, equations (6.6) take the form

$$\tilde{\mathcal{P}}_\mu\bigl(\tilde{\phi}_i^\pm(k)(\tilde{\mathcal{X}}_p))\bigr) = (p_\mu \pm k_\mu)\,\tilde{\phi}_i(k)(\tilde{\mathcal{X}}_p)) \qquad \mathcal{P}_\mu\bigl(\phi_i^\pm(k)(\mathcal{X}_p))\bigr) = (p_\mu \pm k_\mu)\,\phi_i(k)(\mathcal{X}_p)). \tag{6.8}$$

Thus, if $k_0 \geq 0$, we can interpret $\tilde{\phi}_i^+(k)$ and $\phi_i^+(k)$ (resp. $\tilde{\phi}_i^-(k)$ and $\phi_i^-(k)$) as operators creating (resp. annihilating) a particle with 4-momentum k_μ.

If the relations (3.2) (resp. (3.3)) hold, similar considerations are (resp. partially) valid with respect to state vectors with fixed charge (resp. total angular momentum).

As we see, the description of a quantum field theory in momentum representation is quite similar in Heisenberg picture, via the operators $\tilde{\phi}_i(k)$, and in momentum picture, via the operators $\phi_i(k)$. This similarity will be investigated deeper on concrete examples in forthcoming paper(s). The particular form of the operators $\tilde{\phi}_i(k)$ and $\phi_i(k)$ can be found by solving the field equations, respectively (2.2) and (5.3), in momentum representation, but the analysis of the so-arising equations is out of the subject of the present work.

7. The Momentum Picture as 4-Dimensional Analogue of the Schrödinger One

We have introduced the momentum picture and explored some its aspects on the base of the Heisenberg one, i.e. the latter picture was taken as a ground on which the former one was defined and investigated; in particular, the conditions (4.28) turn to be important from this view-point. At that point, a question arises: can the momentum picture be defined independently and to be taken as a base from which the Heisenberg one to be deduced? Below is presented a partial solutions of that problem for theories in which the equations (4.28) hold.

First of all, it should be decided which properties of the momentum picture, considered until now, characterize it in a more or less unique way and then they or part of them to be incorporated in a suitable (axiomatic) definition of momentum picture. As a guiding idea, we shall follow the understanding that the momentum picture is (or should be) a 4-dimensional analogue of the Schrödinger picture in non-relativistic quantum mechanics. Recall, [2, 18, 19], the latter is defined as a representation of quantum mechanics in which: (i) the operators, corresponding to the dynamical variables, are time-independent; (ii) these operators are taken as predefined (granted) in an appropriate way; and (iii) the wavefunctions ψ are, generally, time-dependent and satisfy the Schrödinger equation

$$\frac{\partial \psi}{\partial t} = \frac{1}{i\hbar}\mathcal{H}(\psi), \tag{7.1}$$

with $\mathcal{H}$ being the system's Hamiltonian acting on the system's Hilbert space of states. A 4-dimensional generalization of (i)–(iii), adapted for the needs of quantum field theory, will result in an independent definition of the momentum picture. Since in that theory the operators of the dynamical variables are constructed form the field operators φ_i, the latter should be used for the former ones when the generalization mentioned is carried out. Besides, the field operators satisfy some equations, which have no analogues in quantum mechanics, which indicates to a nontrivial generalization of item (ii) above.

Following these ideas, we define the *momentum picture* of quantum field theory as its representation in which:

(a) The field operators φ_i are spacetime-independent,
$$\partial_\mu(\varphi_i) = 0. \tag{7.2}$$

(b) The state vectors χ are generally spacetime-dependent and satisfy the following first order system of partial differential equations
$$\partial_\mu(\chi) = \frac{1}{i\hbar}\mathcal{P}_\mu(\chi), \tag{7.3}$$
where $\mathcal{P}_\mu$ are the components of the system's momentum operator (constructed according to point (c) below – see (7.8)). If $\chi_0 \in \mathcal{F}$ and $x_0 \in M$ are fixed, the system (7.3) is supposed to have a unique solution satisfying the initial condition
$$\chi|_{x=x_0} = \chi_0. \tag{7.4}$$

(c) If $\tilde{\mathcal{D}}(\tilde{\varphi}_i, \partial_\mu \tilde{\varphi}_j)$ is the density current of a dynamical variable in (ordinary) Heisenberg picture, which is supposed to be polynomial or convergent power series in $\tilde{\varphi}_i$ and $\partial_\mu \tilde{\varphi}_j$, then this quantity in momentum picture is defined to be
$$\mathcal{D} = \mathcal{D}(\varphi_i) := \tilde{\mathcal{D}}\Big(\varphi_i, \frac{1}{i\hbar}[\varphi_j, \mathcal{P}_\mu]_-\Big). \tag{7.5}$$
The corresponding spacetime conserved operator is defined as
$$\mathsf{D} := \frac{1}{c}\int_{x_0=\text{const}} \mathcal{U}^{-1}(x, x_0) \circ \mathcal{D}(\varphi_i) \circ \mathcal{U}(x, x_0) \, d^3 x, \tag{7.6}$$
where $\mathcal{U}(x, x_0)$ is the evolution operator for (7.3)–(7.4), i.e. the unique solution of the initial-value problem
$$\frac{\partial \mathcal{U}(x, x_0)}{\partial x^\mu} = \frac{1}{i\hbar}\mathcal{P}_\mu \circ \mathcal{U}(x, x_0) \tag{7.7a}$$
$$\mathcal{U}(x_0, x_0) = \text{id}_\mathcal{F} \tag{7.7b}$$
with $\mathcal{P}_\mu$ corresponding to (7.6) with the energy-momentum tensor $\mathcal{T}_{\mu\nu}$ for $\mathcal{D}$,
$$\mathcal{P}_\mu := \frac{1}{c}\int_{x_0=\text{const}} \mathcal{U}^{-1(x,x_0)} \circ \mathcal{T}_{0\mu}(\varphi_i) \circ \mathcal{U}(x, x_0) \, d^3 x. \tag{7.8}$$

(d) The field operators φ_i are solutions of the (algebraic) field equations, which (in the most cases) are identified with the Euler-Lagrange equations
$$\left\{\frac{\partial \tilde{\mathcal{L}}(\varphi_j, y_{l\nu})}{\partial \varphi_i} - \frac{1}{i\hbar}\Big[\frac{\partial \tilde{\mathcal{L}}(\varphi_j, y_{l\nu})}{\partial y_{i\mu}}, \mathcal{P}_\mu\Big]_-\right\}\Big|_{y_{j\nu}=\frac{1}{i\hbar}[\varphi_j, \mathcal{P}_\nu]_-} = 0, \tag{7.9}$$
with $\tilde{\mathcal{L}}(\varphi_j, \frac{1}{i\hbar}[\varphi_j, \mathcal{P}_\nu]_-)$ being the system's Lagrangian (in momentum picture, defined according to (7.5)).

A number or comments on the conditions (a)–(d) are in order.

The transition from momentum to Heisenberg picture is provided by the inversion of (4.3) and (4.3) with $\mathcal{U}(x, x_0)$ given via (7.7), i.e.

$$\mathcal{X} \mapsto \tilde{\mathcal{X}} = \mathcal{U}^{-1}(x, x_0)(\tilde{\mathcal{X}}(x)) \tag{7.10}$$

$$\mathcal{A}(x) \mapsto \tilde{\mathcal{A}}(x) = \mathcal{U}^{-1}(x, x_0) \circ (\mathcal{A}(x)) \circ \mathcal{U}(x, x_0). \tag{7.11}$$

Since (7.7) implies

$$\mathcal{H}_\mu(x, x_0) = \frac{1}{i\hbar} \mathcal{P}_\mu \tag{7.12}$$

for the quantities (4.7), the replacement (4.9) is valid. In particular, we have

$$\partial_\mu \tilde{\varphi}_i \mapsto y_{j\mu} = \frac{1}{i\hbar} [\varphi_j, \mathcal{P}_\mu]_-, \tag{7.13}$$

by virtue of (7.2), which justifies the definition (7.5) and the equation (7.9). The Heisenberg relations (4.28b) follow from this replacement:

$$[\tilde{\varphi}_i(x), \tilde{\mathcal{P}}_\mu]_- = \mathcal{U}^{-1}(x, x_0) \circ [\varphi_i, \mathcal{P}_\mu]_- \circ \mathcal{U}(x, x_0) = i\hbar \partial_\mu(\tilde{\varphi}_i).$$

Since the integrability conditions for (7.3) are

$$0 = \partial_\nu \circ \partial_\mu(\chi) - \partial_\mu \circ \partial_\nu(\chi) = \frac{1}{i\hbar} \{\partial_\nu(\mathcal{P}_\mu(\chi)) - \partial_\mu(\mathcal{P}_\nu(\chi))\}$$

$$= \frac{1}{i\hbar} \{(\partial_\nu(\mathcal{P}_\mu) - \partial_\mu(\mathcal{P}_\nu))(\chi) + \mathcal{P}_\mu(\partial_\nu(\chi)) - \mathcal{P}_\nu(\partial_\mu(\chi))\},$$

where (7.3) was applied, the existence of a unique solution of (7.3)–(7.4) implies (use (7.2) again; cf. footnote 3)

$$\partial_\nu(\mathcal{P}_\mu) - \partial_\mu(\mathcal{P}_\nu)) + \frac{1}{i\hbar} [\mathcal{P}_\mu, \mathcal{P}_\nu]_- = 0. \tag{7.14}$$

As $\partial_\nu \tilde{\mathcal{P}}_\mu = 0$, due to the conservation of $\tilde{\mathcal{P}}_\mu$, the replacement (4.6), with $\mathcal{P}_\nu$ for $\mathcal{A}(x)$, together with (7.12) entails $\partial_\mu(\mathcal{P}_\nu)) + \frac{1}{i\hbar}[\mathcal{P}_\nu, \mathcal{P}_\mu]_- = 0$, which, when inserted into (7.14), gives

$$\partial_\nu(\mathcal{P}_\mu) = 0. \tag{7.15}$$

The substitution of (7.15) into (7.14) results in

$$[\mathcal{P}_\mu, \mathcal{P}_\nu]_- = 0, \tag{7.16}$$

which immediately implies (4.28a).

As a result of (7.16) and (7.7), we obtain

$$\mathcal{U}(x, x_0) = e^{\frac{1}{i\hbar}(x^\mu - x_0^\mu)\mathcal{P}_\mu}, \tag{7.17}$$

so that

$$[\mathcal{U}(x, x_0), \mathcal{P}_\mu]_- = [\mathcal{U}^{-1}(x, x_0), \mathcal{P}_\mu]_- = 0 \tag{7.18}$$

and, consequently

$$\tilde{\mathcal{P}}_\mu = \mathcal{U}^{-1}(x, x_0) \circ \mathcal{P}_\mu \circ \mathcal{U}(x, x_0) = \mathcal{P}_\mu, \tag{7.19}$$

which implies the coincidence of the evolution operators given by (4.1) and (7.7). The last conclusion leads to the identification of the momentum picture defined via the conditions (a)–(d) above and by (4.3), (4.4) and (4.28) in Sect. 4.

What regards the conditions (c) and (d) in the definition of the momentum picture, they have no analogues in quantum mechanics. Indeed, equations (7.5)–(7.9) form a closed system for determination of the field operators (via the so-called creation and annihilation operators) and, correspondingly, they provide a method for obtaining explicit forms of the dynamical variables (via the same operators). On the contrary, in quantum mechanics there is no procedure for determination of the operators of the dynamical variables and they are defined by reasons external to this theory.

Thus, we see that a straightforward generalization of the Schrödinger picture in quantum mechanics to the momentum picture in quantum field theory (expressed first of all by (7.2) and (7.3)) is possible if and only if the equations (4.28) are valid for the system considered.

8. Conclusion

In the present paper, we have summarized, analyzed and developed the momentum picture of motion in (Lagrangian) quantum field theory, introduced in [1]. As it was shown, this picture is (expected to be) useful when the conditions (4.28) are valid in (or compatible with) the theory one investigates. If this is the case, the momentum picture has properties that allow one to call it a '4-dimensional Schrödinger picture' as the field operators (and functions which are polynomial in them and their derivatives) in it became spacetime-constant operators and the state vectors have a simple, exponential, dependence on the spacetime coordinates/points. This situation is similar to the one in quantum mechanics in Schrödinger picture, when time-independent Hamiltonians are employed [2], the time replacing the spacetime coordinates in our case.

As we said in Sect 4, there are evidences that the conditions (4.28) should be a part of the basic postulates of quantum field theory (see also [16, § 68]). In the ordinary field theory, based on the Lagrangian formalism to which (anti)commutation relations are added as additional conditions [6,7,14], the validity of (4.28) is questionable and should be checked for any particular Lagrangian [16]. The cause for this situation lies in the fact that (4.28) and the (anti)commutation relations are additional to the Lagrangian formalism and their compatibility is a problem whose solution is not obvious. The solution of that problem is known to be positive for a lot of particular Lagrangians [16], but, in the general case, it seems not to be explored. For these reasons, one may try to 'invert' the situation, i.e. to consider a Lagrangian formalism, to which the conditions (4.28) are imposed as subsidiary restrictions, and then to try to find (anti)commutation relations that are consistent with the so-arising scheme.

This program is realized in the papers [3–5] for free scalar, spinor and vector, respectively, vields. In them, it is demonstrated that the proposed method reproduces most of the known results, reveals ways for their generalizations at different stages of the theory, and also gives new results, such as a (second) quantization of electromagnetic field in Lorentz gauge, imposed directly on the field's operator-valued potentials, and a 'natural' derivation of the paracommutation relations.

Acknowledgments

This research was partially supported by the National Science Fund of Bulgaria under Grant No. F 1515/2005.

References

[1] Bozhidar Z. Iliev. Pictures and equations of motion in Lagrangian quantum field theory. In Charles V. Benton, editor, *Studies in Mathematical Physics Research*, pages 83–125. Nova Science Publishers, Inc., New York, 2004.
http://arXiv.org e-Print archive, E-print No. hep-th/0302002, February 2003.

[2] A. M. L. Messiah. *Quantum mechanics*, volume I and II. Interscience, New York, 1958. Russian translation: Nauka, Moscow, 1978 (vol. I) and 1979 (vol. II).

[3] Bozhidar Z. Iliev. Lagrangian quantum field theory in momentum picture. In O. Kovras, editor, *Quantum Field Theory: New Researcn*, pages 1–66. Nova Science Publishers, Inc., New York, 2005.
http://arXiv.org e-Print archive, E-print No. hep-th/0402006, February 1, 2004.

[4] Bozhidar Z. Iliev. Lagrangian quantum field theory in momentum picture. II. Free spinor fields.
http://arXiv.org e-Print archive, E-print No. hep-th/0405008, May 1, 2004.

[5] Bozhidar Z. Iliev. Lagrangian quantum field theory in momentum picture. III. Free vector fields.
http://arXiv.org e-Print archive, E-print No. hep-th/0505007, May 1, 2005.

[6] N. N. Bogolyubov and D. V. Shirkov. *Introduction to the theory of quantized fields*. Nauka, Moscow, third edition, 1976. In Russian. English translation: Wiley, New York, 1980.

[7] J. D. Bjorken and S. D. Drell. *Relativistic quantum mechanics*, volume 1 and 2. McGraw-Hill Book Company, New York, 1964, 1965. Russian translation: Nauka, Moscow, 1978.

[8] Paul Roman. *Introduction to quantum field theory*. John Wiley&Sons, Inc., New York-London-Sydney-Toronto, 1969.

[9] Lewis H. Ryder. *Quantum field theory*. Cambridge Univ. Press, Cambridge, 1985. Russian translation: Mir, Moscow, 1987.

[10] A. I. Akhiezer and V. B. Berestetskii. *Quantum electrodynamics*. Nauka, Moscow, 1969. In Russian. English translation: Authorized English ed., rev. and enl. by the author, Translated from the 2d Russian ed. by G.M. Volkoff, New York, Interscience Publishers, 1965. Other English translations: New York, Consultants Bureau, 1957; London, Oldbourne Press, 1964, 1962.

[11] Pierre Ramond. *Field theory: a modern primer*, volume 51 of *Frontiers in physics*. Reading, MA Benjamin-Cummings, London-Amsterdam-Don Mills, Ontario-Sidney-Tokio, 1 edition, 1981. 2nd rev. print, Frontiers in physics vol. 74, Adison Wesley Publ. Co., Redwood city, CA, 1989; Russian translation from the first ed.: Moscow, Mir 1984.

[12] N. N. Bogolubov, A. A. Logunov, and I. T. Todorov. *Introduction to axiomatic quantum field theory*. W. A. Benjamin, Inc., London, 1975. Translation from Russian: Nauka, Moscow, 1969.

[13] N. N. Bogolubov, A. A. Logunov, A. I. Oksak, and I. T. Todorov. *General principles of quantum field theory*. Nauka, Moscow, 1987. In Russian. English translation: Kluwer Academic Publishers, Dordrecht, 1989.

[14] C. Itzykson and J.-B. Zuber. *Quantum field theory*. McGraw-Hill Book Company, New York, 1980. Russian translation (in two volumes): Mir, Moscow, 1984.

[15] Bozhidar Z. Iliev. On operator differentiation in the action principle in quantum field theory. In Stancho Dimiev and Kouei Sekigava, editors, *Proceedings of the 6th International Workshop on Complex Structures and Vector Fields, 3–6 September 2002, St. Knstantin resort (near Varna), Bulgaria*, "Trends in Complex Analysis, Differential Geometry and Mathematical Physics", pages 76–107. World Scientific, New Jersey-London-Singapore-Hong Kong, 2003.
http://arXiv.org e-Print archive, E-print No. hep-th/0204003, April 2002.

[16] J. D. Bjorken and S. D. Drell. *Relativistic quantum fields*, volume 2. McGraw-Hill Book Company, New York, 1965. Russian translation: Nauka, Moscow, 1978.

[17] Y. Ohnuki and S. Kamefuchi. *Quantum field theory and parafields*. University of Tokyo Press, Tokyo, 1982.

[18] P. A. M. Dirac. *The principles of quantum mechanics*. Oxford at the Clarendon Press, Oxford, fourth edition, 1958. Russian translation in: P. Dirac, Principles of quantum mechanics, Moscow, Nauka, 1979.

[19] V. A. Fock. *Fundamentals of quantum mechanics*. Mir Publishers, Moscow, 1978. Russian edition: Nauka, Moscow, 1976.

Chapter 2

The Dynamics of a IIA String Theory D0-brane Probing a Four Dimensional Black Hole

A. Chenaghlou[*]
Physics Department, Faculty of Science, Sahand University of Technology,
P O Box 51335-1996, Tabriz, Iran

Abstract

In this paper, we study the motion of the four dimensional black hole for the case of a IIA D0-brane probe. It is shown that the dynamics without angular momentum is reduced to the radial motion of the system. However in presence of the angular momentum, the dynamics has angular motion in addition to the radial motion. It is also noticed that instead of a spiral motion, the D-brane will have uniform motion in the asymptotic region.

1. Introduction

The Dirac-Born-Infeld action describes the dynamic of D-branes [1]. In recent years the study of tachyon condensation by using an effective action of the Dirac-Born-Infeld has been a very powerful tool to investigate various aspect of the D-brane dynamics [2–22]. Study of classical solutions for tachyon dynamics in open string theories has found an important attraction in the context of D-branes. The real time tachyon dynamics reveals many prospects of rolling tachyon solutions of the full open string theory by means of the effective Dirac-Born-Infeld action [10, 16–18]. In fact in Ref [10] a DBI-like effective field theory was proposed to study the tachyon dynamics on the non-BPS brane. Then, the effective field theory described the tachyon condensation successfully [16] which, in turn, leads to a new kind of open-closed string duality [6,7]. Then for example, Kutasov used the Dirac-Born-Infeld action to study the real time dynamics of a Dp-brane propagating in the vicinity of NS5-branes [23]. This problem is closely related to tachyon condensation on an unstable D-brane. The DBI action [24, 25] is

---
[*]E-mail address: a.chenaghlou@sut.ac.ir

$$S_p = -\tau_p \int d^{p+1}X e^{-(\Phi-\Phi_0)}\sqrt{-\det(G_{\mu\nu}+B_{\mu\nu})}, \tag{1}$$

where τ_p is the tension of the Dp-brane. Note that one may lable the worldvolume of the D-brane by X^μ, $\mu = 0, 1, ..., p$. The position of the D-brane in the transverse directions gives rise to scalar fields on the worldvolume of the D-brane $(X^6(X^\mu), ..., X^9(X^\mu))$. $G_{\mu\nu}$ and $B_{\mu\nu}$ are the induced metric and B-field on the D-brane:

$$G_{\mu\nu} = \frac{\partial X^a}{\partial X^\mu}\frac{\partial X^b}{\partial X^\nu}G_{ab}(X),$$
$$B_{\mu\nu} = \frac{\partial X^a}{\partial X^\mu}\frac{\partial X^b}{\partial X^\nu}B_{ab}(X). \tag{2}$$

The indices $a, b = 0, 1, 2, \cdots, 9$ run over the ten dimensional spacetime. G_{ab} and B_{ab} are the metric and B-field in ten dimensions.

2. The Effective Action of a D0-brane Probing a 4D Black Hole

In this section we consider the 4D black hole in IIA string theory. The $D=4$ string frame metric and dilaton are

$$ds^2 = -H_2^{-1/2}H_6^{-1/2}(1+K)^{-1}dt^2 + H_2^{1/2}H_6^{1/2}H_5(dr^2+r^2 d\Omega^2),$$
$$e^{2\phi_4} = \frac{H_5^{1/2}}{1+K}, \tag{3}$$

and the harmonic functions H_2, H_5, H_6 and K are given by

$$H_i = 1 + \frac{r_i}{R}, \quad i=2,5,6 \quad \text{and} \quad K = \frac{r_m}{R}, \tag{4}$$

where

$$r_2 = g_s\frac{N_2 l_s^5}{2V}, \quad r_5 = \frac{N_5 l_s^2}{2R_b}, \quad r_6 = g_s\frac{N_6 l_s}{2}, \quad r_m = g_s^2\frac{N_m l_s^8}{2VR_a^2 R_b}. \tag{5}$$

Note $G_4 = g_s^2 l_s^2/8VR_a R_b = \sqrt{r_2 r_5 r_6 r_m}/2\sqrt{N_2 N_5 N_6 N_m} = l_{pl}^2$. For the RN case $r_2 = r_5 = r_6 = r_m = r_0$ so, $r_0^2 = 2l_{pl}^2\sqrt{N_2 N_5 N_6 N_m}$. Also we have $R_b/l_s = N_5/N_6$, $2V/l_s^4 = N_2/N_6$, $R_a^2/l_s^2 = N_m N_6/N_2 N_5$.

Now we consider a IIA D0-brane probe. In other words, the 4D black hole is embedded in string theory by using D-branes. The position of D0-brane in the 3D X^i-space give rises to 3 scalar fields. The DBI action is

$$S = \tau_0 \int dt\, e^{-\phi_4}\sqrt{-\det G_{\mu\nu}^{(s)}}, \tag{6}$$

where τ_0 and $G_{\mu\nu}$ are the D0-brane tension and the induced metric on the D-brane. Regarding the relation (2), the induced metric takes the form

$$G_{00} = -H_2^{-1/2}H_6^{-1/2}(1+K)^{-1} + H_2^{1/2}H_6^{1/2}H_5 \dot{X}^i \dot{X}^i, \quad i=1,2,3. \tag{7}$$

The DBI action is thus given by

$$S = \int dt \, \frac{\tau_0}{(H_2 H_5 H_6)^{1/4}} \sqrt{1 - H_2 H_5 H_6 (1+K) \dot{X}^2}. \tag{8}$$

This problem can be mapped to the tachyon problem in open string models whose action is

$$S_{\text{tachyon}} = -\int d^{p+1}X \, V(T) \sqrt{-\det G_{\mu\nu}^{(\text{tachyon})}} = -\int d^{p+1}X \, V(T) \sqrt{1 - \dot{T}^2}, \tag{9}$$

where $V(T)$ is the tachyon potential. Comparing (8) and (9) it is seen that one may map one to the other by defining a tachyon field T which is a function of R as

$$\frac{dT}{dR} = \sqrt{H_2 H_5 H_6 (1+K)}. \tag{10}$$

Then

$$S = \int dt \, \frac{\tau_0}{(H_2 H_5 H_6)^{1/4}} \sqrt{1 - \dot{T}^2}. \tag{11}$$

So the tachyon potential $V(T)$ is given by

$$V(T(R)) = \frac{\tau_0}{(H_2 H_5 H_6)^{1/4}}. \tag{12}$$

Note that $dT/dR = \sqrt{(1+r_2/R)(1+r_5/R)(1+r_6/R)(1+r_m/R)} > 0$. For large R, that is, $R \to \infty$ it is seen that $T \simeq R$. For $R \to 0$, we have $dT/dR \sim (r_2 r_5 r_6 r_m)^{1/2}/R^2$ or $T \sim constant - (r_2 r_5 r_6 r_m)^{1/2}/R$. In fact, $R = 0$ corresponds to the black hole horizon.

Now it is better to study the asymptotic behaviour of the tachyon potential. As $R \to \infty$ then $T \to \infty$ and $V(T)/\tau_0 \simeq 1 - (r_2 + r_5 + r_6)/4T$ which look likes $1/R$ potential. Meanwhile, when $R \to 0$ then $T \sim -1/R$ and $V(T) \simeq \tau_0 R^{3/4}/(r_2 r_5 r_6)^{1/4}$.

For the RN black hole one has $dT/dR = (1+r_0/R)^2$. So, $T = R - r_0^2/R + 2r_0 \ln R$. For $R \to 0$, $V(T) \simeq (\tau_0 R^{3/4})/r_0^{3/4}$.

3. The Dynamics of a IIA D0-Brane Probing a 4D Black Hole

In this section, we study the motion of the 4D black hole for the case of a IIA D0-brane probe. We shall show that the dynamics without angular momentum is reduced to the radial motion of the system. However in presence of the angular momentum, the dynamics has angular motion in addition to the radial motion. In what follows we analyze homogeneous dynamics of the DBI action (8) where the corresponding transverse fields of the D-brane depend only on time i.e. $X^i = X^i(t)$, $i=1,2,3$.

The solutions to the equation of motion of the action (8) can be computed by using the conserved charges of the system. The total energy is defined by

$$E = P_i \dot{X}^i - \mathcal{L}, \tag{13}$$

which is conserved due to the time translation invariance. The canonical momentum conjugate to X^i is as follows:

$$P_i = \frac{\partial \mathcal{L}}{\partial \dot{X}^i} = -\frac{\tau_0 (1+K)(H_2 H_5 H_6)^{3/4} \dot{X}^i}{\sqrt{1 - H_2 H_5 H_6 (1+K) \dot{X}^2}}. \tag{14}$$

So, the conserved energy can be obtained by substituting (14) into (13) to yield

$$E = -\frac{\tau_0}{(H_2 H_5 H_6)^{1/4}\sqrt{1 - H_2 H_5 H_6(1+K)\dot{X}^2}}. \tag{15}$$

In what follows, we will require to determine the initial conditions i.e. the values of X and $\dot{X}$ at a specific time. Obviously, these two vectors constitute a plane in R^3. Regarding the transverse $SO(3)$ symmetry, the orbits can be constrained on the (X^1, X^2) plane. Another conserved quantity is the angular momentum of the D-brane which is given by

$$L = X^1 P^2 - X^2 P^1 = -\frac{\tau_0(1+K)(H_2 H_5 H_6)^{3/4}(X^1 \dot{X}^2 - X^2 \dot{X}^1)}{\sqrt{1 - H_2 H_5 H_6(1+K)\dot{X}^2}}. \tag{16}$$

The stress tensor $T_{\mu\nu}$ is another quantity of interest which has only one component and is equal to the energy. Thus it is the expression (15).

Now it is better to work in polar coordinates. So,

$$\begin{aligned} X^1 &= R\cos\theta, \\ X^2 &= R\sin\theta. \end{aligned} \tag{17}$$

Then the energy and angular momentum become

$$E = -\frac{\tau_0}{(H_2 H_5 H_6)^{1/4}\sqrt{1 - H_2 H_5 H_6(1+K)(\dot{R}^2 + R^2\dot{\theta}^2)}},$$
$$L = -\frac{\tau_0(1+K)(H_2 H_5 H_6)^{3/4} R^2 \dot{\theta}}{\sqrt{1 - H_2 H_5 H_6(1+K)(\dot{R}^2 + R^2\dot{\theta}^2)}}. \tag{18}$$

Let us first calculate $\dot{\theta}$ by using the second relation of (18). Therefore, one has

$$\dot{\theta}^2 = \frac{L^2(1 - H_2 H_5 H_6(1+K)\dot{R}^2)}{\tau_0^2(H_2 H_5 H_6)^{3/2}(1+K)^2 R^4 + L^2 H_2 H_5 H_6(1+K)R^2}. \tag{19}$$

Substituting (19) in the first relation of (18) and after some manipulation we find

$$\dot{R}^2 = \frac{1}{H_2 H_5 H_6(1+K)} - \frac{1}{E^2(H_2 H_5 H_6)^2(1+K)^2}\left(\frac{L^2}{R^2} + \tau_0^2(1+K)\sqrt{H_2 H_5 H_6}\right). \tag{20}$$

Then one can obtain the following simplified relation as well

$$\dot{\theta}^2 = \frac{L^2}{E^2(H_2 H_5 H_6)^2(1+K)^2 R^4}. \tag{21}$$

In order to describe the motion of the D-brane we should solve equations (20) and (21). Let us study first the motion when $L=0$. It is evident that in this case θ is constant whereas the radial equation (20) is reduced to

$$\dot{R}^2 = \frac{1}{H_2 H_5 H_6(1+K)} - \frac{\tau_0^2}{E^2(H_2 H_5 H_6)^{3/2}(1+K)}. \tag{22}$$

The above relation imposes a constraint on R due to the fact that the right hand side must be non-negative.

$$\left(1 + \frac{r_2}{R}\right)\left(1 + \frac{r_5}{R}\right)\left(1 + \frac{r_6}{R}\right) \geq \frac{\tau_0^4}{E^4}. \tag{23}$$

This relation has a physical meaning. If the tension of a free BPS D0-brane is smaller than the energy then the D0-brane could escape to infinity. For $\tau_0 > E$, this cannot happen.

We can solve equation (22) exactly provided the orbit stays in the small region $R \ll r_i$ (when the D-brane reaches the horizon region) which occurs when $\tau_0/E \gg 1$. Meanwhile, we would like to study the solutions of the equations of motion in the RN case for which $r_2 = r_5 = r_6 = r_m$. Therefore, we have

$$H_i = \frac{r_i}{R}, i = 2, 5, 6 \quad \text{and} \quad 1 + K = \frac{r_m}{R}. \tag{24}$$

Substituting (24) into (22), the orbit equation is obtained as

$$\dot{R}^2 = \frac{R^4}{r_m r_2 r_5 r_6}\left(1 - \frac{\tau_0^2 R^{3/2}}{E^2 r_2^{1/2} r_5^{1/2} r_6^{1/2}}\right), \tag{25}$$

which has the following solution

$$-\frac{hypergeom([-2/3, 1/2], [1/3], y^{3/2})}{2y} = \pm 1/2\tau, \tag{26}$$

where $y = \left(\tau_0^2/E^2 r_2^{1/2} r_5^{1/2} r_6^{1/2}\right)^{2/3} R$ and the time scaling is $\tau = E^{4/3} t / \left(r_m^{1/2} r_2^{1/6} r_5^{1/6} r_6^{1/6} \tau_0^{4/3}\right)$. The dilaton takes the form

$$e^{2\phi_4} = \sqrt{\frac{R}{r_0}}. \tag{27}$$

Quantum effects are negligible when $e^{2\phi_4} \ll 1$. We notice that for all τ, $e^{2\phi_4} \ll 1$.

The induced metric on the D-brane is

$$G_{00} = -\frac{\tau_0^2 R^{7/2}}{r_2 r_6 r_m r_5^{1/2} E^2}. \tag{28}$$

It is seen that when $\tau \to \pm\infty$ then $G_{00} \to 0$.

Now let us study the motion of the D0-brane with non-vanishing angular momentum. When the D0-brane reaches the horizon region and in the RN case, we have

$$\dot{R}^2 = \frac{R^4}{r_m r_2 r_5 r_6}\left(1 - \frac{\tau_0^2 R^{3/2}}{E^2 r_2^{1/2} r_5^{1/2} r_6^{1/2}} - \frac{L^2 R^2}{r_m r_2 r_5 r_6 E^2}\right), \tag{29}$$

The quantum effect is also small for all τ. For non-zero angular momentum the induced metric on the D-brane is given by

$$G_{00} = -\frac{\tau_0^2 R^{7/2}}{r_2 r_6 r_m r_5^{1/2} E^2} \tag{30}$$

which is the same as (28). Note that when $\tau \to \pm\infty$ then $G_{00} \to 0$. The angular equation of motion is given by

$$\dot{\theta}^2 = \frac{L^2 R^4}{E^2 r_2^2 r_5^2 r_6^2 r_m^2}. \tag{31}$$

It is noticed that when $\tau \to \pm\infty$ then $\dot{\theta} \to 0$ which means instead of a spiral motion, the D-brane will have uniform motion in the asymptotic region.

4. Conclusions

In the last decade, the study of Dirac-Born-Infeld action has attracted much attention. In this paper, the dynamics of a IIA string theory D0-brane probing four dimensional black hole is investigated. We show that the dynamics without angular momentum is reduced to the radial motion of the system. On the other hand, the motion of the D0-brane with non-vanishing angular momentum represents an angular motion in addition to the radial motion. It is also noticed that instead of a spiral motion, the D-brane will have uniform motion in the asymptotic region.

Acknowledgment

We would like to thank Chong-Sun Chu for useful discussions and fruitful comments.

References

[1] J. Dai, R. G. Leigh and J. Polchinski, *Mod. Phys. Lett.* **A4** (1989) 2073.
R. G. Leigh, Dirac-Born-Infeld action from Dirichlet Sigma model, *Mod. Phys. Lett.* **A4** (1989) 2767.

[2] T. Okuda and S. Sugimoto, "Coupling of rolling tachyon to closed strings," *Nucl. Phys. B* **647**, 101 (2002) [arXiv:hep-th/0208196].

[3] N. Lambert, H. Liu and J. Maldacena, "Closed strings from decaying D-branes," *arXiv:hep-th/0303139*.

[4] D. Gaiotto, N. Itzhaki and L. Rastelli, "Closed strings as imaginary D-branes," *arXiv:hep-th/0304192*.

[5] A. Sen, "Field theory of tachyon matter," *Mod. Phys. Lett. A* **17**, 1797 (2002) [*arXiv:hep-th/0204143*].

[6] A. Sen, "Open-closed duality at tree level," *Phys. Rev. Lett.* **91**, 181601 (2003) [*arXiv:hep-th/0306137*].

[7] A. Sen, "Open-closed duality: Lessons from matrix model," *Mod. Phys. Lett. A* **19**, 841 (2004) [*arXiv:hep-th/0308068*].

[8] A. Sen, "Rolling tachyon," *JHEP* **0204**, 048 (2002) [*arXiv:hep-th/0203211*].

[9] A. Sen, "Tachyon matter," *JHEP* **0207**, 065 (2002) [*arXiv:hep-th/0203265*].

[10] A. Sen, "Supersymmetric world-volume action for non-BPS D-branes," *JHEP* **9910**, 008 (1999) [*arXiv:hep-th/9909062*].

[11] A. Sen, "Dirac-Born-Infeld action on the tachyon kink and vortex," *Phys. Rev. D* **68**, 066008 (2003) [*arXiv:hep-th/0303057*].

[12] A. Sen, "Remarks on tachyon driven cosmology," *arXiv:hep-th/0312153*.

[13] G. W. Gibbons, K. Hori and P. Yi, "String fluid from unstable D-branes," *Nucl. Phys. B* **596**, 136 (2001) [*arXiv:hep-th/0009061*].

[14] G. Gibbons, K. Hashimoto and P. Yi, "Tachyon condensates, Carrollian contraction of Lorentz group, and fundamental JHEP **0209**, 061 (2002) [*arXiv:hep-th/0209034*].

[15] H. U. Yee and P. Yi, "Open / closed duality, unstable D-branes, and coarse-grained closed *Nucl. Phys. B* **686**, 31 (2004) [*arXiv:hep-th/0402027*].

[16] M. R. Garousi, "Tachyon couplings on non-BPS D-branes and Dirac-Born-Infeld action," *Nucl. Phys. B* **584**, 284 (2000) [*arXiv:hep-th/0003122*].

[17] E. A. Bergshoeff, M. de Roo, T. C. de Wit, E. Eyras and S. Panda, "T-duality and actions for non-BPS D-branes," *JHEP* **0005**, 009 (2000) [*arXiv:hep-th/0003221*].

[18] J. Kluson, "Proposal for non-BPS D-brane action," *Phys. Rev. D* **62**, 126003 (2000) [arXiv:hep-th/0004106].

[19] D. Kutasov and V. Niarchos, "Tachyon effective actions in open string theory," *Nucl. Phys. B* **666**, 56 (2003) [*arXiv:hep-th/0304045*].

[20] V. Niarchos, "Notes on tachyon effective actions and Veneziano amplitudes," *hep-th/0401066*.

[21] G. W. Gibbons, "Cosmological evolution of the rolling tachyon," *Phys. Lett. B* **537**, 1 (2002) [*arXiv:hep-th/0204008*].

[22] G. Shiu, S. H. H. Tye and I. Wasserman, "Rolling tachyon in brane world cosmology from superstring field theory," *Phys. Rev. D* **67**, 083517 (2003) [*arXiv:hep-th/0207119*].

[23] D. Kutasov, "D-Brane Dynamics Near NS5-Branes", ' [*arXiv:hep-th/0405058*].

[24] J. Polchinski, "String Theory. Vol. 1: An Introduction To The Bosonic String," Cambridge University Press, 1998.

[25] A. A. Tseytlin, "Born-Infeld action, supersymmetry and string theory," *arXiv:hep-th/9908105*.

Chapter 3

THE SEIBERG-WITTEN MAP IN NONCOMMUTATIVE FIELD THEORY: AN ALTERNATIVE INTERPRETATION

Subir Ghosh
Physics and Applied Mathematics Unit,
Indian Statistical Institute,
203 B. T. Road, Calcutta 700108, India.

Abstract

In this article, an alternative interpretation of the Seiberg-Witten map in non-commutative field theory is provided. We show that the Seiberg-Witten map can be induced in a geometric way, by a field dependent co-ordinate transformation that connects noncommutative and ordinary space-times. Furthermore, in continuation of our earlier works, it has been demonstrated here that the above (field dependent co-ordinate) transformation can occur naturally in the Batalin-Tyutin extended space version of the relativistic spinning particle model, (in a particular gauge). We emphasize that the space-time non-commutativity emerges naturally from the particle spin degrees of freedom. Contrary to similarly motivated works, the non-commutativity is not imposed here in an *ad-hoc* manner.

PACS 02.40.Gh; 11.10.Ef; 11.90.+t.

Keywords: Seiberg-Witten map, non-commutative space-time, spinning particle.

Historically the Non-Commutativite (NC) spacetime was introduced by Snyder [1] as a regularization to tame the short distance singularities, inherent in a Quantum Field Theory (QFT). In some sense, this extra structure might appear as a natural generalization of the phase space non-commutativity in quantum mechanics. The softening of the divergence is obviously because NC in spacetime can introduce a lower bound in the continuity of space-time, just as $\hbar$ does in the quantum phase space. The advantage of NC as a regularization is that the computational scheme requires very little changes from the ordinary spacetime and in some forms of NC [1], (for more recent works see [2–4]), manifest Lorentz invariance can be maintained. [1]All the same, the idea of Snyder [1] lost popularity due to the success of renormalization techniques in QFT. Also, now we know [6] that noncommutativity is not of much use as a regulating scheme.

However, in more recent times, NCQFT has matured as an area of intense research activity [6]. It has been established by Seiberg and Witten [7] that the existence of non-commutativity in (open) string boundaries in the presence of a constant two-form Neveu-Schwarz field results in NC D-branes to which the open string endpoints are attached. (Hamiltonian formulation of the above open string boundary noncommutativity is discussed in [8].) This renders QFTs living on the D-brane noncommutative. The authors in [7] also propose an explicit mapping between NC and ordinary space-time dynamical variables. This celebrated map goes by the name of the Seiberg-Witten Map (SWM) [7]. Actually the SWM is the culmination of a much deeper understanding between the connection of QFTs in NC and ordinary space-time. It has been shown in [7] that the appearance of NCQFT is dependent on the choice of regularization and in fact a QFT in ordinary spacetime and an NCQFT both can describe the same underlying QFT. At least to the lowest non-trivial order in $\theta_{\mu\nu}$ [2], the non-commutativity parameter,

$$[x_\mu, x_\nu] = i\theta_{\mu\nu}, \qquad (1)$$

the SWM can be exploited to convert a NCQFT to its counterpart living in ordinary space-time, in which the effects of non-commutativity appear as local interaction terms, supplemented by $\theta_{\mu\nu}$. In the more popular form of NCQFT, $\theta_{\mu\nu}$ is taken to be constant. This can lead to very striking signatures in particle physics phenomenology in the form of Lorentz symmetry breakdown, new interaction vertices etc. [10]

The SWM plays a pivotal role in our understanding of the NCQFT by directly making contact between NCQFT and QFT in ordinary space-time. This has led to new results in the form of axial anomaly [11] in an NC interacting theory of fermions coupled to gauge fields. The previous results [12] are incomplete as they do not conform to the SWM. Furthermore, the SWM is crucial in generalizing [13] to NC space-time, the duality between Maxwell-Chern-Simons and Self-Dual theories in 2+1-dimensions [14]. In the context of bosonization of the massive Thirring model in 2+1-dimensions [15], SWM reveals that the resulting theory is different from the NC self-dual model [16], in contrast to ordinary space-time.

However, as it stands, the SWM is linked exclusively to NC *gauge* theory, since the original derivation of the SWM [7] hinges on the concept of identifying gauge orbits in NC and ordinary space-times. In the explicit form of the SWM [7], the non-commutativity of the *space-time* in which the NC gauge field lives, is not manifest at all since the map is a relation between the NC and ordinary gauge fields and gauge transformation parameters, all having ordinary space-time coordinates as their arguments.

On the other hand, possibly it would have been more natural to consider first a map between NC and ordinary space-time and subsequently to induce the SWM through the change in the space-time argument of the gauge field from the ordinary to NC one. Precisely this type of a geometrical reformulation of the SWM is presented in this paper.

---

[1]The ideas of [3] is discussed in a field theoretic setting in [5].
[2]The SWM for higher orders in θ are discussed in [9].

In the canonical quantization prescription, the Poisson Bracket algebra is elevated to quantum commutator algebra by the replacement

$$\{A, B\} \to \frac{1}{i}[\hat{A}, \hat{B}].$$

But presence of *constraints* may demand a modification in the Poisson Bracket algebra, leading to the Dirac Bracket algebra [17], which are subsequently identified to the commutators,

$$\{A, B\}_{DB} \to \frac{1}{i}[\hat{A}, \hat{B}].$$

However, complications in this formalism can arise, (in particular in case of non-linear constraints), where the Dirac Bracket algebra itself becomes operator valued. To overcome this, Batalin and Tyutin [18] have developed a systematic scheme in which all the physical variables are mapped in an extended canonical phase space, consisting of auxiliary degrees of freedom besides the physical ones, with all of them enjoying canonical free Poisson Bracket algebra. In this formalism, the ambiguity of using (operator valued) Dirac Brackets as quantum commutators does not arise.

In the spinning particle model [19] the canonical $\{x_\mu, x_\nu\} = 0$ Poisson Bracket changes to an operator valued Dirac Bracket,

$$\{x_\mu, x_\nu\}_{DB} = -\frac{S_{\mu\nu}}{M^2}, \tag{2}$$

due to the presence of constraints. In the above, the dynamical variable $S_{\mu\nu}$ represents the spin angular momentum and M is the mass of the particle. This forces us to exploit the Batalin-Tyutin prescription [18].

In a recent paper [3], we have constructed a mapping of the form,

$$\{x_\mu, x_\nu\} = 0 \; , \; x_\mu \to \hat{x}_\mu \; ; \; \{\hat{x}_\mu, \hat{x}_\nu\} = \theta_{\mu\nu}, \tag{3}$$

which bridges the noncommutative and ordinary space-times. Note that $\hat{x}_\mu$ lives in the Batalin-Tyutin [18] extended space and is of the generic form $\hat{x}_\mu = x_\mu + X_\mu$, where X_μ consists of physical and auxiliary degrees of freedom. Explicit expressions for X_μ are to be found later [3].

This space-time map induces in a natural way the following map between noncommutative and ordinary degrees of freedom,

$$\lambda(x) \to \lambda(\hat{x}) \to \hat{\lambda}(x) \; , \; A_\mu(x) \to A_\mu(\hat{x}) \to \hat{A}_\mu(x). \tag{4}$$

Here $\hat{\lambda}$ and $\hat{A}_\mu$ are the NC counterparts of λ and A_μ, the abelian gauge transformation parameter and the gauge field respectively and $\hat{x}_\mu$ and x_μ are the NC and ordinary space-time co-ordinates.

On the other hand, there also exists the SWM [7] which interpolates between noncommutative and ordinary variables,

$$\lambda(x) \to \hat{\lambda}(x) \; , \; A_\mu(x) \to \hat{A}_\mu(x). \tag{5}$$

It is only logical that the above two schemes ((3-4) and 5) can be related. In the present work we have precisely done that. The formulation [3] (3-4) being the more general one, we have explicitly demonstrated how it can be reduced to the SWM [7], in a particular gauge. This incidentally demonstrates the correctness of the procedure. The above idea was hinted in [3]. [3]

In this context, let us put the present work in its proper perspective. Recently a number of works have appeared with the motivation of recovering the SWM in a geometric way, without invoking the gauge theory principles [20]. However, the noncommutative feature of the space-time plays no direct role in the above mentioned re-derivations of the SWM, with non-commutativity just being postulated in an *ad hoc* way. In the present work, we have shown how one can construct a noncommutative sector inside an extended phase space, in a relativistically covariant way. More importantly, we have shown explicitly how this generalized map can be reduced to the SWM under certain approximations. Interestingly, this extended space is physically significant and well studied: It is the space of the relativistic spinning particle [3,19]. Hence it might be intuitively appealing to think that the NC space-time is endowed with spin degrees of freedom, as compared to the ordinary configuration space, since the spin variables directly generate the NC [5]. The analogue of the gauge field is also identified inside this phase space, without any need to consider external fields. This situation is to be contrasted with the NC arising from the background magnetic field in the well known Landau problem [6] of a charge moving in a plane in the presence of a strong, perpendicular magnetic field.

The genesis of the SWM is the observation [7] that the non-commutativity in string theory depends on the choice of the regularization scheme: it appears in *e.g.* point-spitting regularization whereas it does not show up in Pauli Villars regularization. This feature, among other things, has prompted Seiberg and Witten [7] to suggest the map connecting the NC gauge fields and gauge transformation parameter to the ordinary gauge field and gauge transformation parameter. The explicit form of the SWM [7], for abelian gauge group, to the first non-trivial order in the NC parameter $\theta_{\mu\nu}$ is the following,

$$\hat{\lambda}(x) = \lambda(x) + \frac{1}{2}\theta^{\mu\nu}A_\nu(x)\partial_\mu\lambda(x) \,,$$

$$\hat{A}_\mu(x) = A_\mu(x) + \frac{1}{2}\theta^{\sigma\nu}A_\nu(x)F_{\sigma\mu}(x) + \frac{1}{2}\theta^{\sigma\nu}A_\nu(x)\partial_\sigma A_\mu(x). \tag{6}$$

The above relation (6) is an $O(\theta)$ solution of the general map [7],

$$\hat{A}_\mu(A + \delta_\lambda A) = \hat{A}_\mu(A) + \hat{\delta}_{\hat{\lambda}}\hat{A}_\mu(A), \tag{7}$$

which is based on identifying gauge orbits in NC and ordinary space-time.

First let us show that it is indeed possible to re-derive the SWM using geometric objects. We rewrite the SWM (6) in the following way,

$$\hat{\lambda}(x) = \lambda(x) + \frac{1}{2}\{\delta_f[\lambda(x)] - (\lambda(x') - \lambda(x))\} = \lambda(x) + \delta_f[\lambda(x)] \,, \tag{8}$$

$$\hat{A}_\mu(x) = A_\mu(x) + \{\delta_f[A_\mu(x)] - (A_\mu(x') - A_\mu(x))\} = A_\mu(x) + A'_\mu(x) - A_\mu(x'). \tag{9}$$

---
[3]The present analysis being classical, (non)commutativity is to be interpreted in the sense of Poisson or Dirac Brackets.

In the above we have defined,

$$x'_\mu = x_\mu - f_\mu \ , \ A'_\mu(x') = \frac{\partial x^\nu}{\partial x'^\mu} A_\nu(x) \ , \ \lambda'(x') = \lambda(x) \ ,$$

$$f^\mu \equiv \frac{1}{2}\theta^{\mu\nu} A_\nu \ . \tag{10}$$

Here f^μ is the field dependent space-time translation parameter and δ_f constitutes the Lie derivative connected to f^μ,

$$\delta_f[\lambda(x)] = \lambda'(x) - \lambda(x) = -(\lambda(x') - \lambda(x)) = f^i \partial_i \lambda,$$

$$\delta_f[A_\mu(x)] = A'_\mu(x) - A_\mu(x) \ . \tag{11}$$

This shows that the NC gauge parameter ($\hat{\lambda}$) and gauge field ($\hat{A}^\mu$) are derivable from the ordinary one by making a *field dependent* space-time translation f^μ [21]. One can check that the NC gauge transformation of $\hat{A}_\mu(x)$ is correctly reproduced by considering,

$$\hat{\delta}\hat{A}_\mu(x) = \delta(A_\mu(x) + \frac{1}{2}\theta^{\sigma\nu} A_\nu(x) F_{\sigma\mu}(x) + \frac{1}{2}\theta^{\sigma\nu} A_\nu(x) \partial_\sigma A_\mu(x)) \ , \tag{12}$$

where $\delta A_\mu(x) = \partial_\mu \lambda(x)$ is the gauge transformation in ordinary space-time. Hence, if expressed in the form (9), the SWM, (at least to $O(\theta)$), can be derived in a geometrical way, without introducing the original identification (7) obtained from the viewpoint of a matching between NC and ordinary gauge invariant sectors. Also note that the gauge field $A_\mu(x)$ is treated here just as an ordinary vector field, without invoking any gauge theory properties. This constitutes the first part of our result.

Returning to our starting premises, are we justified in making an identification between $\hat{x}_\mu$ in (3)-(4) and x'_μ introduced in (8)-(10), because this relation can connect NC and ordinary space-time. Naively, a relation of the form, $x'_\mu = x_\mu - f_\mu(x)$ can not render the x'-space noncommutative, since the right hand side of the equation apparently comprises of commuting objects only. In our subsequent discussion we will show how this surmise can be made meaningful and will return to this point at the end.

We start by considering a larger space having inherent NC. Such a space, which at the same time is physically appealing, is that of the Nambu-Goto model of relativistic spinning particle [3, 19]. Here the situation is somewhat akin to the open string boundary NC such that the role of Neveu-Schwarz field is played by here by the spin degrees of freedom. The Lagrangian of the model in 2+1-dimensions [3, 19] is,

$$L = [M^2 u^\mu u_\mu + \frac{J^2}{2}\sigma^{\mu\nu}\sigma_{\mu\nu} + MJ\epsilon^{\mu\nu\lambda}u_\mu\sigma_{\nu\lambda}]^{\frac{1}{2}} \ , \tag{13}$$

$$u^\mu = \frac{dx^\mu}{d\tau} \ , \ \sigma^{\mu\nu} = \Lambda^\mu_{\ \rho}\frac{d\Lambda^{\rho\nu}}{d\tau} = -\sigma^{\nu\mu} \ , \ \Lambda^\mu_{\ \rho}\Lambda^{\rho\nu} = \Lambda^\mu_{\ \rho}\Lambda^{\nu\rho} = g^{\mu\nu} \ , \ g^{00} = -g^{ii} = 1. \tag{14}$$

Here (x^μ, $\Lambda^{\mu\nu}$) is a Poincare group element and also a set of dynamical variables of the theory.

In a Hamiltonian formulation, the conjugate momenta are,

$$P^\mu = \frac{\partial L}{\partial u_\mu} = L^{-1}[M^2 u^\mu + \frac{MJ}{2}\epsilon^{\mu\nu\lambda}\sigma_{\nu\lambda}] \ , \ S^{\mu\nu} = \frac{\partial L}{\partial \sigma_{\mu\nu}} = \frac{L^{-1}}{2}[J^2\sigma^{\mu\nu} + MJ\epsilon^{\mu\nu\lambda}u_\lambda]. \tag{15}$$

The Poisson algebra of the above phase space degrees of freedom are,

$$\{P^\mu, x^\nu\} = g^{\mu\nu} \ , \ \{P^\mu, P^\nu\} = 0 \ , \ \{x^\mu, x^\nu\} = 0 \ , \ \{\Lambda^{0\mu}, \Lambda^{0\nu}\} = 0, \tag{16}$$

$$\{S^{\mu\nu}, S^{\lambda\sigma}\} = S^{\mu\lambda}g^{\nu\sigma} - S^{\mu\sigma}g^{\nu\lambda} + S^{\nu\sigma}g^{\mu\lambda} - S^{\nu\lambda}g^{\mu\sigma} \ , \ \{\Lambda^{0\mu}, S^{\nu\sigma}\} = \Lambda^{0\nu}g^{\mu\sigma} - \Lambda^{0\sigma}g^{\mu\nu}. \tag{17}$$

The full set of constraints are,

$$\Psi_1 \equiv P^\mu P_\mu - M^2 \approx 0 \ , \ \Psi_2 \equiv S^{\mu\nu}S_{\mu\nu} - 2J^2 \approx 0, \tag{18}$$

$$\Theta_1^\mu \equiv S^{\mu\nu}P_\nu \ , \ \Theta_2^\mu \equiv \Lambda^{0\mu} - \frac{P^\mu}{M} \ , \ \mu = 0, 1, 2, \tag{19}$$

out of which Ψ_1 and Ψ_2 give the mass and spin of the particle respectively. [4] In the Dirac constraint analysis [17], these are termed as First Class Constraints (FCC), having the property that they commute with *all* the constraints on the constraint surface and generate gauge transformations. The set Θ_2^μ is put by hand [3], to restrict the number of angular co-ordinates.

The non-commuting set of constraints $\Theta_\alpha^\mu, \alpha = 1, 2$, termed as Second Class Constraints (SCC) [17], modify the Poisson Brackets (16) to Dirac Brackets [17], defined below for any two generic variables A and B,

$$\{A, B\}_{DB} = \{A, B\} - \{A, \Theta_\alpha^\mu\}\Delta_{\mu\nu}^{\alpha\beta}\{\Theta_\beta^\nu, B\}, \tag{20}$$

$$\{\Theta_\alpha^\mu, \Theta_\beta^\nu\} \equiv \Delta_{\alpha\beta}^{\mu\nu} \ , \ \alpha, \beta = 1, 2 \ , \ \Delta_{\alpha\beta}^{\mu\nu}\Delta_{\nu\lambda}^{\beta\gamma} = \delta_\alpha^\gamma \delta_\lambda^\mu. \tag{21}$$

$\Delta_{\alpha\beta}^{\mu\nu}$ is non-vanishing even on the constraint surface. The main result, relevant to us, is the following Dirac Bracket [3, 19],

$$\{x_\mu, x_\nu\}_{DB} = -\frac{S_{\mu\nu}}{M^2} \to \{\hat{x}_\mu, \hat{x}_\nu\} = \theta_{\mu\nu}. \tag{22}$$

This is the non-commutativity that occurs naturally in the spinning particle model. Our aim is to express this NC co-ordinate $\hat{x}_\mu$ in the form $\hat{x}_\mu = x_\mu - f_\mu$, with the identification between $\theta_{\mu\nu}$ and $S_{\mu\nu}$. This is indicated in the last equality in (22). In the quantum theory, this will lead to the NC space-time (3).

This motivates us to the Batalin-Tyutin quantization [18] of the spinning particle [3]. For a system of irreducible SCCs, in this formalism [18], the phase space is extended by introducing additional BT variables, ϕ_a^α, obeying

$$\{\phi_\mu^\alpha, \phi_\nu^\beta\} = \omega_{\mu\nu}^{\alpha\beta} = -\omega_{\nu\mu}^{\beta\alpha} \ , \ \omega_{\mu\nu}^{\alpha\beta} = g_{\mu\nu}\epsilon^{\alpha\beta} \ , \ \epsilon^{01} = 1. \tag{23}$$

---
[4] Note that instead of Ψ_2 as above, one can equivalently use $\Psi_2 \equiv \epsilon^{\mu\nu\lambda}S_{\mu\nu}P_\lambda - MJ$, which incidentally defines the Pauli Lubanski scalar.

where the last expression is a simple choice for $\omega_{\mu\nu}^{\alpha\beta}$. The SCCs Θ_α^μ are modified to $\tilde{\Theta}_\alpha^\mu$ such that they become FCC,

$$\{\tilde{\Theta}_\alpha^\mu(q,\phi), \tilde{\Theta}_\beta^\nu(q,\phi)\} = 0 \ ; \ \tilde{\Theta}_\alpha^\mu(q,\phi) = \Theta_\alpha^\mu(q) + \Sigma_{n=1}^\infty \tilde{\Theta}_\alpha^{\mu(n)}(q,\phi) \ ; \ \tilde{\Theta}^{\mu(n)} \approx O(\phi^n), \quad (24)$$

with q denoting the original degrees of freedom. Let us introduce the gauge invariant variables $\tilde{f}(q)$ [18] corresponding to each $f(q)$, so that $\{\tilde{f}(q), \tilde{\Theta}_\alpha^\mu\} = 0$

$$\tilde{f}(q,\phi) \equiv f(\tilde{q}) = f(q) + \Sigma_{n=1}^\infty \tilde{f}(q,\phi)^{(n)}, \quad (25)$$

which further satisfy [18],

$$\{q_1, q_2\}_{DB} = q_3 \ \rightarrow \ \{\tilde{q}_1, \tilde{q}_2\} = \tilde{q}_3 \ , \ \tilde{0} = 0. \quad (26)$$

It is now clear that our target is to obtain $\tilde{x}_\mu$ for x_μ. Explicit expressions for $\tilde{\Theta}^{\mu(n)}$ and $\tilde{f}^{(n)}$ are derived in [18].

Before we plunge into the BT analysis, the reducibility of the SCCs Θ_1^μ (i.e. $P_\mu \Theta_1^\mu = 0$) [3,19] is to removed [22] by introducing a canonical pair of auxiliary variables ϕ and π that satisfy $\{\phi, \pi\} = 1$ and PB commute with the rest of the physical variables. The modified SCCs that appear in the subsequent BT analysis are as shown below:

$$\Theta_1^\mu \equiv S^{\mu\nu} P_\nu + k_1 P^\mu \pi \ ; \ \Theta_2^\mu \equiv (\Lambda^{0\mu} - \frac{P^\mu}{M}) + k_2 (\Lambda^{0\mu} + \frac{P^\mu}{M}) \phi \, , \quad (27)$$

where k_1 and k_2 denote two arbitrary parameters. Since the computations are exhaustively done in [3] they are not repeated here. The results are the following:

$$\tilde{x}_\mu = x_\mu + [S_{\nu\mu} + 2k_1 \pi g_{\nu\mu}](\phi^1)^\nu + \mathcal{R}_{1\mu\nu}(\phi^2)^\nu + higher - \phi - terms \, , \quad (28)$$

$$\{\tilde{x}_\mu, \tilde{x}_\nu\} = -\frac{\tilde{S}_{\mu\nu}}{M^2} \, , \ \tilde{S}_{\mu\nu} = S_{\mu\nu} + \mathcal{R}_{2(\alpha)\mu\nu\lambda} \phi^{(\alpha)\lambda} + higher - \phi - terms \, , \quad (29)$$

where the expressions for $\mathcal{R}$ are straightforward to obtain [3] but are not needed in the present order of analysis. Only it should be remembered that the $\mathcal{R}_1$-term in (28) is responsible for the $(\phi^\alpha)^\mu$-free term $-S_{\mu\nu}/(M^2)$ in the $\{\tilde{x}_\mu, \tilde{x}_\nu\}$ bracket in (29). Thus the problem that we had set out to solve has been addressed successfully in (28), which expresses the NC $\tilde{x}_\mu$ in terms of ordinary x_μ and other variables [3].

Now comes the crucial part of identification of the present map with the SWM [7]. This means in particular that we have to connect (28) to (10), since as we have shown before, (10) is capable of generating the SWM [7]. We exploit the freedom of choosing gauges according to our convenience, since in the BT extended space $\tilde{\Theta}_\alpha^\mu$ are FCCs. For instance, the so called unitary gauge, $\phi_1^\mu = 0, \phi_2^\mu = 0$, trivially converts the system back to its original form before the BT extension. Let us choose the following non-trivial gauge,

$$\phi_1^\mu = \frac{M^2}{2} A^\mu(x) \, , \ \phi_2^\mu = 0, \quad (30)$$

where $A^\mu(x)$ is some function of x_μ, to be identified with the gauge field. Let us also work with terms linear in $A^\mu(x)$. Identifying $\tilde{S}_{\mu\nu}/(M)^2 = \theta_{\mu\nu}$ we end up with the cherished mapping,

$$\tilde{x}_\mu = x_\mu - \frac{1}{2} \theta_{\mu\nu} A^\nu(x) + higher - A(x) - terms \, , \quad (31)$$

$$\{\tilde{x}_\mu, \tilde{x}_\nu\}_{DB} = \theta_{\mu\nu} + higher - A(x) - terms\ . \tag{32}$$

Note that in the above relations (31,32), we have dropped the terms containing k_1, an arbitrary parameter [3], considering it to be very small. Also in (32) Dirac Bracket reappears since the system is gauge fixed and hence has SCCs. This constitutes the second part of our result.

Finally, two points are to be noted. Firstly, the non-commutativity present here does *not* break Lorentz invariance since there appear no constant parameter with non-trivial Lorentz index to start with. The violation will appear only in the identification of $\tilde{S}_{\mu\nu}$ with (constant) $\theta_{\mu\nu}$. Secondly, (28) truly expresses the NC space-time $\tilde{x}_\mu$ in terms of ordinary space-time x_μ. But x_μ becomes NC owing to the Dirac brackets induced by the particular gauge that we fixed in order to reduce our results to the SWM. Obviously, in general, there is no need to fix this particular gauge. This refers to the comment below (12).

In conclusion, let us summarize our work. We have shown that the (abelian $O(\theta)$) Seiberg-Witten map can be viewed as a co-ordinate transformation, albeit with field dependent parameters. The duality concept between gauge orbits in noncommutative and ordinary space-times, which was crucial in the original derivation [7], is not applied here. It has been explicitly demonstrated that a noncommutative space-time sector can be constructed in the Batalin-Tyutin extension of the relativistic spinning particle model [3]. Finally, the above mentioned transformation and subsequently a direct connection with the Seiberg-Witten map is also generated in this model. The present work reveals that noncommutative space-time is endowed with spin degrees of freedom, as compared to the ordinary space-time [5].

Acknowledgement

It is a pleasure to thank Professor R. Jackiw for helpful correspondence.

References

[1] H.S.Snyder, *Phys.Rev.* **71** 38(1947).

[2] S.Doplicher, K.Fredengagen and J.E.Roberts, *Phys.Lett.* B331 39(1994); C.E.Carlson, C.D.Carrone and N.Zobin, *Non-commutative gauge theory without Lorentz violation*, *Phys.Rev.* **D66** (2002) 075001 (*HEP-TH/0206035*).

[3] S.Ghosh, *Phys.Rev.* D66 045031(2002).

[4] H.Kase, K.Morita, Y.Okumura and E.Umezawa, *Lorentz-Invariant Non-Commutative Space-Time Based On DFR Algebra, HEP-TH/0212127*.

[5] S.Ghosh, *Modelling a noncommutative two-brane, hep-th/0212123*.

[6] For a review see M.R.Douglas and N.A.Nekrasov, *Rev.Mod.Phys.* **73** 977(2002) (arXiv: *HEP-TH/ 0106048*); R.J.Szabo, *HEP-TH/0109162*; A.Connes, *Noncommutative Geometry*, Academic Press, 1994.

[7] N.Seiberg and E.Witten, *JHEP* **09**(1999)032.

[8] R.Banerjee, B.Chakraborty and S.Ghosh, *Phys.Lett.* **B537** 340(2002).

[9] Y.Okawa and H.OOguri, *Phys.Rev.* **D64** 046009(2001); S.Fidanza, *JHEP* **0206** 016(2002).

[10] X.Calmet et. al., *HEP-PH/0111115*; H. Bozkaya, P. Fischer, H. Grosse, M. Pitschmann, V. Putz, M. Schweda, R. Wulkenhaar, *Space/time noncommutative field theories and causality*, HEP-TH/0209253; M.Chaichian, K.Nishijima and A.Tureanu, *Spin-Statistics and CPT Theorems in Noncommutative Field Theory*, HEP-TH/0209008; T.Tamaki, T.Harada, U.Miyamoto and T.Torii, *Particle velocity in noncommutative space-time*, *Phys.Rev.* **D66** (2002) 105003 (gr-qc/0208002); J.L.Cortes, J.Gamboa, F.Mendez, *Noncommutativity in Field Space and Lorentz Invariance Violation*, J.M.Carmona, *HEP-TH/0207158*.

[11] R.Banerjee and S.Ghosh, *Phys.Lett.* **B533** 162(2002).

[12] J.M.Gracia-Bondia and C.P.Martin, *Phys.Lett.* **B479** 321(2000); L.Bonora, M.Schnabl and A.Tomasiello, *Phys.Lett.* **B485** 311(2000); F.Ardalan and N.Sadooghi, *Int.J.Mod.Phys.* **A16** 3157(2001).

[13] S.Ghosh, *Gauge invariance and duality in the noncommutative plane*, hep-th/0210107, (to appear in *Phys.Lett.* B); M.B.Cantcheff and P.Minces, *Duality between noncommutative Yang-Mills-Chern-Simons and non-abelian self-dual models*, hep-th/0212031, (to appear in *Phys.Lett.* B); O.F.Dayi, *Noncommutative Maxwell-Chern-Simons theory, duality and a new noncommutative Chern-Simons theory in d=3*, hep-th/0302074.

[14] S.Deser, R.Jackiw and S.Templeton, *Phys.Rev.Lett.* **48**(1982)975; *Ann.Phys.* **140** (1982)372; S.Deser and R.Jackiw, *Phys.Lett.* **139B**(1984)371.

[15] E.Fradkin and F.A.Schaposnik, *Phys.Lett.* **338B**(1994)253; R.Banerjee, *Phys.Lett.* **358B** (1995)297.

[16] S.Ghosh, *Bosonization in the noncommutative plane*, hep-th/0303022.

[17] P.A.M.Dirac, *Lectures on Quantum Mechanics* (Yeshiva University Press, New York, 1964).

[18] I.A.Batalin and I.V.Tyutin, *Int.J.Mod.Phys.* **A6** 3255(1991).

[19] S.Ghosh, *Phys.Lett.* **B338** 235(1994); *J.Math.Phys.* **42** 5202(2001). For the original works see, A.J.Hanson and T.Regge, *Ann.Phys. (N.Y.)* **87** 498(1974); see also A.J.Hanson, T.Regge and C.Teitelboim, *Constrained Hamiltonian System*, Roma, Accademia Nazionale Dei Lincei, (1976).

[20] L.Cornalba, *D-brane physics and noncommutative Yang-Mills theory*, HEP-TH/9909081; A.A.Bichl et. al., *Noncommutative Lorentz symmetry and the origin of the Seiberg-Witten map*, HEP-TH/0108045; B.Jurco and P.Schupp, *Euro.Phys.J.* **C14** 367(2000); B.Jurco and P.Schupp and J.Wess, *Nucl.Phys.* **B584** 784(2001);

H.Liu, *Nucl.Phys.* **B614** 305(2001); R.Jackiw, S.-Y.Pi and A.P.Polychronakos, *Noncommutating gauge fields as a Lagrange fluid*, *Phys.Rev.Lett.* **88** 111603(2002) (*HEP-TH/0206014*).

[21] Field dependent transformations have appeared in *eg* R.Jackiw, *Phys.Rev.Lett.* **41**, 1635 (1978); R. Jackiw and S.-Y. Pi, *HEP-TH/0111122*.

[22] R.Banerjee and J.Barcelos-Neto, *Ann.Phys.* **265** 134(1998).

Chapter 4

HIGHER DIMENSIONAL MODEL SPACES AND DEFECTS

*Dominic G.B. Edelen**
3503 Avenue P, Galveston TX 77550

Abstract

Higher dimensional model spaces are used to obtain general solutions of the kinematic equations of defects; that is, for dislocations and disclinations that result from local action of the Euclidean group SO(3) ◁ T(3) as a gauge group. Appropriate choices of the Lie algebra valued generating matrices are used to obtain distortion 1-forms that satisfy the kinematic equations of defects and the equations of equilibrium of linear elasticity for both dislocations (local action of T(3)) and disclinations (local action of the full Euclidean group). These solutions lead to an improved understanding of the classical Volterra solutions, and to new classes of solutions that are self screening. It is shown that the undetermined functions in these solutions for screw dislocation problems can be determined by use of the field equations of the full gauge theory of defects, and explicit solutions are obtained.

1. Introduction

The discussion is confined to static problems in the interests of simplicity. The interested reader can extend our results to dynamic problems by use of the methods given in [1, 2]. Presentation of the theoretical foundations will be *via* the exterior calculus (see [3]), while explicit problems will use the more customary formulation.

Let $\mu = dX \wedge dY \wedge dZ$ be the volume element of the reference manifold, E_3, with the standard Cartesian coordinate cover (X, Y, Z). The corresponding oriented boundary elements are

$$\mu_A = \partial_A \,\rfloor\, \mu, \quad \mu_1 = dY \wedge dZ, \quad \mu_2 = -dX \wedge dZ, \quad \mu_3 = dX \wedge dY. \tag{1}$$

Linear elasticity is predicated on the exact distortion 1-forms

$$\beta^i = \beta^i_A dX^A = d(X^A \delta^i_A + u^i(X^A)) = (\delta^i_A + \partial_A u^i) dX^A. \tag{2}$$

---

*E-mail address: domod@earthlink.net

that necessarily satisfy the integrability conditions

$$d\beta^i = 0. \tag{3}$$

Set $\beta = ((\beta_A^i))$, so that the linear elastic strain, ϵ, is given by

$$\epsilon = \frac{1}{2}(\beta + \beta^T) - E, \tag{4}$$

where E is the identity matrix. Linear elasticity is also predicated upon the assumption that the stress is a linear tensor-valued function of the strain. For an isotropic material, the corresponding linear elastic stress, $\sigma = ((\sigma_i^A))$, is given by

$$\sigma = \lambda Tr(\epsilon)E + 2\hat{\mu}\epsilon, \tag{5}$$

This stress can also be derived from the elastic energy function

$$\mathcal{E}_e = \frac{1}{2}Tr(\sigma\epsilon) = \frac{1}{2}\lambda Tr(\epsilon)^2 + \hat{\mu}Tr(\epsilon^2), \tag{6}$$

If we define the traction 2-forms by $\mathcal{T}_i = \sigma_i^A \mu_A$, then the equations of equilibrium take the form

$$d\mathcal{T}_i = 0; \tag{7}$$

that is, $\partial_A \sigma_i^A = 0$, because $dX^B \wedge \mu_A = \delta_A^B \mu$.

When defects (dislocations and disclinations) are present, as consequences of minimal replacement engendered by local action of the translation group T(3) and the rotation group SO(3) (see [2]), the integrability conditions (1.3) take the form

$$d\beta^i + \Gamma_j^i \wedge \beta^j = \alpha^i = \alpha^{iA}\mu_A, \quad d\Gamma_j^i + \Gamma_k^i \wedge \Gamma_j^k = R_j^i, \tag{8}$$

where α^i are the *dislocation density 2-forms* and R_j^i are the *disclination density 2-forms*. The first of (1.8) thus shows that $\alpha^i = \Gamma_j^i \wedge \beta^j$ are the *obstructions* to satisfaction the kinematic equations, $d\beta^i = 0$, of elastic bodies. The system (1.8) entails the integrability conditions

$$d\alpha^i + \Gamma_k^i \wedge \alpha^k = R_k^i \wedge \beta^k, \quad dR_j^i + \Gamma_k^i \wedge R_j^k = R_k^i \wedge \Gamma_j^k. \tag{9}$$

Under these circumstances, the equations of equilibrium go over into

$$d\mathcal{T}_i - \Gamma_i^k \wedge \mathcal{T}_k = 0, \quad \Rightarrow \quad \partial_A \sigma_i^A - \Gamma_{Ai}^k \sigma_k^A = 0, \tag{10}$$

where $\Gamma_i^k = \Gamma_{Ai}^k dX^A$.

These equations, in their full generality, were first obtained by Kossecka & de Wit [4] on the basis of careful physical reasoning. Kossecka & de Wit and subsequent authors took $R_k^i \wedge \beta^k - \Gamma_k^i \alpha^k = d\alpha^i$ to be the disclination density 3-forms of the material. This is wrong because when $R_k^i = 0$ there would still appear to be disclination density 3-forms $-\Gamma_k^i \wedge \alpha^k$. This is the basis for numerous difficulties and confusions in the literature, with the claim [5] that the translation group, T(3), can generate disclinations. If we introduce matrix notation, then $R = d\Gamma + \Gamma \wedge \Gamma = 0$ are integrability conditions for the existence of an SO(3)-valued matrix L (recall that both R and F are so (3) valued functions) such $dL + \Gamma L = 0$.

These latter equations are just the group equations for SO(3)-valued matrix functions. The matrix L thus satisfies (see [3], Chapter 5) the system of Riemann-Graves integral equations $L = E - H[\Gamma L]$. We then have $\Gamma = -(dL)L^{-1}$, and the gauge transformation $\beta = L\hat{\beta}$ results in $\hat{\Gamma} = 0$; simply note that $\Gamma = L\hat{\Gamma}L^{-1} - (dL)L^{-1} = -(dL)L^{-1}$ implies $\hat{\Gamma} = 0$. Such "disclinations" can thus be transformed away by completely integrable SO(3) gauge transformations! Further, the states characterized by β and $\hat{\beta}$ are *kinetically equivalent* because they give the same evaluation of the right Cauchy-Green strain tensor (i.e., $C_{AB} = \beta^i_A \delta_{ij} \beta^j_B = \hat{\beta}^i_A \delta_{ij} \hat{\beta}^j_B$ because L belongs to SO(3)). It is for this reason that we now take R^i_j as the disclination density 2-forms (i.e., $R^j_i \neq 0$ guarantee that the disclinations can not be removed by a completely integrable SO(3) gauge transformation).

2. Higher Dimensional Model Spaces

Classic differential geometry tells us that, no matter how complicated a differentiable manifold may be, there is always a higher dimensional flat space in which it can be immersed. An explicit calculation algorithm for this immersion problem has been given in Appendix B of [6]. The structure of these results is such that it provides direct access to problems of material manifolds with defects. Let us therefore consider a flat -dimensional manifold En with the 8I??R?8standard Cartesian coordinate cover $N = \{N^1, N^2, \ldots, N^n\}^T$ with $n > 3$. The space E_n has the natural basis dN for 1-forms. Any other basis takes the form

$$\beta = A dN, \tag{11}$$

where A is a GL(n)-valued function on E_n. The Cartan equations of structure,

$$d\beta + \Gamma \wedge \beta = 0, \quad d\Gamma + \Gamma \wedge \Gamma = \Theta = 0, \quad \Gamma = -(dA)A^{-1}. \tag{12}$$

are then satisfied *identically*, as the reader can easily check. The first of these says that E_n is torsion-free, while the second says that E_n is curvature-free. Further, and of particular importance, if we introduce the gauge variables y on E_n by $y = AN$, the system (2.2) has the first integrals

$$\beta = dy + \Gamma y. \tag{13}$$

In general, we will require A to be an SO(n)-valued function on E_n, or an element of GL(n) that contains SO(n-1) and leaves the nth-direction in E_n invariant (i.e., it accommodates both rotations and translations). Calculation of such matrices is usually accomplished by exponentiation of the corresponding $\mathfrak{a}(n)$-valued functions, which is nearly an impossible task if explicit evaluations are called for. A generalization (see the Appendix of [7]) of an old representation theorem of Cayley for SO(3), that was first brought to my attention by Professor Junkins [8], gives the following *algebraic* evaluations. Let B be $\mathfrak{a}(n)$-valued function on E_n with none of its eigenvalues equal to 1. Then A has the algebraic evaluation $A = (E - B)(E + B)^{-1}$, with the reciprocal evaluation $B = (E - A)(E + A)^{-1}$, and the matrix of connection 1-forms has the evaluation

$$\Gamma = -(dA)A^{-1} = 2(E + B)^{-1} dB (E - B)^{-1}. \tag{14}$$

The purpose of all of this can now be made clear. If we use the index sets $(i, j, k) \subset (1, 2, 3)$, $(a, b, c) \subset (4, 5, \ldots, n)$, then the first of (2.2) gives

$$d\beta^i + \Gamma^i_j \wedge \beta^j = -\Gamma^i_a \wedge \beta^a = -\Gamma^i_4 \wedge \beta^4 - \Gamma^i_5 \wedge \beta^5 - \ldots - \Gamma^i_n \wedge \beta^n = \alpha^i. \tag{15}$$

and the subsidiary relations.

$$d\beta^a + \Gamma^a_i \wedge \beta^i + \Gamma^a_b \wedge \beta^b = 0. \tag{16}$$

The first of (1.8) thus shows that there is a mode of dislocation density that is associated with each of the additional dimension of the model space! Further, $\Theta^i_j = d\Gamma^i_j + \Gamma^i_k \wedge \Gamma^k_j + \Gamma^i_a \wedge \Gamma^a_j = 0$ and the second of (1.8) gives the *algebraic evaluation*

$$R^i_j = -\Gamma^i_a \wedge \Gamma^a_j = -\Gamma^i_4 \wedge \Gamma^4_j - \Gamma^i_5 \wedge \Gamma^5_j - \ldots - \Gamma^i_n \wedge \Gamma^n_j. \tag{17}$$

There is therefore a mode of disclination density that is likewise associated with each additional dimension of the model space!

The evaluations (2.5) and (2.6) show that there are no defects if $\{\Gamma^i_a = 0 | 1 \leq i \leq 3 < a \leq n\}$. The evaluation (2.4) thus shows that the generating matrix B must be such that at least one of the quantities $\{B^i_a | 1 \leq i \leq 3 < a \leq n\}$ must be nonzero on a region of the reference manifold in order for defects to be present. This means that defects are put into a problem by specifying the quantities $\{B^i_a | 1 \leq i \leq 3 < a \leq n\}$.

When the general solution of these kinematic equations of defects, that is given by (2.3), is used, we obtain the essential results of this paper:

$$\beta^i = dy^i + \Gamma^i_k y^k + \Gamma^i_a y^a, \quad \beta^a = dy^a + \Gamma^a_i y^i + \Gamma^a_b y^b. \tag{18}$$

When A is a constant matrix, which can be Emapped onto the identity matrix by a trivial gauge transformation, (2.7) reduces to $\beta^i = dy^i$ Since $B = 0$ in this case, elasticity theory gives the identification

$$y^i = X^A \delta^i_A + u^i. \tag{19}$$

Accordingly, (2.7) gives the evaluation

$$\beta^i = d(X^A \delta^i_A + u^i) + \Gamma^i_k (X^A \delta^k_A + u^k) + \Gamma^i_a u^a. \tag{20}$$

In fact, the structure of the β^i given by (2.9) is exactly the same as that obtained in [1,2], by use of the minimal replacement construct of gauge theory, if we use the identifications $\chi^i = X^A \delta^i_A + u^i$, $\phi^i = \Gamma^i_a y^a$. The free variables of a defect problem are therefore $\{u^i, y^a\}$, for any determination of the defect generating quantities Γ that result from a nontrivial map from the reference configuration into the Lie algebra valued matrix B (see (2.4)). Put another way, a choice of the quantities $\{B, u^i, y^a\}$ serves to determine the quantities $\{\beta^i, \alpha^i, \Gamma^i_j, R^i_j\}$ such that all of the kinematic equations of defects are satisfied. In this regard, the entries of B determine the nature and the placement of the defects, the quantities $\{y^4, y^5, \ldots, y^n\}$ determine the amplitudes of the defects, while $\{u^1, u^2, u^3\}$ determine the elastic displacements so that the equations of equilibrium are satisfied.

There are certain observations that can be made at this point. First, the upper left-hand 3×3 submatrix of B can be set to zero. This simply suppresses those "disclinations" that can be transformed away by integrable SO(3)-gauge transformations. Next, (2.4) and (2.5) show that we can have disclination free dislocation problems only when $\Gamma_i^a = 0$, which is the case only when the lower left-hand $(n-3) \times 3$ submatrix of B is set to zero. This means that B only generates translations, so it is sufficient to restrict to E_4 in this case! On the other hand, dislocation free disclinations can occur only when $\beta^a = 0$. These latter conditions, with B necessarily restricted to (n), are extremely restrictive; their integrability conditions are satisfied only when $y^i = 0$, $y^a = k^a$, which would appear to eliminate such a possibility for defect theory. Put another way, any disclination of a material will require supporting dislocations. The restrictions $y^i = 0$, $y^a = k^a$ are precisely those for immersions of *curved* 3-dimensional spaces in flat spaces of higher dimension; that is, for the classic immersion problem. We know, however, that the current configuration of a material body occurs in a flat 3-dimensional space of current configurations, so there is no curved 3-dimensional space that requires immersion.

Those that are interested in the geometry of defects can proceed as follows. Once a problem has been solved, we will have the quantities $y(X^A)$ and $A(X^A)$ gauge transformation $y = AN$ on E_n thus gives the graph of the defect solution as the 3-dimensional surface

$$N(X^A) = A(X^A)^{-1} y(X^A), \tag{21}$$

in E_n. Since $\beta = AdN$, the *first fundamental form* on this surface is given by

$$dS^2 = \beta^T A^{T-1} A^{-1} \beta, \tag{22}$$

which reduces to $dS^2 = \beta^T \beta$ is A is SO(n)-valued function. In this latter instance, we have

$$dS^2 = (\beta_A^i \delta_{ij} \beta_B^j + \beta_A^a \delta_{ab} \beta_B^b) dX^A dX^B, \tag{23}$$

and only the first sum is the minimal replacement of the right Cauchy-Green strain tensor. This surface has the curvature 2-forms R_j^i, R_j^a, R_b^i and R_b^a. Thus, even if the disclination density 2-forms R_j^i all vanish, the surface will still be a surface with nonzero extrinsic curvature if any of the dislocation density 2-forms are nonzero. Once a problem has been solved, by using the complete solution of the kinematic equations of defects given above to obtain a solution of the equations of equilibrium, there will usually be arbitrary functions that still appear in the solution. Such solutions are obtained for a few easy problems of materials with dislocations and with disclinations in the discussions to follow. These arbitrary functions can then be evaluated by using the gauge theory of defects presented in [1, 2] with appropriate choices of the minimally coupled Lagrangian function for the defects and the null Lagrangian function. Such evaluations are obtained for the gauge-theoretic generalization of the screw dislocation and for the "self screening" screw-like dislocations.

There is one further process that must be executed before identification of the actual physical quantities can be made. Any solution of a problem with defects, with use of the representation (2.9), means that a particular choice of gauge has been made. Other choices of gauge will result in different representations. In particular, we can always transform to the antiexact gauge in the manner discussed in [2]. Let H be the linear homotopy operator with center at the origin in the reference configuration (see [3] Chapter 5). The following identity is then satisfied by any 1-form on a star shaped region:

$$\beta^i = dH[\beta^i] + H[d\beta^i]. \tag{24}$$

When this is applied to (2.9), we obtain

$$H[\beta^i] = X^A\delta_A^i + u^i + H[\Gamma_j^i(X^A\delta_A^j + u^j) + \Gamma_a^i y^a] = X^A\delta_A^i + u_{\ddagger}^i. \quad (25)$$

where

$$u_{\ddagger}^i = u^i + H[\Gamma_j^i(X^A\delta_A^j + u^j) + \Gamma_a^i y^a]. \quad (26)$$

are the *total displacement functions* (in the antiexact gauge) of the material body. The reason for this assignment is that (2.9) and (2.12) give

$$\beta^i = d(X^A\delta_A^i + u_{\ddagger}^i) + H[d\beta^i]. \quad (27)$$

and $H[d\beta^i]$ are antiexact 1-forms (i.e., they are evaluated in the antiexact gauge) that are determined solely by the obstructions to satisfaction of the kinematic equations, $d\beta^i = 0$, of elastic materials. The relations (2.14) thus show that the quantities u^i, that occur in (2.9), are only partial displacement functions that obtain with the choice of gauge in which the representation (2.9) is obtained!

3. Dislocations without Disclinations: The Class of "Screw-Like" Defects

Screw-like defects are engendered by the requirement that the configuration be invariant under translations along the Z-axis and under rotations about the Z-axis. The only scalar valued function with these properties is of the form $f(R)$, and only basis 1-forms with this property are dR and $d\theta$, where

$$R = \sqrt{X^2 + Y^2}, \quad \theta = ArcTan\left(\frac{Y}{X}\right). \quad (28)$$

Dislocations arise from allowing the translation group, T(3), to act locally. We therefore restrict the model space to be E_4 and take the only nonzero element of the generating matrix B to be $B_4^3 = \theta$; that is

$$B = \{\{0,0,0,0\}, \{0,0,0,0\}, \{0,0,0,\theta\}, \{0,0,0,0\}\}. \quad (29)$$

Here, we have used block-row notation (i.e. $\{0,0,0,0\}$ is the first row of B, etc.). This means that only local translations parallel to the Z-axis are considered. Edge-like dislocations can be treated in a similar fashion with the choice $B_4^1 = \theta$. Calculation of solutions for edge-like dislocations are of such a length that they will not be presented.

For the purposes of this Section, we take

$$y^1 = X, \quad y^2 = Y, \quad y^3 = Z, \quad y^4 = \frac{a}{2}M(R), \quad (30)$$

where α is a constant. Due to the linearity+ of the problem, we can go back later and replace $y^i = X^i$ by $y^i = X^i + u^i$, where $\{u^i | 1 \leq i \leq 3\}$ constitute the solution of any displacement boundary value problem of linear elasticity The substitutions of (3.2) into (2.9) give

$$\beta^1 = dX, \quad \beta^2 = dY, \quad \beta^3 = dZ + aMd\theta, \quad \beta^4 = \frac{a}{2}dM = \frac{a}{2}M'(R)dR, \quad (31)$$

because the only nonzero element of Γ is

$$\Gamma_4^3 = 2d\theta. \tag{32}$$

We therefore have

$$R_j^i = 0, \quad \Gamma_j^i = 0, \tag{33}$$

so that there are no disclinations present. Calculation of the dislocation densities can now be made, either from the evaluations of the β's given by (3.3), or from (2.5). This gives the evaluations $\alpha^1 = \alpha^2 = 0$, $\alpha^3 = -\Gamma_4^3 \wedge \beta^4 = \beta^4 \wedge \Gamma_4^3$, and hence

$$\alpha^3 = adM(R) \wedge d\theta = aM'(R)dR \wedge d\theta = a\frac{M'(R)}{R}dR \wedge Rd\theta = a\frac{M'(R)}{R}dX \wedge dY. \tag{34}$$

The only nonzero dislocation density, with $\alpha^i = \alpha^{iA}\mu_A$, is therefore

$$\alpha^3 = a\frac{M'(R)}{R}\mu_3 = am(R)\mu_3. \tag{35}$$

If the dislocation density amplitude is taken to be the physically assigned quantity (i.e., $m(R)$ is assigned), then (3.6) gives

$$M(R) = \int_0^R \xi\mu(\xi)d\xi. \tag{36}$$

Noting that $R^2 d\theta = XdY - YdX$, (3.3) gives us the explicit evaluation

$$\beta^3 = dZ + a\frac{M(R)}{R^2}(XdY - YdX) = dZ + a\frac{M(R)}{R}(\frac{X}{R}dY - \frac{Y}{R}dX). \tag{37}$$

The important datum to note here, is that use of (3.7) shows that $M(R)/R^2$ is *bounded* at the origin, and that $M(R)/R$ *vanishes* at the origin if $m(R)$ is bounded at the origin. It is therefore useful to define the dislocation amplitude, *Adis*, by

$$Adis(R) = \frac{M(R)}{R}. \tag{38}$$

Accordingly, (3.8) shows that β^3 is a *globally bounded 1-form* if the dislocation density amplitude, *m(R)*, is globally bounded. This result is in sharp contrast with the classical Volterra solution that obtains from $aM(R) = b/2\pi$, where b is the Burgers' vector for the screw dislocation In fact, (3.7) show that the $m(R)$ must have the distribution evaluation $m(R) = \delta'(R)$, where $\delta(R)$ is the Dirac distribution, and $a = b/2\pi$.

It is now a simple matter to use the results given above to calculate ϵ and σ:

$$\epsilon = \frac{aM(R)}{2R^2}\{\{0,0,-Y\},\{0,0,X\},\{-Y,X,0\}\}.$$

$$\sigma = \frac{\hat{\mu}aM(R)}{R^2}\{\{0,0,-Y\},\{0,0,X\},\{-Y,X,0\}\}. \tag{39}$$

Since $dM(R) = M'(R)(XdX + YdY)/R$, we see that *the equations of equilibrium,* $[\partial_X, \partial_Y, \partial_Z]\sigma = 0$, *are satisfied for all piecewise smooth choices of $M(R)$!* The solutions obtained in this section are thus exact solutions the kinematic equations of defects and of the equations of equilibrium, and they obtain for any choice of the function $M(R) = \int_0^R \xi m(\xi) d\xi$ for which the dislocation density amplitude $m(R)$ is finite at the origin. Further, if we write $\beta^3 = dX + \hat{\beta}^3$, then (3.8) shows that $\hat{\beta}^3$ is an antiexact 1-form (see [3]) with respect to the linear homotopy operator H with center at the origin. The antiexact gauge [2] is therefore in use, and hence (3.3) shows that $H[\beta^i] = X^A \delta_A^i$. Accordingly,(i.e., *these solutions obtain with vanishing total elastic displacements* (i.e., $u_\ddagger = v_\ddagger = w_\ddagger = 0$). These screw-like dislocations simply are; they are not a response to the choice of badly behaved elastic displacement functions.

There is an alternative derivation of these results that shows the usefulness of higher dimensional model spaces. Here, we take

$$B = \{\{0,0,0,0\}, \{0,0,0,0\}, \{0,0,0, \frac{Y}{X}\}, \{0,0,-\frac{Y}{X},0\}\}, \quad (40)$$

This B generates the local element

$$A = \{\{1,0,0,0\}, \{0,1,0,0\}, \{0,0, \frac{X^2-Y^2}{X^2+Y^2}, -\frac{2XY}{X^2+Y^2}\}, \{0,0, \frac{2XY}{X^2+Y^2}, \frac{X^2-Y^2}{X^2+Y^2}\}\}, \quad (41)$$

of SO(4) that gives position dependent rotations in the (N^3, N^4)-plane, rather than a local element of the translation group T(4). The nonzero entries of Γ are now

$$\Gamma_4^3 = -\Gamma_3^4 = \frac{2(XdY - YdX)}{X^2 + Y^2}. \quad (42)$$

A direct calculation with the same $y = \{X, Y, Z, M(R)\}^T$ shows that all of the physical quantities, β^1, β^2, β^3, α^1, α^2, α^3, Γ_j^i, R_j^i are the same as before, even though β^4 is now given by

$$\beta^4 = \frac{a}{2} dM - \frac{Z(XdY - YdX)}{X^2 + Y^2}. \quad (43)$$

Again, all quantities of physical interest are bounded, even though B is now singular on the hyperplane $X = 0$. This derivation seems preferable to that given previously because it does not involve the multiple-valued transcendental function $\theta = ArcTan(Y/X)$.

4. Screw-Like Dislocations with Compact Dislocation Densities

In general, dislocations are found to have core regions of finite extent due to the nature of the inter-mollecular forces that are involved. These circumstances are modeled by problems for which the dislocation density, $m(R)$, vanishes for $R > r$. A direct use of (3.7) gives

$$M(R) = \begin{pmatrix} a \int_0^R \xi m(\xi) d\xi, \text{ for } R < r \\ a \int_0^r \xi m(\xi) d\xi, \text{ for } R > r \end{pmatrix}. \quad (44)$$

There are obviously two cases.

Case I. $M(R) \neq 0$.

For this case we may choose $a = b/(2\pi \int_0^r \xi m(\xi) d\xi)$ so as to obtain agreement with the classical Volterra screw dislocation for $R > r$. We thus obtain

$$M(R) = \begin{pmatrix} \frac{b}{2\pi \int_0^r \xi m(\xi) d\xi} \int_0^R \xi m(\xi) d\xi, \text{ for } R < r \\ \frac{b}{2\pi}, \text{ for } R > r \end{pmatrix}. \tag{45}$$

This can be written in the more convenient form

$$M(R) = \begin{pmatrix} \frac{b}{2\pi} \hat{M}(R), \text{ for } R < r \\ \frac{b}{2\pi}, \text{ for } R > r \end{pmatrix}, \tag{46}$$

with $\hat{M}(R) = \int_0^R \xi m(\xi) d\xi / \int_0^r \xi m(\xi) d\xi$ for $R < r$, so that

$$\hat{M}(R) = 1, \tag{47}$$

Case II. $M(R) \neq 0$: **Self Screening Dislocations.**

This case means that the dislocation density has the property $\int_0^r \xi m(\xi) d\xi$, so that $m(R)$ must take on both positive and negative values on the interval $0 \leq R \leq r$. When this evaluation is substituted into (3.11), we obtain

$$M(R) = \begin{pmatrix} a \int_0^R \xi m(\xi) d\xi, \text{ for } R < r \\ 0, \text{ for } R > r \end{pmatrix}. \tag{48}$$

Since $M(R) = 0$ for $R > r$, the outside world can not detect the presence of such dislocations; they are self. It is therefore convenient to write

$$M(R) = \begin{pmatrix} a\hat{M}(R), \text{ for } R < r \\ 0, \text{ for } R > r \end{pmatrix}, \tag{49}$$

with $\hat{M}(R) = \int_0^R \xi m(\xi) d\xi$ for $R < r$ and

$$\hat{M}(R) = 0. \tag{50}$$

I have not been able to find any previous references to self screening screw dislocations, so self screening screw dislocations appear to be new. These solutions may have relevance in problems where the material has a global pattern that can exhibit a very localized disruption of that pattern to become another pattern.

5. Screw -Like Dislocations with Terminations

If we go back to (3.2) and make the alternative choice

$$y^4 = \frac{a}{2} q(Z) M(R), \tag{51}$$

then

$$\beta^3 = dZ + aq(Z) M(R) d\theta \tag{52}$$

and
$$\alpha^3 = aq(Z)\frac{M'(R)}{R} + aq(Z)M(R)dR \wedge Rd\theta + aq'(Z)M(R)dZ \wedge d\theta. \tag{53}$$

When this is translated into (X, Y) format, we obtain

$$\alpha^3 = aq(Z)\frac{M'(R)}{R} + aq(Z)M(R)dX \wedge dY + aq'(Z)\frac{M(R)}{R^2}(YdX \wedge dZ - XdY \wedge dZ)$$

$$= aq(Z)\frac{M'(R)}{R}\mu_3 - aq'(Z)\frac{M(R)}{R^2}(Y\mu_2 + X\mu_1). \tag{54}$$

This gives

$$\alpha^{31} = -aXq'(Z)\frac{M(R)}{R^2}, \quad \alpha^{32} = -aYq'(Z)\frac{M(R)}{R^2}, \quad \alpha^{33} = aq(Z)\frac{M'(R)}{R}. \tag{55}$$

These evaluations show that all quanatities are globally bounded if $q'(Z)$ and $M'(R)/R$ are globally bounded. We have thus calculated the required dislocation densities in order for there to be Z-dependence.

Such solutions of the kinematic equations of defects have relevance in the termination problem for screw-like dislocations. For this problem we take

$$q(Z) = 1 \text{ for } Z < Z_1, q(Z) = 0 \text{ for } Z > Z_2 \tag{56}$$

with $Z_1 < Z_2$, and a smooth transition between these values for $Z_1 \leq Z \leq Z_2$. The evaluation (5.4) shows that $\alpha^3 = 0$ for $Z > Z_2$, while $\alpha^3 = a\frac{M'(R)}{R}\mu_3$ for $Z < Z_1$. Indeed, α^{31} and α^{32} are nonzero only in the "transition" region, $Z_1 \leq Z \leq Z_2$ they generate a boundary layer of dislocation densities, because $q'(Z)$ is only nonzero in this boundary layer. Such situations would appear to have particular relevance for self screening dislocations where the boundary layer will also have compact support. In particular, (5.5) shows that the boundary layer is not compact if $M(r) \neq 0$ because $M(R) = M(r)$ for $R > r$.

The reason why we have been able to obtain these results is that we have models with a distributed core region (i.e., a region where $M'(R) \neq 0$), while the classical Volterra solutions dismiss the core region as a mysterious place that must be excluded because they are places where the stresses go to infinity. We note in passing that the techniques presented here, when extended to include dynamics, allows us to take $q = q(R, t)$ rather than just $q = q(R)$. In this event, (5.2) becomes

$$\beta^3 = dZ + aq(R, t)M(R)d\theta \tag{57}$$

and hence $d\beta^3$ will contain the additional term

$$a\partial_t q(R, t)M(R)dt \wedge d\theta = a\partial_t q(R, t)\frac{M(R)}{R^2}(Xdt \wedge dY - Ydt \wedge dX), \tag{58}$$

as well as the term $a\partial_R q(R, t)M(R)dR \wedge d\theta$. Since we must use the dynamic formulation, rather than just the static one, the $\partial_t q(R, t)$-term will serve to determine the required dislocation current 2-forms! Thus, for instance, we could study the propagation of a termination boundary layer for a screw dislocation.

The equations of motion will no longer be satisfied without compensating displacement fields. This will require the altered evaluation

$$y^1 = X + u, \quad y^2 = Y + v, \quad y^3 = Z + w \tag{59}$$

before the strain and stress matrices are calculated. Satisfaction of the equations of equilibrium for such problems entails huge calculations that are best left to the reader if the need should arise. Such calculations are indeed possible because the equations of linear elasticity posses a Green's tensor for problems with imposed body force fields, and the defects lead to equivalent body force fields that are nonzero only in the transition region $Z_1 \leq Z \leq Z_2$.

6. Determination of the Function $M(R)$ by Use of Gauge Field Theory

The full gauge theory of defects is predicated on stationarizing an action integral with suitable choice of the Lagrangian function. We use this theory to obtain evaluations of the only free remaining function, $M(R)$. This Lagrangian function is made up of several parts [1, 2].

The first part of the Lagrangian is the minimally replaced strain energy of the body. This has the evaluation

$$L_e = \frac{1}{2} Tr(\sigma \epsilon) = \frac{\hat{\mu}}{2R^2} M(R)^2 \tag{60}$$

when (3.10) are used. Here we have absorbed the constant a in the undetermined function $M(R)$.

The second part of the Lagrangian accounts for the energy in the dislocation fields. It has the evaluation $L_g = \frac{s_1}{2} \alpha^{iA} \alpha_{iA}$ is a coupling constant. This is the so called minimal coupling of the elasticity fields to the defect fields. Since only α^{33} is nonzero for screw-like dislocations, the evaluation (3.6) gives

$$L_g = \frac{s_1}{2R^2} (M'(R))^2. \tag{61}$$

The third part of the Lagrangian consists of a *null Lagrangian* null Lagrangian is included so as to guarantee that the minimally replaced stress field and the stress field of the corresponding elasticity problem lead to the same evaluations of any boundary tractions. It has the evaluation $L_n = -Tr(\sigma_e \epsilon)$, where σ_e is the stress tensor for the classic Volterra solution. We therefore obtain the evaluation

$$L_n = -\frac{b\hat{\mu}}{2\pi R^2} M(R). \tag{62}$$

The total Lagrangian for the screw like dislocations thus has the evaluation

$$L = \frac{\hat{\mu}}{2R^2} M(R)^2 + \frac{s_1}{2R^2} (M'(R))^2 - \frac{b\hat{\mu}}{2\pi R^2} M(R). \tag{63}$$

The Euler-Lagrange equation that is determined by this Lagrangian has the form

$$\frac{d}{dR}\left(\frac{M'}{R^2}\right) = \frac{\hat{\mu}}{s_1} \frac{M}{R^2} - \frac{b\hat{\mu}}{2\pi s_1} \frac{1}{R^2}. \tag{64}$$

It is useful to introduce the constants

$$\kappa^2 = \frac{\hat{\mu}}{s_1}, \quad L^2 = \kappa^2 \frac{b}{2\pi}. \tag{65}$$

so that κ has the dimension of a reciprocal, length. The gauge field equation (6.5) thus assumes the standard form

$$\frac{R}{dR}\left(\frac{M'(R)}{R^2}\right) = \kappa^2 \frac{M(R)}{R^2} - L^2 \frac{1}{R^2}. \tag{66}$$

This equation obviously has the particular solution $L^2/kappa^2 = b/2\pi$, so we have

$$M(R) = \frac{L^2}{\kappa^2} + \hat{M}(R), \tag{67}$$

where $\hat{M}(R)$ satisfies the associated homogeneous equation

$$\frac{d}{dR}\left(\frac{M'(R)}{R^2}\right) = \kappa^2 \frac{M(R)}{R^2}. \tag{68}$$

This latter equation can be integrated in closed form to obtain

$$M(R) = \frac{L^2}{\kappa^2} + \frac{1}{4\kappa^3}e^{-\kappa R}(1+\kappa R)C_1 + \frac{6}{4\kappa^3}\left\{e^{-\kappa R}(1+\kappa R) + e^{\kappa R}(-1+\kappa R)\right\}C_2. \tag{69}$$

Here, the integration constants C_1, C_2, are to be determined by the conditions imposed by the various cases considered above.

For the general screw dislocation problem, we have the conditions $M(0) = 0$, $M(R)$ bounded as $R \mapsto \infty$. Under these circumstances, (6.8) gives

$$M(R) = \frac{b}{2\pi}(1 - e^{-\kappa R}(1 + \kappa R)). \tag{70}$$

We make particular note of the simplicity of the result (6.9), in contrast to the usually reported solutions that involve Bessel functions of the second kind. The reason for the increased simplicity of (6.9) is that this solution was obtained only after the kinematic equations of defects and the equations of equilibrium were satisfied. The evaluation (6.9) gives the dislocation density amplitude $m(R) = M'(R)/R = \frac{b}{2\pi}\kappa^2 e^{-\kappa R}$, so that there is exponential decay of the dislocation density for this problem. For $s_1 \ll \mu$, which is the elasticity limit by (6.4), (6.6) shows that K becomes large and the corresponding characteristic length becomes small. In this case, the solution obtained from (6.9) approaches the classical Volterra solution for $R \geq 10/\kappa$, but remains bounded everywhere.

The exponential decay of the dislocation density can be overcome by the brute force method of requiring the dislocation density to have compact support, as presented in Section 4. For screw dislocation problems with dislocation densities of compact support and $M(r) \neq 0$, the appropriate boundary conditions are $M(0) = 0$, $M(r) = L^2/\kappa^2 = b/2\pi$, and $M(R) = M(r)$ for $R > r$. Here it is useful to introduce the new variables

$$\rho = \kappa R, \quad s = \kappa r, \quad M(R) = \hat{M}(\rho). \tag{71}$$

When these boundary conditions and changes of variables are used with (6.8), we obtain

$$\frac{2\pi}{b}\hat{M}(\rho) = 1 - \epsilon^{-\rho}(1+\rho) + e^{-\rho}(1+s)\frac{1+\rho+e^{2\rho}(\rho-1)}{1+s+e^{2s}(s-1)}. \qquad (72)$$

For the self screening case, the appropriate boundary conditions are $M(0) = M(r) = 0$, and $M(R) = 0$ for $R > r$. Use of the new variables (6.10) with (6.8) thus gives

$$\frac{2\pi}{b}\hat{M}(\rho) = 1 - \epsilon^{-\rho}(1+\rho) + e^{-\rho}(1+s-e^s)\frac{1+\rho+e^{2\rho}(\rho-1)}{1+s+e^{2s}(s-1)}. \qquad (73)$$

7. A Class of States of Material Bodies with Nontrivial Disclinations

Gauge theory teaches us that disclinations are generated by local action of the rotation group. For the purpose of this discussion, we confine attention to the model space E_4. We know that the upper left-hand 3×3 submatrix of the generating matrix $B \subset \mathfrak{so}(4)$ can be set to zero because its effects can be removed by an integrable gauge transformation. Next, we note that $R^i_j = -\Gamma^i_4 \wedge \Gamma^4_j \neq 0$ only if at least two of the remaining generators of $\mathfrak{so}(4)$ are nonzero. Thus, by analogy with plane strain problems, we take

$$B = \{\{0,0,0,\frac{X}{L}\}, \{0,0,0,\frac{Y}{L}\}, \{0,0,0,0\}, \{-\frac{X}{L}, -\frac{X}{L}, 0, 0\}\}. \qquad (74)$$

Here, L is a scale factor with the dimension of a length, which is required because entries in generating matrices must be dimensionless quantities. When (7.1) is substituted into (2.4), the nonzero connection 1-forms are given by

$$\Delta\Gamma^1_2 = -\Delta\Gamma^2_1 = 2(X\,dY - Y\,dX), \qquad (75)$$

which are antiexact 1-forms (i.e., they can not be removed by an SO(3) gauge transformation), and

$$\Delta\Gamma^1_4 = -\Delta\Gamma^4_1 = 2L\,dX, \quad \Delta\Gamma^2_4 = -\Delta\Gamma^4_2 = 2L\,dY. \qquad (76)$$

Here, we have set

$$R^2 = X^2 + Y^2, \quad \Delta = L^2 + R^2, \qquad (77)$$

so that Δ^{-1} is globally bounded. A direct calculation using (2.6) gives the following nonzero *disclination density 2-forms*

$$\Delta^2 R^1_2 = -\Delta^2 R^2_1 = 4L^2\,dX \wedge dY = \Delta^2 d\left(\frac{2L^2 R^2}{\Delta}d\theta\right). \qquad (78)$$

A combination of (7.5) and Stokes' theorem shows that the integral of R^1_2 over a disk of radius R_0 in the (X,Y)-plane with center on the Z axis has the evaluation $4\pi L^2 R_0^2/(L^2 + R_0^2)$.

All of these quantities are invariant under translation along the Z-axis. We therefore take

$$u = U(R)X, \quad v = U(R)Y, \quad w = 0, \qquad (79)$$

so that the elastic displacement vector is perpendicular to the Z-axis and radially directed. This leads naturally to the choice

$$y = \{(1+U(R))X, (1+U(R)Y, z, N(R))\}^T. \tag{80}$$

From now on we will simply write U and N, where the R dependence is understood. When (2.9) is used, with $\beta^i = \beta^i_A dX^A$, we obtain the following nonzero components of the distortion 1-forms

$$\beta^1_1 = 1 + \frac{2LN}{\Delta} + U - \frac{2Y^2(1+U)}{\Delta} + \frac{X^2 U'}{R},$$

$$\beta^1_2 = \beta^2_1 - \frac{2XY(1+U)}{\Delta} + \frac{XYU'}{R},$$

$$\beta^2_2 = 1 + \frac{2LN}{\Delta} + U - \frac{2X^2(1+U)}{\Delta} + \frac{Y^2 U'}{R}, \quad \beta^3_3 = 1, \tag{81}$$

and

$$R\Delta\beta_4 = (XdX + YdY)(\Delta N' - 2LR(1+U)). \tag{82}$$

Since $\alpha^i = -\Gamma^i_4 \wedge \beta^4$, the dislocation density 2-forms that are engendered by the disclination density 2-forms (7.5) are given by

$$R\Delta^2\alpha^1 = 2LYF dX \wedge dY, \quad R\Delta^2\alpha^2 = -2LXF dX \wedge dY, \quad \alpha^3 = 0, \tag{83}$$

where

$$F = 2L(1+U)R - \Delta N'. \tag{84}$$

Accordingly, (7.10) and (7.11) show that there are no accompanying dislocations if and only if $U(R)$ and $N(R)$ are chosen so that $F = 0$. We note that all of these evaluations result in *globally bounded* quantities if $U(R)$ and $N(R)$ are globally bounded. All of the kinematic equations of defects have now been satisfied.

Now that we know the distortion 1-forms (i.e., $\beta = ((\beta^i_A))$, which is a symmetric matrix), we can compute ϵ and σ. Since some of the entries of Γ^i_j are not zero, the resulting equilibrium equations take the form (see (1.10))

$$0 = Eqt1 = \partial_1\sigma^1_1 + \partial_2\sigma^2_1 - \frac{2Y}{\Delta}\sigma^1_2 + \frac{2X}{\Delta}\sigma^2_2,$$

$$0 = Eqt2 = \partial_1\sigma^1_2 + \partial_2\sigma^2_2 + \frac{2Y}{\Delta}\sigma^1_1 - \frac{2X}{\Delta}\sigma^2_1. \tag{85}$$

A lengthy calculation shows that all of the X's and the Y's combine so that only functions of R remain, with the exception of a leading coefficient, and that

$$YEqt1 = XEqt2. \tag{86}$$

There is thus only one resulting equation to be satisfied. Set

$$K = \frac{\lambda}{2\hat{\mu}}. \tag{87}$$

These calculations show that the equilibrium equations, (7.11), are satisfied if and only if $N(R)$ and $U(R)$ stand in the relation

$$2(1+2K)L\Delta N' = 2(2K-1)L^2R + 2(1+2K)R^3$$
$$-4L^2RU - 3(1+K)\Delta^2 U' - (1+K)R\Delta U''. \tag{88}$$

When (7.12) is satisfied, the kinematic equations of defects and the equations of equilibrium are satisfied. These exact solutions of disclination problems may dispel the myth that elastic materials can not support disclinations.

There is a two parameter (c, $L \neq 0$) family of disclinated states without elastic displacements (i.e., with $U(R) = 0$ in (7.12) when $N(R)$ is given by

$$2L(1+2K)N(R) = c + (1+2K)R^2 - 2L^2 ln(\Delta). \tag{89}$$

All quantities of interest are globally bounded in this case because only the combination $N(R)/\Delta$ is present in the evaluations of the distortion 1-forms, the dislocation density 2-forms, and the disclination density 2-forms.

An externally imposed screw dislocation field can be combined with this disclination field by going to E_5 with the assigned generating matrix

$$B = \{\{0,0,0,X/L,0\},\{0,0,0,Y/L,0\},\{0,0,0,0,\theta\}\},$$
$$\{\{-X/L,-Y/L,0,0,0\},\{0,0,0,0,0\}\}; \tag{90}$$

that is, we adjoin a local translation in the z-direction through use of the added dimension for the model space. With y given by $y = \{x,y,z,N,\frac{a}{2}M\}^T$, $\Delta = L^2 + X^2 + Y^2$, and $\{x,y,z\} = \{X+u, Y+v, Z+w\}$, the disclination density 2-forms R^i_j, and Γ^i_j, β^1, β^2, β^4 are the same as before if we take $u = XU(R)$, $v = YU(R)$, $w = 0$, and $N = N(R)$. The rest of the β's are given by

$$\beta^3 = aMd\theta + dz, \quad \beta^5 = \frac{a}{2}dM. \tag{91}$$

This shows that the plane strain disclinations and the anti-plane strain screw dislocations combine by superposition if we set $M = M(R)$ and use the information developed in the discussions of screw dislocations given above. The solutions of the equations of equilibrium for the two problems can thus be taken over directly because the engineering stress tensor is a linear tensor values function of the linear engineering strain tensor and the self-evident separation of effects between plane strain and antiplane strain problems. This superposition does not hold if we adjoin a local translation in the x-direction (edge dislocation) rather than in the z-direction.

References

[1] Kadic A & Edelen DGB, *A Gauge Theory of Dislocations and Disclinations, Lecture Notes in Physics*, **174**, (1983), Springer, Berlin.

[2] Edelen DGB & Lagoudas D, *NorthGauge Theory and Defects In Solids*, (2005), North Holland, Amsterdam.

[3] Edelen DGB, *Dover, New York Applied Exterior Calculus*, **174**, (2005), Dover, New York.

[4] Kosseccka E & de Wit R, *Arch. Mech. Stos.*, **174**, 749, (1977).

[5] Lazar M, *Phys. Let. A*, **311**, 416, (2003).

[6] Edelen DGB, *Clas. Quantum Grav.*, **20**, 3661, (2003).

[7] Edelen DGB, *Clas. Quantum Grav.*, **17**, 4715, (2000).

[8] Schaub H, Tsiotras P & Junkins JL, *Int. J. Engng. Sci.*, **33**, 2277, (1995).

In: Theoretical Physics and Nonlinear Optics
Editors: Thomas F. George et al
ISBN 978-1-61122-939-4
© 2012 Nova Science Publishers, Inc.

Chapter 5

ONE-MAGNON SYSTEMS IN AN ANISOTROPIC NON-HEISENBERG FERROMAGNETIC IMPURITY MODEL

S.M. Tashpulatov*
Institute of Nuclear Physics of Uzber Academy of Science,
Tashkent, Uzbekistan

Abstract

We consider a one-magnon system in an anisotropic non-Heisenberg impurity model with an arbitrary spin s and investigate the spectrum and the localized impurity states of the system on the $\nu-$ dimensional integer lattice Z^ν. We show that there are at most four types of localized impurity states (not counting the degeneracy multiplicities of their energy levels) in this system. We find the domains of these states and calculate the degeneracy multiplicities of their energy levels.

Keyword: impurity states, lattice, interaction, creation operator, annihilation operator, anisotropic model.

The use of films in various areas of physics and technology arouses great interest in studying a localized impurity state (LIS) of a magnet. The LISs in a Heisenberg ferromagnet with ferromagnetic and antiferromagnetic impurities were investigated in many papers (see, e. g., [1-6]), where the situations with linear and cubic lattices were considered in detail. It was shown that there are two LIS types in the linear case and three types in the cubic case.

In [7], the case of a $\nu-$ dimensional lattice was considered, and it was proved that there are at most three types of LISs (not counting the degeneracy multiplicities of their energy levels) in the ν- dimensional case. It was shown that the number of types of LISs in the system changes with varying parameters of the Hamiltonian, and the LIS domains were found. In this case, it turns out that the three types of LISs in the system are respectively nondegenerate, ν-fould degenerate, and $(\nu - 1)$-fould degenerate.

---
*E-mail address: toshpul@mail.ru, toshpul@rambler.ru

In theoretical investigations of magnetically ordered systems and in the interperatation of experimental data, the starting point was usually the Heisenberg exchange Hamiltonian (for an arbitrary spin s),

$$H = J \sum_{m,\tau} (\vec{S}_m \vec{S}_{m+\tau}), \qquad (1)$$

where J is the parameter of the bilinear exchange interaction between the nearest-neighbor atoms, $\vec{S}_m$ is the atomic spin operator for the spin s at the mth lattice site in the ν-dimensional lattice Z^ν, and the summation over τ ranges the nearest neighbors.

For an arbitrary spin s, the isotropic spin exchange Hamiltonian in fact has the form [8]

$$H = \sum_{m,\tau} \sum_{n=1}^{2s} J_n (\vec{S}_m \vec{S}_{m+\tau})^n, \qquad (2)$$

where J_n are the parameters of the multipole exchange interactions between the nearest-neighbor atoms. Hamiltonian (2) coincides with (1) only for $s = 1/2$, whereas if $s > 1/2$, then some terms containing higher degrees of $(\vec{S}_m \vec{S}_{m+\tau})$ appear, which must be taken into consideration in studying magnets with spins $s > 1/2$. Expression (2) is called a non-Heisenberg Hamiltonian.

For many decades, magnetic theory has been developing based on the Heisenberg model. We note that there are not very many magnets close to ideal Heisenberg magnets. For magnets with spins $s > 1/2$, the Heisenberg model gives a somewhat inadequate description of their properties. The magnetic properties of non-Heisenberg magnets strongly differ from those of Heisenberg magnets, and calculating the higher-order exchange terms with respect to the spin is incomparably more complex for them that in the Heisenberg case.

In the paper [9], the considered one-magnon system in an isotropic non-Heisenberg ferromagnetic impurity model with an arbitrary spin values s. The spectrum and the LISs of the system on the ν-dimensional integer lattice Z^ν are investigated. Number of the LISs, their energies, the degeneracy multiplicities of the related energy levels, and the LIS domains of existence change with varying parameters of the Hamiltonian are finded.

In this paper, we consider one-magnon states in an anisotropic (XXZ-model) non-Heisenberg ferromagnetic impurity model with an arbitrary spin s. We find how the number of the LISs, their energies, the degeneracy multiplicities of the related energy levels, and the LIS domains of existence change with varying parameters of the Hamiltonian. So far as we know, the spectrum and the LISs of one-magnon systems in an anisotropic non-Heisenberg ferromagnetic imputity model with an arbitrary spin s regarding the multipole exchange interactions included have not been discussed in the literature.

The Hamiltonian of the system in question has the form

$$H = -\sum_{m,\tau} \sum_{n=1}^{2s} J_n \{ S_m^z S_{m+\tau}^z + \frac{\alpha}{2}(S_m^+ S_{m+\tau}^- + S_m^- S_{m+\tau}^+) \}^n -$$

$$-\sum_{m,\tau}\sum_{n=1}^{2s}(J_n^0 - J_n)\{S_0^z S_\tau^z + \frac{\alpha}{2}(S_0^+ S_\tau^- + S_0^- S_\tau^+)\}^n. \quad (3)$$

Here $J_n > 0$ are the parameters of the multipole exchange interaction between the nearest-neighbor atoms in the lattice, J_n^0 are the atom-impurity multipole exchange interaction parameters, α- are the anisotropic parameter, $\tau = \pm e_j, j = 1, 2, ..., \nu$, where e_j are unit coordinate vectors, i. e., the summation over τ ranges the nearest neighbors, and $\vec{S}_m = (S_m^x; S_m^y; S_m^z)$ is the atomic spin operator for the spin s at the lattice site m. We set $S_m^\pm = S_m^x \pm i S_m^y$, where S_m^- and S_m^+ are the corresponding magnon creation and annihilation operators at the site m.

Hamiltonian (3) acts in the summetric Fock space $\mathcal{H}$. We let φ_0 denote the so-called vacuum vector uniquely determined by the conditions $S_m^+ \varphi_0 = 0$, $S_m^z \varphi_0 = s\varphi_0$, $||\varphi_0|| = 1$. The vector $S_m^- \varphi_0$ describes the state of a system of one magnon with the spin s located at the site m. The vectors $\{\frac{1}{\sqrt{2s}} S_m^- \varphi_0\}$ form an orthonormal system. The closure of the space spanned by these vectors is denoted by $\mathcal{H}_1$. It is a Euclidian space with respect to the natural inner product and is called a space of one-magnon states. The following proposition holds.

Proposition 1. *The space $\mathcal{H}_1$ is invariant with respect to the operator H. The operator $H_1 = H/\mathcal{H}_1$ is a bounded self-adjoint operator generating a bounded self-adjoint operator $\overline{H}_1$ acting in the space $l_2(Z^\nu)$ according to the formula*

$$(\overline{H}_1 f)(p) = \sum_{p,\tau}\sum_{n=1}^{2s}(-1)^{n+1} J_n s^n [2(1 + \binom{n}{2}\alpha^2 + \binom{n}{4}\alpha^4 + ... + \alpha^n) f(p) -$$

$$-(\binom{n}{1}\alpha + \binom{n}{3}\alpha^3 + ... + \binom{n}{n-1}\alpha^{n-1})[f(p+\tau) + f(p-\tau)]] +$$

$$+ \sum_{p,\tau}\sum_{n=1}^{2s}(-1)^{n+1} 2s^n [(1 + \binom{n}{2}\alpha^2 + \binom{n}{4}\alpha^4 +$$

$$+ ... + \alpha^n)(\delta_{p,\tau} + \delta_{p,0}) f(p) - (\binom{n}{1}\alpha + \binom{n}{3}\alpha^3 + \binom{n}{5}\alpha^5 + ... +$$

$$+ \binom{n}{n-1}\alpha^{n-1})(\delta_{p,0} f(\tau) + \delta_{p,\tau} f(0))], \quad (4)$$

where $\delta_{k,j}$ is the Kronecker delta and the summation over τ ranges the nearest neighbors. The operator H_1 itself acts on the vector $\Psi = \frac{1}{\sqrt{2s}}\sum_p f(p) S_p^- \varphi_0$, $\Psi \in \mathcal{H}_1$, by the formula

$$H_1 \Psi = \sum_p (\overline{H}_1 f)(p) \frac{1}{\sqrt{2s}} S_p^- \varphi_0. \quad (5)$$

Proposition 1 is proved using the well known commutation relations for the operators S_m^+, S_p^-, and S_q^z.

Lemma 1. *The spectra of the operators H_1 and $\overline{H_1}$ coinside, i. e., $\sigma(H_1) = \sigma(\overline{H_1})$.*

Proof. Because H_1 and $\overline{H_1}$ are bounded self-adjoint operators, it follows that it $\lambda \in \sigma(H_1)$, then the Weyl criterion (see [10]) implies that there is a sequence $\{\Psi_n\}_{n=1}^\infty$ such that $\|\Psi_n\| = 1$ and $\lim_{n\to\infty} \|(H_1 - \lambda)\Psi_n\| = 0$. We set $\Psi_n = \frac{1}{\sqrt{2s}}(\sum_p f_n(p) S_p^- \varphi_0)$. Then $\|(H_1 - \lambda)\Psi_n\|^2 = ((H_1 - \lambda)\Psi_n, (H_1 - \lambda)\Psi_n) = \sum_p \|(\overline{H_1} - \lambda) f_n(p)\|^2 (\frac{1}{\sqrt{2s}} S_p^- \varphi_0, \frac{1}{\sqrt{2s}} S_p^- \varphi_0) = \|(\overline{H_1} - \lambda) F_n\|^2 (\frac{1}{2s} S_p^+ S_p^- \varphi_0, \varphi_0) = \|(\overline{H_1} - \lambda) F_n\|^2 \to 0$ as $n \to \infty$, where $F_n = \sum_p f_n(p)$. It follows that $\lambda \in \sigma(\overline{H_1})$. Consequently, $\sigma(H_1) \subset \sigma(\overline{H_1})$. Conversely, let $\overline{\lambda} \in \sigma(\overline{H_1})$. Then, by the Weyl criterion, there is a sequence $\{F_n\}_{n=1}^\infty$ such that $\|F_n\| = 1$ and $\lim_{n\to\infty} \|(\overline{H_1} - \overline{\lambda}) F_n\| = 0$. Setting $F_n = \sum_p f_n(p)$, $\|F_n\| = \sqrt{\sum_p |f_n(p)|^2}$, we conclude that $\|\Psi_n\| = \|F_n\| = 1$ and $\|(\overline{H_1} - \overline{\lambda}) F_n\| = \|(H_1 - \overline{\lambda})\Psi_n\| \to 0$ as $n \to \infty$. Thus means that $\overline{\lambda} \in \sigma(H_1)$ and hence $\sigma(\overline{H_1}) \subset \sigma(H_1)$. These two relations imply $\sigma(H_1) = \sigma(\overline{H_1})$.

As is seen, if vector (5) is an eigenfunction of H_1 with the eigenvalue $z \notin G_\nu$, then $F = \sum_p f(p)$ is an eigenfunction of the operator $\overline{H_1}$ with the same eigenvalue $z \notin G_\nu$, and this eigenvalue has the same multiplicity. Therefore, to investigate the spectrum of the operator H_1, it suffices to consider that of the operator $\overline{H_1}$ acting in $l_2(Z^\nu)$ by formula (4).

The spectrum and the LISs of the operator H_1 can be studed easily in its quasimomentum representation. Let $\mathcal{F}$ denote the Fourier transformation,

$$\mathcal{F}: l_2(Z^\nu) \Rightarrow L_2(T^\nu) \equiv \mathcal{H}_1,$$

where T^ν is the ν-dimensional torus endowed with the normalized Lebesgue measure $d\lambda$: $\lambda(T^\nu) = 1$. We write

$$\mathcal{F}: \mathcal{H}_1 \Rightarrow \tilde{\mathcal{H}}_1 \equiv L_2(T^\nu).$$

Proposition 2. *The Fourier transform of $\overline{H}_1$ is an operator $\tilde{H}_1 = \mathcal{F}\overline{H}_1\mathcal{F}^{-1}$ acting in the space $\mathcal{H}_1$ by the formula*

$$(\tilde{H}_1 f)(x) = h_\alpha(x) f(x) + \int_{T^\nu} h_{1\lambda}(x;t) f(t) dt, \tag{6}$$

where $h_\alpha(x) = \sum_{n=1}^{2s} (-1)^{n+1} 4 J_n s^n [\nu \varphi_n(\alpha) - \Psi_n(\alpha) \sum_{i=1}^\nu \cos x_i]$,

$$h_{1\alpha}(x;t) = \sum_{n=1}^{2s} (-1)^{n+1} 2 s^n (J_n^0 - J_n) \sum_{i=1}^\nu \varphi_n(\alpha)[1 + \cos(x_i - t_i)] - \Psi_n(\alpha)[\cos t_i + \cos x_i],$$

and

$$\varphi_n(\alpha) = \begin{cases} 1 + \binom{n}{2}\alpha^2 + \binom{n}{4}\alpha^4 + \ldots + \alpha^n & \text{if} \quad n = 2k, \\ 1 + \binom{n}{2}\alpha^2 + \binom{n}{4}\alpha^4 + \ldots + \binom{n}{n-1}\alpha^{n-1} & \text{if} \quad n = 2k+1. \end{cases}$$

and
$$\Psi_n(\alpha) = \begin{cases} \binom{n}{1}\alpha + \binom{n}{3}\alpha^3 + \binom{n}{5}\alpha^5 + ... + \binom{n}{n-1}\alpha^{n-1} & \text{if} \quad n = 2k, \\ \binom{n}{1}\alpha + \binom{n}{3}\alpha^3 + \binom{n}{5}\alpha^5 + ... + \binom{n}{n-2}\alpha^{n-2} + \alpha^n & \text{if} \quad n = 2k+1. \end{cases}$$

To prove Proposition 2, the Fourier transform of (4) should by considered directly.

In is clear that the continuous spectrum of $\tilde{H}_1$ is independent of $h_{1\alpha}(x;t)$ and fills the whole closed interval $\sigma_{cont.}(\tilde{H}_1) = G_\nu = [m_\nu; M_\nu]$, where $m_\nu = min_{x \in T^\nu} h_\alpha(x)$, $M_\nu = max_{x \in T^\nu} h_\alpha(x)$.

An eigenfunction $\varphi \in L_2(T^\nu)$ of $\tilde{H}_1$ corresponding to an eigenvalue $z \notin G_\nu$ is called a LIS of operator $\tilde{H}_1$, and z is called the energy of this state.

We consider the operator $K_\nu(z)$ acting in the space $\tilde{\mathcal{H}}_1$ according to the formula

$$(K_\nu(z)f)(x) = \int_{T^\nu} \frac{h_{1\alpha}(x;t)}{h_\alpha(t) - z} f(t)dt, \, x, t \in T^\nu.$$

It is a completely continuous operator in $\tilde{\mathcal{H}}_1$ for $z \notin G_\nu$.

Let
$$\Delta_\nu(z) = (1 + A\int_{T^\nu} \frac{\varphi_n(\alpha)sin^2 t_1 dt}{h_\alpha(t) - z})^\nu \times (1 + 1/2 A \varphi_n(\alpha) \int_{T^\nu} \frac{(cost_1 - cost_2)^2 dt}{h_\alpha(t) - z})^{\nu-1} \times$$
$$\times [\{1 + A \int_{T^\nu} \frac{cost_1[\varphi_n(\alpha) \sum_{i=1}^{\nu} cost_i - \nu \Psi_n(\alpha)]}{h_\alpha(t) - z} dt\} \times$$
$$\times \{1 + A \int_{T^\nu} \frac{\nu \varphi_n(\alpha) - \Psi_n(\alpha) \sum_{i=1}^{\nu} cost_i}{h_\alpha(t) - z} dt\} -$$
$$- A^2 \int_{T^\nu} \frac{\varphi_n(\alpha) \sum_{i=1}^{\nu} cost_i - \nu \Psi_n(\alpha)}{h_\alpha(t) - z} dt \times \int_{T^\nu} \frac{cost_1[\nu \varphi_n(\alpha) - \Psi_n(\alpha) \sum_{i=1} \nu cost_i]}{h_\alpha(t) - z} dt],$$
(7)

where $dt = dt_1 dt_2 ... dt_\nu$ and $A = \sum_{n=1}^{2s}(-1)^{n+1} 2s^n (J_n^0 - J_n)$.

Lemma 2. *A number $z_0 \notin G_\nu$ is an eigenvalue of operator $\tilde{H}_1$ if and only if it is a zero of the function $\Delta_\nu(z)$, i. e., $\Delta_\nu(z_0) = 0$.*

Proof. In the case under consideration, the equation for the eigenvalues is an integral equation with a degenerate kernel. Therefore, it is equivalent to a homogeneous linear system of equations. As is known, a homogeneous linear system of algebraic equations has a nontrivial solution if and only if the determinant of the system is zero, and this determinant is the function $\Delta_\nu(z)$ here.

At first we consider a some particular case and detail investigate spectrum and LISs of operator $\tilde{H}_1$ in this special case.

A. Let $\alpha = 1$ (isotropic case).

When $\varphi_n(\alpha) = 1 + \binom{n}{2}\alpha^2 + \binom{n}{4}\alpha^4 + ... + \alpha^n = 1 + \binom{n}{2} + \binom{n}{4} + ... + \binom{n}{n-2} + 1 = 2^{n-1}$ and $\Psi_n(\alpha) = \binom{n}{1}\alpha + \binom{n}{3}\alpha^3 + \binom{n}{5}\alpha^5 + ... + \binom{n}{n-1}\alpha^{n-1} = \binom{n}{1} + \binom{n}{3} + \binom{n}{5} + ... + \binom{n}{n-1} = 2^{n-1}$;
And in this case equation for eigenvalues has the form:

$$\{-2\sum_{n=1}^{2s}(-2s)^n J_n[\nu - \sum_{i=1}^{\nu} cosx_i] - z\}f(x) -$$

$$-2\sum_{n=1}^{2s}(-2s)^n(J_n^0 - J_n)\int_{T^\nu}\{\nu + \sum_{i=1}^{\nu}[\cos(x_i - t_i) - \cos x_i - \cos t_i]\}f(t)dt = 0 \quad (8).$$

Let $p(s) = -2\sum_{n=1}^{2s}(-2s)^n J_n$, $q(s) = -2\sum_{n=1}^{2s}(-2s)^n(J_n^0 - J_n)$ and

$$\Delta_\nu(z) = (1 + q(s))\int_{T^\nu}\frac{(1 - \cos t_1)(\nu - \sum_{i=1}^{\nu}\cos t_i)dt}{p(s)[\nu - \sum_{i=1}^{\nu}\cos t_i] - z}) \times$$

$$\times (1+q(s)\int_{T^\nu}\frac{\sin^2 t_1 dt}{p(s)[\nu - \sum_{i=1}^{\nu}\cos t_i] - z})^\nu \times (1+\frac{q(s)}{2}\int_{T^\nu}\frac{(\cos t_1 - \cos t_2)^2 dt}{p(s)[\nu - \sum_{i=1}^{\nu}\cos t_i] - z})^{\nu-1}, \quad (9)$$

where $dt = dt_1 dt_2 ... dt_\nu$.

We let Ω denote the range of all pairs $\omega = (p(s); q(s))$ and introduce the following subsets in Ω for $\nu = 1$:

$$A_1 = \{\omega : p(s) > 0, -p(s) \leq q(s) < 0\}, A_2 = \{\omega : p(s) > 0, q(s) > -p(s)\},$$

$$A_3 = \{\omega : p(s) < 0, q(s) > p(s)\}, A_4 = \{\omega : p(s) > 0, p(s) < q(s)\},$$

$$A_5 = \{\omega : p(s) > 0, 0 < q(s) \leq p(s)\}, A_6 = \{\omega : p(s) < 0, q(s) \geq p(s)\},$$

$$A_7 = \{\omega : p(s) < 0, 0 < q(s) < -p(s)\}, A_8 = \{\omega : p(s) < 0, q(s) > -p(s)\}.$$

We write $z_1 = -\frac{[p(s)+q(s)][p(s)-3q(s)+\sqrt{D}]}{4q(s)}$, $z_2 = \frac{[p(s)+q(s)]^2}{2q(s)}$, $z_3 = -\frac{[p(s)+q(s)][p(s)-3q(s)-\sqrt{D}]}{4q(s)}$, where $D = [p(s) + q(s)][p(s) + 9q(s)]$.

The following theorem discribes the variation of the energy spectrum of $\tilde{H}_1$ in the one-dimensional case.

Theorem 1.

A. If $\omega \in A_2 \bigcup A_2$ ($\omega \in A_4 \bigcup A_8$), then the operator $\tilde{H}_1$ has exactly two LISs φ_1 and φ_2 with the respective energies z_1 and z_2 (z_2 and z_3), satisfying the inequalities $z_1 < z_2$ ($z_2 < z_3$) and $z_i < m_1, i = 1, 2$ ($z_j > M_1, j = 2, 3$).

B. If $\omega \in A_6$ ($\omega \in A_5$), then the operator $\tilde{H}_1$ has a single LIS φ with the energy $z = z_1$ ($z = z_3$) satisfying the inequality $z_1 < m_1$ ($z_3 > M_1$).

C. If $\omega \in A_1 \bigcup A_7$, then the operator $\tilde{H}_1$ has no LIS.

We sketch the proof of Theorem 1. In the one-dimensional case, the equation $\Delta_1(z) = 0$ is equivalent to the system of the two equations

$$1 + q(s)\int_T \frac{(1 - \cos t)^2 dt}{p(s)(1 - \cos t) - z} = 0 \quad (10)$$

and

$$1 + q(s)\int_T \frac{\sin^2 t\, dt}{p(s)(1 - \cos t) - z} = 0. \quad (11)$$

Integrating by quadratures in (10) and (11) with $z < m_1$ ($z > M_1$), whence the proof of Theorem 1 immediately follows in view of the existence conditions for the solutions.

In the case of the dimension $\nu = 2$, we introduce the following ranges for the pairs ω:

$$Q_1 = \{\omega : p(s) > 0, -p(s) \leq q(s) < 0\}, Q_2 = \{\omega : p(s) < 0, 0 < q(s) \leq -p(s)\},$$

$$Q_3 = \{\omega : p(s) > 0, -\frac{100}{27}p(s) \leq q(s) < -p(s)\}, Q_4 = \{\omega : p(s) < 0, \frac{25}{9}p(s) \leq q(s) < 0\},$$

$$Q_5 = \{\omega : p(s) > 0, 0 < q(s) < \frac{25}{9}p(s)\}, Q_6 = \{\omega : p(s) < 0, -p(s) \leq q(s) < -\frac{25}{9}p(s)\},$$

$$Q_7 = \{\omega : p(s) > 0, -\frac{100}{27}p(s) \leq q(s) < -p(s)\}, Q_8 = \{\omega : p(s) < 0, \frac{100}{27}p(s) \leq q(s) < \frac{25}{9}p(s)\},$$

$$Q_9 = \{\omega : p(s) > 0, \frac{25}{9}p(s) \leq q(s) < \frac{100}{27}p(s)\}, Q_{10} = \{\omega : p(s) < 0, -\frac{25}{9}p(s) \leq q(s) < -\frac{100}{27}p(s)\},$$

$$Q_{11} = \{\omega : p(s) > 0, -\frac{25}{9}p(s) < q(s)\}, Q_{12} = \{\omega : p(s) < 0, \frac{100}{27}p(s) < q(s)\},$$

$$Q_{13} = \{\omega : p(s) > 0, \frac{100}{27}p(s) \leq q(s)\}, Q_{14} = \{\omega : p(s) < 0, -\frac{100}{27}p(s) < q(s)\}.$$

The next theorem describes the variation of the energy spectrum of the operator $\tilde{H}_1$ in the two-dimensional case.

Theorem 2.
A. If $\omega \in Q_1 \bigcup Q_2$, then the operator $\tilde{H}_1$ has no LIS.
B. If $\omega \in Q_3 \bigcup Q_4$ ($\omega \in Q_5 \bigcup Q_6$), then $\tilde{H}_1$ has a single LIS φ with the energy z_1 (z_2) satisfying the inequality $z_1 < m_2$ ($z_2 > M_2$). Then related energy level is non-degenerate.
C. If $\omega \in Q_7 \bigcup Q_8$ ($\omega \in Q_9 \bigcup Q_10$), the the operator $\tilde{H}_1$ has exactly two LISs φ_1 and φ_2 with the respective energies z_1 and z_2, $z_i < m_2, i = 1, 2$ (z_3 and z_4, $z_j > M_2, j = 3, 4$), and the related energy levels are nondegenerate.
D. If $\omega \in Q_{11} \bigcup Q_{12}$ ($\omega \in Q_{13} \bigcup Q_{14}$), then $\tilde{H}_1$ has three LISs φ_1, φ_2 and φ_3 with the respective energies z_1, z_2 and z_3 (z_4, z_5 and z_6)) satisfying the inequalities $z_i < m_2, i = 1, 2, 3$ ($z_j > M_2, j = 4, 5, 6$). The energy levels z_1 and z_3 (z_4 and z_6) are nondegenerate, while $z_2(z_5)$ is two-fold degenerate.

Proof. The expressions

$$\varphi(z) = \int_{T^2} \frac{(1 - \cos t_1)(2 - \cos t_1 - \cos t_2)}{p(s)(2 - \cos t_1 - \cos t_2) - z} dt, \Psi(z) = \int_{T^2} \frac{\sin^2 t_1}{p(s)(2 - \cos t_1 - \cos t_2) - z} dt,$$

$$\theta(z) = \int_{T^2} \frac{(\cos t_1 - \cos t_2)^2}{p(s)(2 - \cos t_1 - \cos t_2) - z} dt$$

are increasing functions z for $z \notin [m_2; M_2]$. Their values can be calculated exactly at the points $z = m_2$ and $z = M_2$. For $z < m_2$ and $p(s) > 0$, the function $\varphi(z)$ increases from 0 to $\frac{1}{p(s)}$, $\Psi(z)$ increases from 0 to $\frac{27}{50p(s)}$. For $z > M_2$ and $p(s) > 0$, these functions also accordingly increase from $-\infty$ to 0, from $-\frac{9}{25p(s)}$ to 0, and from $-\frac{27}{50p(s)}$ to 0. If $p(s) < 0$ and $z < m_2$, then they respectively increase from 0 to ∞, from 0 to $-\frac{9}{25p(s)}$, and from 0 to $-\frac{27}{50p(s)}$. For $p(s) < 0$ and $z > M_2$, the functions $\varphi(z), \Psi(z)$ and $\theta(z)$ increase from $\frac{1}{p(s)}$ to 0, from $\frac{9}{25p(s)}$ to 0, and from $\frac{27}{50p(s)}$ to 0. Investigating the equation $\Delta_2(z) = 0$ outside the domain of the continuum spectrum, we immediately prove the assertion of the theorem.

In the case $\nu = 3$, we introduce the notation

$$a = \int_{T^3} \frac{\sin^2 t_1 dt_1 dt_2 dt_3}{3 - \cos t_1 - \cos t_2 - \cos t_3} = \int_{T^3} \frac{\sin^2 t_1 dt_1 dt_2 dt_3}{3 + \cos t_1 + \cos t_2 + \cos t_3},$$

$$b = \int_{T^3} \frac{(\cos t_1 - \cos t_2)^2 dt_1 dt_2 dt_3}{3 - \cos t_1 - \cos t_2 - \cos t_3} = \int_{T^3} \frac{(\cos t_1 - \cos t_2)^2 dt_1 dt_2 dt_3}{3 + \cos t_1 + \cos t_2 + \cos t_3}.$$

As is seen, we have $0 < a < b < 1$ and $2a < b$. We now consider the following subsets in Ω for the case $\nu = 3$:

$$Q_1 = \{\omega : p(s) > 0, -p(s) < q(s) < 0\}, Q_2 = \{\omega : p(s) > 0, 0 < q(s) < \frac{p(s)}{3}\},$$

$$Q_3 = \{\omega : p(s) < 0, \frac{p(s)}{3} < q(s) < 0\}, Q_4 = \{\omega : p(s) < 0, 0 < q(s) < -p(s)\},$$

$$Q_5 = \{\omega : p(s) > 0, -\frac{2p(s)}{b} < q(s) \le -p(s)\}, Q_6 = \{\omega : p(s) < 0, \frac{2p(s)}{b} < q(s) \le \frac{p(s)}{3}\},$$

$$Q_7 = \{\omega : p(s) > 0, \frac{p(s)}{3} < q(s) \le \frac{2p(s)}{b}\}, Q_8 = \{\omega : p(s) < 0, -p(s) < q(s) \le -\frac{2p(s)}{b}\},$$

$$Q_9 = \{\omega : p(s) > 0, -\frac{p(s)}{a} \le q(s) < -\frac{2p(s)}{b}\}, Q_{10} = \{\omega : p(s) < 0, \frac{p(s)}{a} < q(s) \le \frac{2p(s)}{b}\},$$

$$Q_{11} = \{\omega : p(s) > 0, \frac{2p(s)}{b} \le q(s) < \frac{p(s)}{a}\}, Q_{12} = \{\omega : p(s) < 0, -\frac{2p(s)}{b} \le q(s) < -\frac{p(s)}{a}\},$$

$$Q_{13} = \{\omega : p(s) > 0, -\frac{p(s)}{a} \le q(s)\}, Q_{14} = \{\omega : p(s) < 0, \frac{p(s)}{a} \le q(s)\},$$

$$Q_{15} = \{\omega : p(s) > 0, \frac{p(s)}{a} \le q(s)\}, Q_{16} = \{\omega : p(s) < 0, -\frac{p(s)}{a} \le q(s)\}.$$

Theorem 3.
A. If $\omega \in Q_1 \bigcup Q_2 \bigcup Q_3 \bigcup Q_4$, then the operator $\tilde{H}_1$ has no LIS.

B. If $\omega \in Q_5 \bigcup Q_6$ ($\omega \in Q_7 \bigcup Q_8$), then the operator $\tilde{H}_1$ has a single LIS φ with the energy $z < m_3$ ($z > M_3$). This energy level is nondegenerate.

C. If $\omega \in Q_9 \bigcup Q_{10}$ ($\omega \in Q_{11} \bigcup Q_{12}$), then $\tilde{H}_1$ has two LISs φ_1 and φ_2 with the respective energies z_1 and z_2 (z_3 and z_4) satisfying the inequalities $z_i < m_3, i = 1, 2$ ($z_j > M_3, j = 3, 4$). Furtheremore, the energy level z_1 (z_3) is nondegenerate, while z_2 (z_4) is two-fold degenerate.

D. If $\omega \in Q_{13} \bigcup Q_{14}$ ($\omega \in Q_{15} \bigcup Q_{16}$), then the operator $\tilde{H}_1$ accordingly has exactly three LISs φ_1, φ_2 and φ_3 with the energies z_1, z_2, and z_3 (z_4, z_5 and z_6) satisfying the inequalities $z_i < m_3, i = 1, 2, 3$ ($z_j > M_3, j = 4, 5, 6$). Moreover, the level z_1 (z_4) is nondegenerate, z_2 (z_5) is two-fold degenerate, and z_3 (z_6) is three-fold degenerate.

Theorem 3 is proved based on the monotonicity of the functions

$$\varphi(z) = \int_{T^3} \frac{(1 - \cos t_1)(3 - \cos t_1 - \cos t_2 - \cos t_3) dt}{p(s)h(t) - z}, \Psi(z) = \int_{T^3} \frac{\sin^2 t_1 dt}{p(s)h(t) - z},$$

$$\theta(z) = \int_{T^3} \frac{(\cos t_1 - \cos t_2)^2 dt}{p(s)h(t) - z}$$

for $z \notin [m_3; M_3]$. In what follows, we use the values of the Watson integrals [11]. It is necessary here to take into account that the measure is normalized in the case under consideration.

It can be similarly proved that in the $\nu-$ dimensional lattice, the system has at most three types of LISs (not counting the degeneracy multiplicities of their energy levels) with the energies $z_i \notin [m_\nu; M_\nu]$. Furthermore, for $i = 1, 2, 3$, the corresponding energy levels are nondegenerate, $\nu-$ fold degenerate, and $(\nu - 1)-$ fold degenerate. The domains of these LISs can also be found.

We now consider the case $p(s) \equiv 0$. If $p(s) \equiv 0$ and $J_n \neq 0, n = 1, 2, ..., 2s$, then $\Delta_\nu(z)$ assumes the form

$$\Delta_\nu(z) = det A . det B,$$

where

$$A = \begin{pmatrix} a_1 & b_1 & b_1 & b_1 ...... & b_1 \\ a_2 & b_2 & 0 & 0...... & 0 \\ a_2 & 0 & b_2 & 0...... & 0 \\ a_2 & 0 & 0 & b_2...... & 0 \\ \vdots & \vdots & \vdots & \vdots & \vdots \\ a_2 & 0 & 0 & 0...... & b_2 \end{pmatrix},$$

$$B = \begin{pmatrix} b_2 & 0 & 0...... & 0 \\ 0 & b_2 & 0...... & 0 \\ 0 & 0 & b_2...... & 0 \\ \vdots & \vdots & \vdots & \vdots \\ 0 & 0 & 0...... & b_2 \end{pmatrix},$$

where A is a $(\nu + 1) \times (\nu + 1)$ matrix, B is a diagonal $\nu \times \nu$ matrix, $a_1 = 1 - \nu q(s)/z$, $a_2 = q(s)/(2z)$, $b_1 = q(s)/z$, and $b_2 = 1 - q(s)/(2z)$.

Theorem 4. If $p(s) \equiv 0$ and $J_n \neq 0, n = 1, 2, ..., 2s$, then the operator $\tilde{H}_1$ has exactly two LISs (not counting the degeneracy multiplicities of their energy levels) φ_1 and φ_2 with the respective energies $z_1 = q(s)/2$ and $z_2 = (2\nu + 1)q(s)/2$. The energy level z_1 is $(2\nu - 1)-$ fold degenerate, while z_2 is nondegenerate.

Proof. The equation $\Delta_\nu(z) = 0$ is equivalent to the system of two equations

$$b_2^{2\nu-1} = 0 \tag{12}$$

and

$$a_1 b_2 - \nu a_2 b_1 = 0. \tag{13}$$

Equation (12) has a root equal to $z = q(s)/2$, and it is clear that its multiplicity is $2\nu - 1$, while Eq. (13) has a solution $z = z_2$. Consequently, for arbitrary values of ν, the system has at most three types of LISs.

B. Let $\alpha = 0$ (anisotropic case).

When $\varphi_n(\alpha) = 1$ and $\Psi_n(\alpha) = 0$; And in this case equation for eigenvalues has the form

$$\{\sum_{n=1}^{2s}(-1)^{n+1}4s^n J_n \nu - z\}f(x) + \sum_{n=1}^{2s}(-1)^{n+1}2s^n(J_n^0 - J_n)\int_{T^\nu}\{\nu + \sum_{i=1}^{\nu}\cos(x_i - t_i)\}f(t)dt = 0$$

$$\tag{14}.$$

Let $a(s) = \sum_{n=1}^{2s}(-1)^{n+1}4s^n J_n \nu$, $b(s) = \sum_{n=1}^{2s}(-1)^{n+1}2s^n(J_n^0 - J_n)$ and

$$\Delta_\nu(z) = (1 + \frac{\nu b(s)}{a(s) - z}) \times \{1 + \frac{b(s)}{2[a(s) - z]}\}^{2\nu}. \quad (15)$$

and $z_1 = a(s) + \nu b(s)$, $z_2 = a(s) + \frac{b(s)}{2}$.

The following theorem describes the variation of the energy spectrum of the operator $\tilde{H}_1$ in the case $\alpha = 0$.

Theorem 5. If $\alpha = 0$, then the operator $\tilde{H}_1$ has exactly two LISs φ_1 and φ_2 with the respective energies z_1 and z_2, satisfying the inequalities $z_i < m_\nu$ ($z_i > M_\nu$), if $b(s) < 0$ ($b(s) > 0$). The energy level z_1 are nondegenerate, while z_2 is $2\nu-$ fold degenerate.

C. Let $\alpha = -1$ (anisotropic case).

When $\varphi_n(\alpha) = 1 + \binom{n}{2} + \binom{n}{4} + ... + \binom{n}{n-k} + 1 = 2^{n-1}$ and $\Psi_n(\alpha) = -(\binom{n}{1} + \binom{n}{3} + \binom{n}{5} + ... + \binom{n}{n-1}) = -2^{n-1}$; And in this case equation for eigenvalue has the form:

$$\{\sum_{n=1}^{2s}(-1)^{n+1}2^{n+1}J_n s^n[\nu - \sum_{i=1}^{\nu}\cos x_i] - z\}f(x_1; x_2; ...; x_\nu) + \sum_{n=1}^{2s}(-1)^{n+1}(2s)^n(J_n^0 - J_n) \times$$

$$\times \int_{T^\nu}[\nu + \sum_{i=1}^{\nu}\{\cos(x_i - t_i) + \cos x_i + \cos t_i\}]f(t_1; t_2; ...; t_\nu)dt_1 dt_2...dt_\nu = 0 \quad (16).$$

Let $\varepsilon(s) = 2\sum_{n=1}^{2s}(-2s)^n(J_n^0 - J_n)$ and $I(s) = \sum_{n=1}^{2s}(-2)^{n+1}J_n s^n$, and

$$\Delta_\nu(z) = (1 - \frac{\varepsilon(s)}{2}\int_{T^\nu}\frac{(\cos t_1 - \cos t_2)^2 dt}{h(t) - z})^{\nu-1} \times (1 - \varepsilon(s)\int_{T^\nu}\frac{(1 + \cos t_1)(\nu + \sum_{i=1}^{\nu}\cos t_i)dt}{h(t) - z}) \times$$

$$\times (1 - \varepsilon(s)\int_{T^\nu}\frac{\sin^2 t_1 dt}{h(t) - z})^\nu;$$

We let $\mathcal{P}$ denote the range of all pairs $P = (I(s); \varepsilon(s))$ and introduce the following subsets in $\mathcal{P}$ for $\nu = 1$:

$G_1 = \{P : I(s) > 0, 0 < \varepsilon(s) \leq I(s)\}, G_2 = \{P : I(s) < 0, I(s) \leq \varepsilon(s) < 0\},$

$G_3 = \{P : I(s) < 0,) < \varepsilon(s) \leq -I(s)\}, G_4 = \{P : I(s) > 0, -I(s) \leq \varepsilon(s) < 0\},$

$G_5 = \{P : I(s) > 0, \varepsilon(s) > I(s)\}, G_6 = \{P : I(s) < 0, \varepsilon(s) > -I(s)\},$

$G_7 = \{P : I(s) > 0, \varepsilon(s) < -I(s)\}, G_8 = \{P : I(s) < 0, \varepsilon(s) < I(s)\}.$

We write

$$z_1 = \frac{I^2(s) + 6I(s)\varepsilon(s) - 3\varepsilon^2(s) - \sqrt{D}}{4\varepsilon(s)}, z_2 = -\frac{(I(s) - \varepsilon(s))^2}{2\varepsilon(s)}, z_3 = \frac{I^2(s) + 6I(s)\varepsilon(s) - 3\varepsilon^2(s) + \sqrt{D}}{4\varepsilon(s)};$$

where $D = 9\varepsilon^4(s) + I^4(s) + 12I^3(s)\varepsilon(s) + 30I^2(s)\varepsilon^2(s) + 28I(s)\varepsilon^3(s);$

The following theorem describes the variation of the energy spectrum of the operator $\tilde{H}_1$ in the one-dimensional case.

Theorem 6. A. if $P \in G_3 \bigcup G_4$, then the operator $\tilde{H}_1$ has no LIS.

B. If $P \in G_1$ $(P \in G_2)$, then the operator $\tilde{H}_1$ has a single LIS φ with the energy $z = z_1$ $(z = z_3)$ satisfying the inequality $z_1 < m_1$ $(z_3 > M_1)$.

C. If $P \in G_5 \cup G_6$ $(P \in G_7 \cup G_8)$, then the operator $\tilde{H}_1$ exactly two LISs φ_1 and φ_2 with the respective energies z_1 and z_2 (z_2 and z_3), satisfying the inequalities $z_1 < z_2$ ($z_2 > z_3$) and $z_i < m_1, i = 1, 2$ $(z_j > M_1, j = 2, 3)$.

We sketch the proof of Theorem 6. In the one-dimensional case, the equation $\Delta_\nu(z) = 0$ is equivalent to the system of the two equations

$$1 - \varepsilon(s) \int_{T^\nu} \frac{(1 + cost)^2 dt}{I(s)(1 - cost) - z} = 0 \qquad (17)$$

and

$$1 - \varepsilon(s) \int_{T^\nu} \frac{sin^2 t dt}{I(s)(1 - cost) - z} = 0 \qquad (18).$$

Integrating by quadratures in (17) and (18) with $z < m_1$ $(z > M_1)$, we obtain two algebraic equations for $z < m_1$ $(z > M_1)$, whence the proof of Theorem 6 immediately folloows in view of the existence conditions for the solutions.

In the case of the dimension $\nu = 2$, we introduce the following ranges for the pairs P:

$$Q_1 = \{P : I(s) > 0, 0 < \varepsilon(s) \le \frac{I(s)}{3}\}, Q_2 = \{P : I(s) < 0, \frac{I(s)}{3} \le \varepsilon(s) < 0,\}$$

$$Q_3 = \{P : I(s) > 0, \frac{I(s)}{3} < \varepsilon(s) \le I(s)\}, Q_4 = \{P : I(s) < 0, 0 < \varepsilon(s) \le -\frac{I(s)}{3}\},$$

$$Q_5 = \{P : I(s) > 0, -\frac{I(s)}{3} \le \varepsilon(s) < 0\}, Q_6 = \{P : I(s) < 0, I(s) \le \varepsilon(s) < \frac{I(s)}{3}\},$$

$$Q_7 = \{P : I(s) > 0, \varepsilon(s) > I(s)\}, Q_8 = \{P : I(s) < 0, \varepsilon(s) > -\frac{I(s)}{3}\},$$

$$Q_9 = \{P : I(s) > 0, \varepsilon(s) < -\frac{I(s)}{3}\}, Q_{10} = \{P : I(s) < 0, \varepsilon(s) < I(s)\}.$$

The next theorem describes the variation of the energy spectrum of $\tilde{H}_1$ in the two-dimensional case.

Theorem 7.

A. If $P \in Q_2 \cup Q_5$, then the operator $\tilde{H}_1$ has no LIS.

B. If $P \in Q_1 \cup Q_4$ $(P \in Q_3 \cup Q_6)$, then the operator $\tilde{H}_1$ has a single LIS φ with the energy $z = \tilde{z}_1$ $(z = \tilde{z}_2)$ satisfying the inequality $\tilde{z}_1 < m_2$ $(\tilde{z}_2 > M_2)$. Moreover, the energy level $\tilde{z}_1$ ($\tilde{z}_2$) is nondegenerate, if $P \in Q_1$ $(P \in Q_3)$. If $P \in Q_4$ $(P \in Q_6)$, then this energy level is two-fold degenerate.

C. If $P \in Q_7 \cup Q_8$ $(P \in Q_9 \cup Q_{10})$, then the operator $\tilde{H}_1$ accordingly has exactly three LISs φ_1, φ_2, and φ_3 with the energy levels z_1, z_2, and z_3 satisfying the inequalities $z_i < m_3$ $(z_i > M_3), i = 1, 2, 3$; Moreover, the level z_1 and z_2 is nondegenerate, z_3 is two-fold degenerate.

Theorem 7 is proved analoguesly to theorem 2.

It can be similarly proved that in the $\nu-$ dimensional ($\nu \le 3$) lattice, the system has at most three types of LISs (not counting the degeneracy multiplicities of their energy levels) with the energies $z_i \notin [m_\nu; M_\nu]$. Furtheremore, for $i = 1, 2, 3$, the corresponding energy levels are nondegenerate, $\nu-$ fold degenerate, and $(\nu-1)-$ fold degenerate. The domains of these LISs can also be found.

D. We now consider the general case, i. e., α is arbitrary number, $\alpha > 0$.

In this case equation $\Delta_\nu(z) = 0$ is equivalent to system of three equations:

$$(1 + A \int_{T^\nu} \frac{\varphi_n(\alpha) \sin^2 t_1 dt}{h_\alpha(t) - z})^\nu = 0 \tag{19}$$

and

$$(1 + \frac{A}{2} \int_{T^\nu} \frac{(\cos t_1 - \cos t_2)^2 dt}{h_\alpha(t) - z})^{\nu-1} = 0, \tag{20}$$

and

$$\{1 + A \int_{T^\nu} \frac{\cos t_1 [\varphi_n(\alpha) \sum_{i=1}^\nu \cos t_i - \nu \Psi_n(\alpha)] dt}{h_\alpha(t) - z}\} \times \{1 + a \int_{T^\nu} \frac{\nu \varphi_n(\alpha) - \Psi_n(\alpha) \sum_{i=1}^\nu \cos t_i}{h_\alpha(t) - z} dt\} -$$
$$- A_2 \int_{T^\nu} \frac{\varphi_n(\alpha) \sum_{i=1}^\nu \cos t_i - \nu \Psi_n(\alpha)}{h_\alpha(t) - z} dt \times \int_{T^\nu} \frac{\cos t_1 [\nu \varphi_n(\alpha) - \Psi_n(\alpha) \sum_{i=1}^\nu \cos t_i]}{h_\alpha(t) - z} dt = 0, \tag{21}$$

where $dt = dt_1 dt_2 ... dt_\nu$, and $a = \sum_{n=1}^{2s} (-1)^{n+1} 2s^n (J_n^0 - J_n)$.

Equation (19) has a single solution and this solutions is $\nu-$ fold degenerate, while Eq. (20) has a also single solution, and this solution is $(\nu-1)-$ fold degenerate.

Expressing all integrals in the equation (21) through the integral $J^\star(z) = \int_{T^\nu} \frac{dt_1 dt_2 ... dt_\nu}{h_\alpha(t_1, t_2, ..., t_\nu) - z}$, we find that the equation (21) is equivalent to the equation

$$\eta_\alpha(z) J^\star(z) = \theta_\alpha(z), \tag{22}$$

where

$$\eta_\alpha(z) = -\nu^2 \varepsilon(s) I^2(s) \varphi_n(\alpha) \Psi_n^2(\alpha) - 2\nu \varepsilon(s) I(s) \Psi_n^2(\alpha) [z - \nu I(s) \varphi_n(\alpha)] - \varepsilon(s) \varphi_n(\alpha) [z - \nu I(s) \varphi_n(\alpha)]^2 +$$
$$+ \nu \varepsilon^2(s) \Psi_n(\alpha) [z - \nu I(s) \varphi_n(\alpha)] - \nu \varepsilon^2(s) \varphi_n^2(\alpha) [z - \nu I(s) \varphi_n(\alpha)],$$

and

$$\theta_\alpha(z) = -\nu I^2(s) \Psi_n^2(\alpha) + 2\nu \varepsilon(s) I(s) \Psi_n^2(\alpha) + \varepsilon(s) \varphi_n(\alpha) [z - \nu I(s) \varphi_n(\alpha)] + \nu \varepsilon^2(s) \Psi_n^2(\alpha) + \nu \varepsilon^2(s) \varphi_n^2(\alpha).$$

In turn, for $\eta_\alpha(z) \neq 0$, the latter equation is equivalent to the equation of the form

$$J^\star(z) = \frac{\theta_\alpha(z)}{\eta_\alpha(z)}. \tag{23}$$

Analyzing Eq. (23) outside the continuous spectrum of the operator $\tilde{H}_1$, and taking into account that the function $J^\star(z)$ is monotonic for $z \notin \sigma_{cont.}(\tilde{H}_1)$, we can easily verify that the equation has no more than two solutions outside the continuous spectrum.

Consequently, in the general case, the system has at most four types of LISs (not counting the degeneracy multiplicities of their energy levels) with the energies $z_i \notin \sigma_{cont.}(\tilde{H}_1)$. Furtheremore, the corresponding energy levels them two LISs are nondegenerate, the third one is $\nu-$ fold degenerate, and, finally, the fourth LIS is $\nu-1-$ fold degenerate. We can found conditions of the existence and regions of the existence their LISs.

References

[1] Yu. A. Izyumov, *JETP*, **21**, 381-388 (1965);

[2] V.V. Gann and L.G. Zazunov, *Fiz. Tverd. Tela*, **15**, 3535-3569 (1973);

[3] Y.-L. Wang and H. Callen, *Phys. Rev.*, **160**, 358-363 (1967);

[4] T.Oguchi and I.Ono, *J. Phys. Soc. Japan*, **26**, 32-42 (1969);

[5] T. Wolfram and J. Callawy, *Phys. Rev.*, **130**, 2207-2217 (1963);

[6] I.Ono and Y.Endo, *Phys. Rev. Lett. A,* **41**, 440-442 (1972);

[7] S.M.Tashpulatov, *Theor. Math. Phys.*, **126**, 403-408 (2001).

[8] E. Schrodinger, *Pros. Roy. Irish. Acad. A,* **48**, 39, (1941).

[9] S.M.Tashpulatov, *Theor. Math. Phys.*, **142**, 71-78 (2005).

[10] M.Reed and B. Simon, *Methods of Modern Mathematical Physics*, Vol. 1, Functional Analysis, Acad. Press, New York (1972).

[11] V.V.Val'kov, S.G. Ovchinnikov, and O.P. Petrakovskii, *Fiz. Tverd. Tela,* **30**, 3044-3047 (1988).

In: Theoretical Physics and Nonlinear Optics
Editors: Thomas F. George et al
ISBN 978-1-61122-939-4
© 2012 Nova Science Publishers, Inc.

Chapter 6

GROUND STATE MASS SPECTRA OF LOW LYING BARYONS IN AN EQUALLY MIXED SCALAR AND VECTOR POTENTIAL MODEL

S.N. Jena[1,*], *T.C. Tripathy*[2,†] *and M.K. Muni*[3,]
[1]Department of Physics, Berhampur University,
Berhampur-760 007, Orissa, India
[2]Department of Physics, Roland Institute of Technology,
Surya Vihar, Berhampur-761008, Orissa, India
[3]Department of Mathematics and Science, SMIT,
Ankuspur, Ganjam, Orissa, India

Abstract

Assuming the baryons as an assembly of independent quarks, confined in a first approximation by an equally mixed scalar and vector linear potential which presumably represents the non-perturbative multi-gluon interactions including gluon self-coupling, the ground state mass spectra of low-lying baryons are studied, taking into account perturbatively the contribution of the quark-gluon coupling due to one-gluon exchange and that of the Goldstone boson (π, η and k) exchange interaction arising from spontaneous breaking of chiral symmetry over and above that of the center-of-mass motion. The results obtained for the masses of the ground state baryons with a suitable choice of the strong coupling constant $\alpha_c = 0.16$ agree reasonably well with the corresponding experimental values.

1. Introduction

In earlier works an equally mixed scalar and vector linear potential of the form

$$V_q(r) = \frac{1}{2}\left(1 + \gamma^0\right)(a^2 r + V_0) \tag{1}$$

---

[*]E-mail address: snjena@rediffmail.com
[†]E-mail address: tarinitripathy@yahoo.com

with $a > 0$, has been found quite successful in the study of static baryon properties[1], the weak-electric and -magnetic form factors for the semileptonic baryon decays[2]. Then incorporating chiral symmetry in the $SU(2)$ flavour sector in the usual manner, this model has been used to study the electromagnetic properties of nucleons[3] and the magnetic moments of the baryons in the nucleon octet[4] in reasonable agreement with the experimental data. This model has also been adopted to study successfully the mass and decay constant of the $(q\bar{q})$ pion [5], S-state mass splittings of the mesons of $s\bar{s}$, $c\bar{c}$ and $b\bar{b}$ systems, ground state mass splittings of the heavy non-self-conjugate mesons in strange, charm and bottom flavour sector [6] and electromagnetic decays of mesons[6,7]. This model has also been employed to explain reasonably well the mass spectrum of octet baryons taking into account the contributions due to the colour-electric and magnetic energies arising out of the residual OGE interaction alongwith that due to the residual quark-pion coupling arising out of the requirement of the chiral symmetry and the neccessary centre-of-mass(c.m) motion. But in this work the interactions between the constituent quarks arising out of the Goldstone-boson exchange (GBE) which are considered to play an important role in contributing to the energy of the baryon core were not taken into account. Therefore in the present work we wish to take into account the GBE contributions in a perturbative manner to study the ground state mass spectrum of octet baryons.

In the past several non-relativistic quark models have been used [8] for the study of light and strange baryons. Recently, Glozman et al [9] have shown that a chiral constituent quark model with a Q-Q interaction relying solely on GBE is capable of providing a unified description not only of the N and Δ spectra but also of all strange baryons in good agreement with phenomenology. They have also presented a constituent quark model with the confinement potential in linear and harmonic [9] forms for the light and strange baryons providing a unified description of their ground states and excitation spectra. Their model which relies on constituent quarks and Goldstone bosons arising as effective degrees of freedom of low energy QCD from the spontaneous breaking of chiral symmetry (SBCS) has been found to be quite proficient in reproducing the spectra of the three quark systems from a precise variational solution of a Schroedinger equation with a semi-relativistic Hamiltonian.

Thus several papers based on non-relativistic quark models have appeared in the literature in connection with the study of the mass spectrum of light and strange baryons. Although the phenomenological picture is reasonable at the non-relativistic level, a relativistic approach is quite indispensable on this account in view of the fact that the baryonic mass splittings are of the same order as the constituent quark masses. Of course, the chiral constituent quark model which has been constructed by Glozman et al [9] in a semi-relativistic framework is a step in this direction and shows an essential improvement over non-relativistic approaches. On the other hand, the MIT bag model [10] has also been found to be relatively successful in this respect. In its improved versions, the Chiral Bag Model (CBM) [11] have included the effect of pion self-energy due to baryon-pion coupling at the vertex to provide a better understanding of the baryon masses. Nevertheless, such models still contain some dubious phenomenological elements which are objectionable. The sharp spherical bag boundary, the zero point energy, the exclusion of pions from within the bag or adhoc inclusion of pions within it, are a few such points to be noted in this context. Furthermore it is somewhat difficult to believe that the static spherical bag remains unperturbed even after the creation of a pion. However, the sharp spherical bag boundary in the bag confinement, which is at the root of all objections and difficulties encountered by the otherwise successful CBM, is nonetheless arbitrary and phenomenological in nature and can therefore be replaced by an alternative and suitable phenomenological average potential for individual quarks, presuming at the same time its good features together with its successful predictions in the study of light baryons in their ground states.

The chiral potential models [12] which are comparatively more straight-forward in the above respects are obviously attempts in this direction. In such models the confining potentials which basically represent the interaction of quarks with the gluon field are usually assumed phenomenologically as Lorentz scalars in harmonic and cubic forms. Potentials of a different type of Lorentz structure with equally mixed scalar and vector parts in harmonic [13], non-Coulombic power-law[14], square-root [15], and logarithmic [16] form are also used in this context.

In the present work we prefer to work in an alternative, but similar scheme based on Dirac equation with a purely phenomenological individual quark potential in equation(1). In this work we take the Lorentz structure of the potential as an equal admixture of scalar and vector parts because of the fact that both the scalar and vector parts in equal proportions at every point render the solvability of the Dirac equation for independent quarks by reducing it to the form of a $Schr\ddot{o}dinger$-like equation. This Lorentz structure of the potential also has an additional advantage of generating no spin-orbit splittings, as observed in the experimental baryon spectrum.

In the present work, we assume that the constituent quarks in a baryon core are independently confined in a first approximation by a relativistic potential of the form in equation(1) which presumably represents the nonperturbative multigluon interactions. Here we take into account perturbatively the contributions of OGE alongwith that of GBE arising from the SBCS over and above the c.m corrections. In the present model we consider the constituent quarks of flavours u, d, s with masses considerably larger than the corresponding current quark masses so that the underlying chiral symmetry of QCD is spontaneously broken. As a consequence of SBCS, at the same time Goldstone bosons appear, which couple directly to the constituent quarks[9]. Hence beyond the scale of SBCS one is left with constituent quarks with dynamical masses related to $<\bar{q}q>$ condensates and with Goldstone bosons as the effective degrees of freedom. This feature, that in the Nambu-Goldstone mode of chiral symmetry constituent quark and Goldstone boson fields prevail together, is well supported, e.g., by the σ model[17] or the NJL model[18]. In the same framework also with the spin and flavor content of the nucleon are naturally resolved[19].

The present work is organised as follows. In section-II we outline the potential model with the solutions for the relativistic bound states of the individually confined quarks in the ground states of baryons.

Section-III provides a brief account of the energy correction due to the spurious c.m motion. This section also provides an account of a further correction to the baryon mass due to color-electric and -magnetic interaction energies originating from the hopefully weak residual OGE interaction, treated perturbatively. A brief account of the correction due to GBE interactions between the constituent quarks arising out of SBCS is also presented in this section. Finally, in Section-IV we present the results of the present model for the ground state masses of light and strange baryons which are in very good agreement with the corresponding experimental values.

2. Basic Framework

In baryonic dimensions corresponding to light quark configuration, the long distance confining part of the interaction arising out of the non-perturbative multigluon mechanism including gluon self-couplings is presumed to be dominant in comparision with any other residual interactions such as the quark-gluon interactions originating from OGE at short distances and also the GBE interactions arising from SBCS. Therefore, to a first approximation, the confining part of the interaction represented here by an average flavour independent potential of the form in equation(1) is believed to provide zeroth order constituent quark dynamics inside such baryons. We further assume that the constituent independent quarks obey the Dirac equation

$$\left[\gamma^0 E_q - \vec{\gamma} \cdot \vec{p} - m_q - V_q(r)\right] \psi_q(\vec{r}) = 0 \tag{2}$$

implying thereby a Lagrangian density of zeroth order

$$\mathcal{L}_q^0(x) = \overline{\psi}_q(x) \left[\frac{i}{2}\gamma^\mu \overleftrightarrow{\partial}_\mu - m_q - V_q(r)\right] \psi_q(x) \tag{3}$$

Assuming that all the quarks in a baryon core are in their ground $1S_{\frac{1}{2}}$ states, the normalised quark wave function $\psi_q(\vec{r})$ satisfying the Dirac equation (2) can be written in two component form as

$$\psi_q(\vec{r}) = N_q \begin{pmatrix} \phi_q(\vec{r}) \\ \frac{\vec{\sigma} \cdot \vec{p}}{\lambda_q} \phi_q(r) \end{pmatrix} \chi \uparrow \tag{4}$$

where $\phi_q(\vec{r})$ is the radial angular part of the upper component $\psi_q(\vec{r})$ and is given by

$$\phi_q(\vec{r}) = \frac{i}{\sqrt{4\pi}} f_q(r)/r \tag{5}$$

Finally the reduced radial part $f_q(r)$ can be found to satisfy a *Schrödinger*-type equation

$$f_q''(r) + \left[\lambda_q(E_q - m_q - a^2 r - V_0)\right] f_q(r) = 0 \tag{6}$$

which can be transformed into a convenient dimensionless form

$$f_q''(\rho) + (\epsilon_q - \rho) f_q(\rho) = 0 \tag{7}$$

where ($\rho = r/r_{0q}$) is a dimensionless variable with $r_{0q} = (\lambda_q a^2)^{-1/3}$ and

$$\epsilon_q = \left(\frac{\lambda_q}{a^4}\right)^{\frac{1}{3}} (E_q - m_q - V_0) \tag{8}$$

The equation(7) provides the basic eigen value equation, which can be solved as follows:
With $z = \rho - \epsilon_q$, equation(7) reduces to the Airy equation

$$f_q''(z) - z f_q(z) = 0 \tag{9}$$

The solution $f_q(z)$ of equation(9) is the Airy function $Ai(z)$. Since at $r = 0$ we require $f_q(r) = 0$, we have $Ai(z) = 0$ at $z = -\epsilon_q$. If z_n are the roots of the Airy function such that $Ai(z_n) = 0$, then we have $z = -\epsilon_q = z_n$. For the ground state of quarks the ϵ_q value is given by the first root z_1 of the Airy function so that

$$\epsilon_q = -z_1 \qquad (10)$$

The value of this root $z_1 = -2.33811$.

Now the individual quark binding energy E_q of zeroth order in the baryon ground state can be obtained from energy eigen value condition given in equation(8) through the relation

$$E_q = m_q - V_0 + ax_q \qquad (11)$$

where x_q is the solution of the root equation obtained through substitution from the equation(8) in the form

$$x_q^4 + bx_q^3 - \epsilon_q^3 = 0 \qquad (12)$$

with $b = \frac{2E_q + V_0}{a}$.

The quark binding energy E_q leads immediately to the energy of the baryon core in zeroth order as

$$M_B^0 = E_B^0 = \sum_q E_q \qquad (13)$$

The over all normalisation constant N_q of $\psi_q(\vec{r})$ appearing in equation(4) is obtained in a simplified form in the present model as

$$N_q^2 = \frac{3(E_q + m_q)}{(4E_q + 2m_q - V_0)} \qquad (14)$$

3. Energy Corrections to Baryon Masses

The contribution of quark-binding energy to the mass M_B^0 of the baryon core needs correction due to the center-of-mass motion, OGE interaction and GBE interaction between the quarks which need to be calculated separately in order to obtain the physical masses of the baryons.

A. Centre of Mass Correction

In this model there would be a sizeable spurious contribution to the energy from the motion of the centre-of-mass of the three-quark system. Unless this aspect is duly accounted for, the concept of the independent motion of quarks inside the baryon core will not lead to a physical baryon state of definite momentum. Although there is still some controversy on this subject, we adopt here the technique adopted by Bartelski et al[20] and E.Eich et al[21], which is just one way of accounting for the c.m motion. Following their prescription a ready estimate of the c.m momentum $\vec{P}_B$ of the baryon core can be obtained as

$$<\vec{P}_B^2> = \sum_q <\vec{p}_q^2> \qquad (15)$$

where $<\vec{p}_q^2>$ is the average value of the square of the individual quark momentum taken over the $1S_{\frac{1}{2}}$ single quark states $\psi_q(\vec{r})$ and is given in this model as

$$<\vec{p}^2>_q = \frac{(E_q+m_q)(E_q-m_q-V_0)(8E_q+2m_q-3V_0)}{5(4E_q+2m_q-V_0)} \qquad (16)$$

Therefore, if E_B^0 is the energy of the baryon core in zeroth order then the centre-of-mass corrected baryon mass is

$$E_B = <E_B^2>^{\frac{1}{2}} = \left[E_B^{0\,2} - <\vec{P}_B^2>\right]^{1/2} \qquad (17)$$

which provides the necessary centre-of-mass correction to the energy of the baryon core as

$$(\Delta E_B)_{c.m} = \left(E_B - E_B^0\right) = \left[\left(E_B^{0\,2} - <\vec{P}_B^2>\right)^{1/2} - E_B^0\right]. \qquad (18)$$

B. One-gluon-exchange Correction

The individual quarks in a baryon-core are considered so far to be experiencing the only force coming from the average effective potential $V_q(r)$ in equation(1) which is assumed to provide a suitable phenomenological description of the non-perturbative gluon interaction including gluon self-couplings. All that remains inside the quark core is the hopefully weak OGE interaction provided by the interaction Lagrangian density

$$\mathcal{L}_I^g = \sum_{\alpha=1}^{8} J_i^{\mu\alpha}(x) A_\mu^\alpha(x) \qquad (19)$$

where $A_\mu^\alpha(x)$ are the eight-vector gluon fields and $J_i^{\mu\alpha}(x)$ is the i-th quark colour current. Since at small distances the quarks should almost be free, it is reasonable to calculate the energy shift in the mass spectrum arising out of the quark interaction energy due to their coupling to the coloured gluons, using a first-order perturbation theory.

If we only keep the terms of order α_c, the problem reduces to evaluating the diagrams shown in (fig.1), where (fig.1(a)) corresponds to the OGE part while (fig.1(b)) implies the quark self-energy that normally contributes to the renormalization of quark masses. If $\vec{E}_i^\alpha$ and $\vec{B}_i^\alpha$ are the colour-electric and -magnetic fields, respectively, generated by the i-th quark colour-current

$$J_i^{\mu\alpha}(x) = g_c \overline{\psi}_i(x) \gamma^\mu \lambda_i^\alpha \psi_i(x) \qquad (20)$$

with λ_i^α being the usual Gell-Mann SU(3) matrices and $\alpha_c = (g_c^2/4\pi)$, then the contribution to the mass due to relevant diagram can be written as a sum of the colour-electric and -magnetic parts as

$$(\Delta E_B)_g = (\Delta E_B)_g^e + (\Delta E_B)_g^m \qquad (21)$$

where

$$(\Delta E_B)_g^e = \frac{1}{8\pi} \sum_{i,j} \sum_{\alpha=1}^{8} \int\int \frac{d^3\vec{r}_i d^3\vec{r}_j}{|\vec{r}_i - \vec{r}_j|} <B|J_i^{0\alpha}(\vec{r}_i) J_j^{0\alpha}(\vec{r}_j)|B> \qquad (22)$$

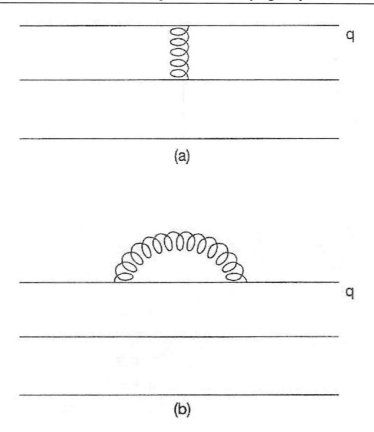

Figure 1. One-gluon-exchange contribution to the energy.

$$(\Delta E_B)_g^m = \frac{-1}{4\pi} \sum_{i<j} \sum_{\alpha=1}^{8} \int \int \frac{d^3\vec{r}_i d^3\vec{r}_j}{|\vec{r}_i - \vec{r}_j|} <B|\vec{J}_i^\alpha(\vec{r}_i) \cdot \vec{J}_j^\alpha(\vec{r}_j)|B> \qquad (23)$$

We have not included the self-energy diagram in the calculation of the magnetic part of the interaction, which contributes to the renormalisation of the quark masses and can possibly be accounted for in the phenomenological quark masses. The exclusion of this diagram, however, requires that each $\vec{B}_i^\alpha$ should satisfy the boundary condition $\hat{r} \times \vec{B}_i^\alpha = 0$, separately at the edge of the confining region, which is a possible case. On the other hand, as the electric field $\vec{E}_i^\alpha$ is necessarily in the radial direction, it is only possible to satisfy the boundary condition $\hat{r} \times (\sum_i E_i^\alpha) = 0$ for a colour singlet state $|B>$ for which $(\sum_i \lambda_i^\alpha = 0)$. Therefore, in order to preserve the boundary conditions we are forced to take into account the self-energy diagrams in (fig.1(b)) in the calculation of the electric part only. Now using equation(4) in equation(20) we find

$$\begin{aligned} J_i^{0\alpha}(\vec{r}_i) &= g_c \lambda_i^\alpha N_i^2 \left[\phi^2(r_i) + \frac{\phi'^2(r_i)}{\lambda_i^2} \right] \\ \vec{J}_i^\alpha(\vec{r}_i) &= -2g_c \lambda_i^\alpha N_i^2 (\vec{\sigma}_i \times \hat{r}_i) \frac{\phi(r_i)\phi'(r_i)}{\lambda_i} \end{aligned} \qquad (24)$$

Again using equation(24) together with the identity

$$\frac{1}{|\vec{r}_i - \vec{r}_j|} = \frac{1}{2\pi^2} \int \frac{d^3\vec{k}}{k^2} e^{-i\vec{k}\cdot(\vec{r}_i-\vec{r}_j)}$$

in equations.(22) and (23), we obtain

$$(\Delta E_B)_g^e = \frac{\alpha_c}{4\pi^2} \sum_{i,j}\langle\sum_\alpha \lambda_i^\alpha \lambda_j^\alpha\rangle N_i^2 N_j^2 \int \frac{d^3\vec{k}}{k^2} F_i^e(k) F_j^e(k) \qquad (25)$$

$$(\Delta E_B)_g^m = \frac{-2\alpha_c}{\pi^2} \sum_{i<j}\langle\sum_\alpha \lambda_i^\alpha \lambda_j^\alpha\rangle \frac{N_i^2 N_j^2}{\lambda_i \lambda_j} \int \frac{d^3\vec{k}}{k^2} \vec{F}_i^m(k) \cdot \vec{F}_j^m(k) \qquad (26)$$

where

$$F_i^e(k) = \frac{1}{\lambda_i^2}\left[\left((2E_i - V_0)\lambda_i - k^2\right) << j_0(|\vec{k}|r_i) >> -\lambda_i a^2 << r_i j_0(|\vec{k}|r_i) >>\right] \qquad (27)$$

$$\vec{F}_i^m(k) = \frac{i}{2} << j_0(|\vec{k}|r_i) >> \left(\vec{\sigma}_i \times \vec{k}\right) \qquad (28)$$

Then equations(25) and (26) can be written as

$$(\Delta E_B)_g^e = \frac{\alpha_c}{\pi} \sum_{i,j}\langle\sum_\alpha \lambda_i^\alpha \lambda_j^\alpha > N_i^2 N_j^2 I_{ij}^e \qquad (29)$$

$$(\Delta E_B)_g^m = \frac{-4\alpha_c}{3\pi} \sum_{i<j}\langle\sum_\alpha \lambda_i^\alpha \lambda_j^\alpha (\vec{\sigma}_i \cdot \vec{\sigma}_j)\rangle \frac{N_i^2 N_j^2}{\lambda_i \lambda_j} I_{ij}^m \qquad (30)$$

where

$$I_{ij}^e = \int_0^\infty dk F_i^e(k) F_j^e(k) \qquad (31)$$

$$I_{ij}^m = \int_0^\infty dk k^2 << j_0(|\vec{k}|r_i) >><< j_0(|\vec{k}|r_j) >> \qquad (32)$$

Finally, by taking into account the specific quark flavour and spin configurations in various states of baryon and using the relations $<\sum_\alpha(\lambda_i^\alpha)^2>=\frac{16}{3}$ and $<\sum_\alpha \lambda_i^\alpha \lambda_j^\alpha>_{i\neq j}=\frac{-8}{3}$ for baryons, in general one can write the energy correction due to OGE as

$$\begin{aligned}(\Delta E_B)_g^e &= \alpha_c\left(a_{uu}T_{uu}^e + a_{us}T_{us}^e + a_{ss}T_{ss}^e\right)\\ (\Delta E_B)_g^m &= \alpha_c\left(b_{uu}T_{uu}^m + b_{us}T_{us}^m + b_{ss}T_{ss}^m\right)\end{aligned} \qquad (33)$$

where a_{ij} and b_{ij} are the numerical coefficients depending on each baryon listed in Table-1 and the terms $T_{ij}^{e,m}$ are

$$T_{ij}^e = \frac{48(E_i + m_i)(E_j + m_j)}{\pi(4E_i + 2m_i - V_0)(4E_j + 2m_j - V_0)} I_{ij}^e \qquad (34)$$

$$T_{ij}^m = \frac{32}{\pi(4E_i + 2m_i - V_0)(4E_j + 2m_j - V_0)} I_{ij}^m \qquad (35)$$

From Table-1 one can note that the colour-electric contribution for the baryon masses vanishes when all the constituent quark masses in a baryon are equal, whereas it is non-zero otherwise. However, even in the case of strange baryons, it would subsequently be seen that the colour-electric contribution is quite small. Therefore, the degeneracy among the baryons is essentially removed through the spin-spin interaction energy in the colour-magnetic part.

Table 1. The coefficients a_{ij} and b_{ij} used in the calculation of the colour-electric and -magnetic energy corrections due to OGE

Baryons	a_{uu}	a_{us}	a_{ss}	b_{uu}	b_{us}	b_{ss}
N	0	0	0	-3	0	0
Δ	0	0	0	+3	0	0
Λ	1	-2	1	-3	0	0
Σ	1	-2	1	1	-4	0
Ξ	1	-2	1	0	-4	1
Σ^*	1	-2	1	1	2	0
Ξ^*	1	-2	1	0	2	1
Ω^-	0	0	0	0	0	+3

C. Goldstone Boson Exchange Correction

The coupling of Goldstone bosons (π, η, k) to the constituent quarks arising from SBCS in QCD can be taken into account over and above the dominant quark confining interactions arising out of the non-perturbative multigluon interactions. Such GBE interactions between the core quarks can be treated in a perturbative manner like it is done in the CBM[11]. Here the fields of Goldstone bosons may be treated independently without any constraint and their interactions with quarks may be assumed to be linear as it is done in case of pion in earlier works[22].

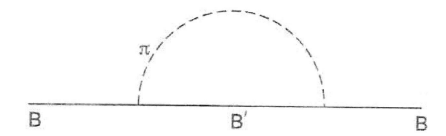

Figure 2. Baryon self-energy due to coupling with pion.

Following the Hamiltonian technique[23] as has been used in the CBM, we can first of all describe the effect of pion coupling in low-order perturbation theory as follows. The pionic self-energy of the baryons can be evaluated with the help of the single-loop self-energy diagram (fig.2) as

$$\sum_B (E_B) = \sum_k \sum_{B'} \frac{V^{\dagger BB'} V^{BB'}}{(E_B - \omega_k - M_{B'}^0)} \qquad (36)$$

where $\sum_k = \sum_j \int \frac{d^3 \vec{k}}{(2\pi)^3}$.

Here j corresponds to the pion-isospin index and B' is the intermediate baryon state. $V^{BB'}(\vec{k})$ is the general baryon-pion absorption vertex function obtained[24] in this model

as
$$V_j^{BB'}(\vec{k}) = i\sqrt{4\pi}\frac{f_{BB'\pi}}{m_\pi}\frac{ku(k)}{\sqrt{2\omega_k}}\left(\vec{\sigma}^{BB'}\cdot\hat{k}\right)\tau_j^{BB'} \tag{37}$$

where $\vec{\sigma}_j^{BB'}$ and $\tau_j^{BB'}$ are spin and isospin matrices and $\omega_k^2 = \vec{k}^2 + m_\pi^2$. The form factor u(k) in this model can be expressed as

$$u(k) = \frac{5N_{nl}^2}{3\lambda_{nl}g_A}[(2m_q + V_0) << j_0(|\vec{k}|)r) >> + a^2 << rj_0(|\vec{k}|r) >>$$
$$+ a^2 << j_1(|\vec{k}|r)/k >>] \tag{38}$$

where $j_0(|\vec{k}|r)$ and $j_1(|\vec{k}|r)$ represent the zeroth and first order spherical Bessel functions, respectively. The double angular brackets stand for the expectation values with respect to $\phi_q(r)$. In this model the axial vector coupling constant g_A for the beta decay of the neutron is given by

$$g_A = \frac{5}{9}\left(\frac{8E_q + 10m_q + V_0}{4E_q + 2m_q - V_0}\right)$$

Now with the vertex function $V_j^{BB'}(\vec{k})$ at hand, it is possible to calculate the pionic self-energy for various baryons with appropriate baryon intermediate states contributing to the process. For degenerate intermediate states on mass shell with $M_B^0 = M_{B'}^0$, the self-energy correction becomes

$$(\delta M_B)_\pi = \sum_B \left(E_B^0 = M_B^0 = M_{B'}^0\right) = -\sum_{k,B'}\frac{V^{\dagger BB'}V^{BB'}}{\omega_k} \tag{39}$$

Now using equation(37), we find

$$(\delta M_B)_\pi = \frac{-I_\pi}{3}\sum_{B'} C_{BB'} f_{BB'\pi}^2 \tag{40}$$

where
$$C_{BB'} = \left(\vec{\sigma}^{BB'}\cdot\vec{\sigma}^{B'B}\right)\left(\vec{\tau}^{BB'}\cdot\vec{\tau}^{B'B}\right) \tag{41}$$

and
$$I_\pi = \frac{1}{\pi m_\pi^2}\int_0^\infty \frac{dk k^4 u^2(k)}{\omega_k^2} \tag{42}$$

For the intermediate baryon states B', we consider only the octet and decuplet ground states. The pionic self-energy $(\delta M_B)_\pi$ for different baryons can be computed by using the values of $f_{BB'\pi}$ and $C_{BB'}$[8] as has been done in our earlier work[22]. The self-energy $(\delta M_B)_\pi$ calculated here contains both the quark self-energy (fig.3(a)) and the one-pion-exchange contributions (fig.3(b)).

The corrections due to η-exchange and kaon exchange interactions between the quarks can be calculated following the same approach as we have used in case of pion. Thus in general, one writes energy corrections due to GBE interactions as

$$(\delta M_B)_\chi = \frac{-I_\chi}{3}\sum_{B'} C_{BB'} f_{BB'\chi}^2 , \tag{43}$$

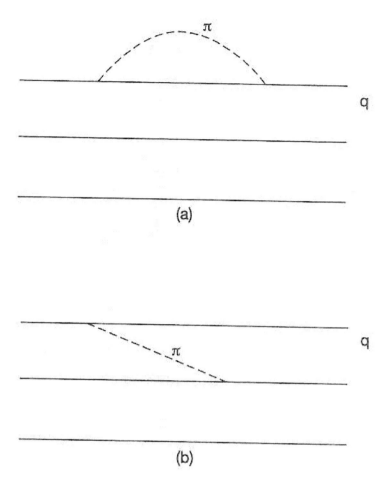

Figure 3. One-pion-exchange contributions to the energy.

where $C_{BB'}$ is given by equation(41) and

$$I_\chi = \frac{1}{\pi m_\chi^2} \int_0^\infty \frac{dk k^4 u^2(k)}{k^2 + m_\chi^2} \ , \qquad (44)$$

where $\chi = \pi, \eta$,k. Following the discussions given in ref.[8] the baryon-meson coupling constant $f_{BB'\chi}$ can be expressed in terms of the nucleon-meson coupling constant $f_{NN\chi}$. Pion exchange interaction acts only between light quarks whereas η-exchange is allowed in all quark pair states. The kaon exchange interaction takes place in u-s and d-s pair states.

4. Results and Discussion

The zeroth-order mass $M_B^0 = E_B^0$ of the ground state of a baryon arising out of the binding energies of the constituent quarks, confined independently by a phenomenological average potential $V_q(r)$ which presumably represents the dominant non-perturbative multigluon interactions is subjected to certain corrections due to the residual quark-gluon interaction and the GBE interaction between the constituent quarks, together with that due to the spurious centre-of-mass motion. All of these corrections can be treated independently, as though they are of the same order of magnitude, so that the physical mass of a baryon can be obtained as

$$M_B = E_B + (\Delta E_B)_g^m + (\Delta E_B)_g^e + (\delta M_B)_\chi \tag{45}$$

where $(\delta M_B)_\chi$, with $(\chi = \pi, \eta, k)$ is the GBE correction [equation(40)] and $[(\Delta E_B)_g^m + (\Delta E_B)_g^e]$ is the colour-magnetic and -electric interaction energies arising out of the residual OGE processes [equation(33)]. Here $E_B = E_B^0 + (\Delta E_B)_{c.m}$ is the c.m corrected baryon mass [equation(18)].

The quantitative evaluation of these terms on the right hand side of equation(45) within the frame-work of the model primarily involves the choice of potential parameters (a, V_0) of the model, the quark masses m_q and the corresponding binding energy E_q alongwith the other relevant model quantities. We find that with a suitable choice of potential parameters

$$(a, V_0) = (0.2960, -0.3649) \; GeV \tag{46}$$

and the constituent quark masses

$$(m_u = m_d, m_s) = (0.32, 0.51) \; GeV \tag{47}$$

the energy eigen-value condition, given in equation(8), yields the individual quark binding energies

$$(E_u = E_d, E_s) = (456.84, 591.68) \; MeV. \tag{48}$$

which are used to compute the corrections due to the OGE and GBE interactions between the quarks alongwith that due to c.m motion. We then evaluate the integral expressions for $I_{ij}^{e,m}$ in equations(31) and (32) with the help of a standard numerical method and calculate the terms $T_{ij}^{e,m}$ from equations(34) and (35) as

$$(T_{uu}^e, T_{us}^e, T_{ss}^e) = (675.386, 660.005, 696.160) \; MeV \tag{49}$$

$$(T_{uu}^m, T_{us}^m, T_{ss}^m) = (171.144, 152.969, 149.730) \; MeV \tag{50}$$

which are necessary for computing $(\Delta E_B)_g^{e,m}$. The integral I_χ is then evaluated from equation(44) as

$$(I_\pi, I_\eta, I_k) = (1988.075, 109.506, 137.823) \; MeV \tag{51}$$

Indeed, in the chiral limit there is only one coupling constant for all Goldstone bosons. Due to explicit chiral symmetry breaking the coupling constant for π, η and k may become different. However, in order to prevent a proliferation of the free parameters we try to keep the number of free parameters as small as possible and assume a single phenomenological pion-nucleon coupling constant $f_{NN\pi}$=0.283 for all mesons (π, η, k) and this value is used here to compute the GBE corrections. The GBE corrections for the ground state of baryons are presented in Table-2. The energy corrections and the results obtained for the mass spectra of light and strange baryons in their ground states are presented in the Table-3. The calculated values of the ground state masses of light and strange baryons are found to agree

Table 2. GBE corrections $(\delta M_B)_\chi$ (where $\chi = \pi, \eta, k$) for the ground state baryons (in MeV)

Baryon	$(\delta M_B)_\pi$	$(\delta M_B)_k$	$(\delta M_B)_\eta^{uu}$	$(\delta M_B)_\eta^{us}$	$(\delta M_B)_\eta^{ss}$	total	$(\delta M_B)_\chi$
N	-185.34	-	10.29	-	-	10.29	-175.05
Δ	-37.06	-	-10.29	-	-	-10.29	-47.35
Λ	-111.21	-47.02	10.29	-	-	10.29	-147.94
Σ	-12.36	-78.36	3.43	-17.21	-	-20.64	-111.36
Σ*	-12.36	-31.35	-3.43	8.61	-	5.18	-38.53
Ξ	-	-78.36	-	-17.21	-5.40	-22.61	-100.97
Ξ*	-	-31.35	-	8.61	-5.40	3.21	-28.14
Ω⁻	-	-	-	-	-16.20	-16.20	-16.20

Table 3. The energy corrections $(\Delta E_B)_g^{m,e}$, $(\delta M_B)_\chi$ and the physical masses (M_B) of ground state baryons (in MeV)

Baryon	E_B	$(\Delta E_B)_g^m$	$(\Delta E_B)_g^e$	$(\delta M_B)_\chi$	M_B Calculation	M_B Expt.
N	1197.20	-82.15	0	-175.05	940	940
Δ	1197.20	82.15	0	-47.35	1232	1232
Λ	1337.84	-82.15	8.25	-147.94	1116	1116
Σ	1337.84	-70.52	8.25	-111.36	1164.21	1193
Σ*	1337.84	76.33	8.25	-38.53	1383.89	1385
Ξ	1477.40	-73.94	8.25	-100.97	1310.74	1321
Ξ*	1477.40	72.90	8.25	-28.14	1530.41	1533
Ω⁻	1616.15	71.87	0	-16.20	1671.82	1672

quite well with the experiment. It is found that the OGE corrections require a value of quark-gluon coupling constant $\alpha_c = 0.16$ which is quite consistent with the idea of treating OGE effects in low-order perturbation theory.

In this model we find that the SU(3) breaking effect due to quark masses $m_u = m_d \neq m_s$ lifts the degeneracy in the baryon masses through the c.m corrected energy term E_B among the groups (N, Δ), $(\Lambda, \Sigma, \Sigma^*)$, (Ξ, Ξ^*) and Ω^-. Then in second step, the GBE corrections $(\delta M_B)_\chi$ arising from $SBCS$ in QCD removes the degeneracy partially between N and Δ; Λ, Σ and Σ*; Ξ and Ξ*. However, energy corrections due to OGE, particularly the colour-magnetic interaction energy removes the mass degeneracy completely among these baryons. It should be pointed out here that the colour-electric interaction energy due to OGE being minimal in this model.

In the present work the meson degree of freedom which is taken into account as in the CBM[11], ignores to a large extent the short-range part of the meson exchange interaction, which is of crucial importance of splittings. Only when the complete infinite set of all radially excited intermediate states B' is taken into account, this method could be adequate[25].

For example, the meson exchange contribution to the N -Δ difference will become much larger. It will also be strongly enhanced when the meson exchange contribution is calculated non-perturbatively. The meson exchange contribution is strongly dependent on the radius of the bare wave function also, i.e., on the type of confinement. This dependence has not been studied in the present work. Thus we find that the present model which (i) includes an appreciable contribution from the color-magnetic interaction and (ii) treats the meson degree of freedom perturbatively in a manner like it is done in the CBM (11), can reproduce the ground state mass spectra of octet baryons.

Acknowledgments

The authors are grateful to Prof.N.Barik, Mayurbhanj Professor of Physics, Utkal University, Bhubaneswar, Orissa, India for his valuable suggestions and useful discussions on this work.

References

[1] P.Leal.Ferreira, *Lett.Nuovo-Cimento*, **20**, 157(1977); A.P.Kobuskin Report No. ITP-76-58E, Kiev. S.N.Jena and S.Panda, *Pramana-J.Phys.* **35**, 21(1990).

[2] S.N.Jena and S.Panda, *Pramana-J.Phys.* **37**, 47(1990).

[3] S.N.Jena and S.Panda, *Int.J.Mod.Phys.* **A7**, 2841(1992), *J.Phys.G:Nucl. Part.Phys.* **18**, 273(1992).

[4] S.N.Jena and S.Panda, *Int.J.Mod.Phys.***A8**, 4563(1993); ibid. A9, 327(1994).

[5] S.N.Jena, M.R.Behera and S.Panda, *Pramana-J.Phys.* **51**, 711(1998).

[6] S.N.Jena, S.Panda and T.C.Tripathy, *Nucl.Phys.* **A658**, 249(1999).

[7] S.N.Jena, S.Panda and J.N.Mohanty, *J.Phys.G:Nucl.Part.Phys.* **24**, 1869(1998).

[8] N.Isgur and G.Karl, *Phys.Rev.* **D18**, 4187(1978); 20, 1191(1979); Y.Nogami and N.Ohtsuka, ibid. **26**, 261(1982).

[9] L.Ya.Glozman, W.Plessas, K.Varga and R.F.Wagenbrunn, *Phys.Rev.* **D58**, 094030(1998); L.Ya.Glozman, D.O.Riska, *Phys.Rep.* **268**, 263-303(1996); L.Ya.Glozman *Nucl.Phys.* **A663**, 103(2000).

[10] A.Chodos, R.L.Jaffe, K.Johnson, C.B.Thorn and V.F.Weisskopf, *Phys.Rev.* **D9**, 1471(1974a); A.Chodos, R.L.Jaffe, K.Johnson and C.B.Thorn *Phys. Rev.***D10**, 2599(1974b); T.De.Grand, R.L.Jaffe, K.Johnson and J.Kiskis, *Phys.Rev.* **D12**, 2060(1975).

[11] A.W.Thomas, *Adv.Nucl.Phys.* **13**, 1(1983).

[12] R.Tegen, R.Brockmann and W.Weise, *Z.Phys.* **A307**, 339(1982); R.Tegen and W.Weise, *Z.Phys.* **A314**, 357(1983); R.Tegen, M.Schedle and W.Weise, *Phys. Lett.* **125B**, 9(1983).

[13] P.Leal.Ferreira and N.Zagury, *Lett.Nuovo-Cimento.* **20**, 511(1977); P.Leal. Ferreira, J.A.Helayel and N.Zagury, *Lett.Nuovo-Cimento.* **A55**, 215(1980); N.Barik, B.K.Dash and M.Das, *Phys.Rev.***D31**, 1652(1985); N.Barik and B.K.Dash, *Phys.Rev.***D33**, 1925(1986).

[14] A.Martin, *Phys.Lett.***93B**, 338(1980); ibid. 100B, 5111(1981); N.Barik and S.N.Jena, *Phys.Lett.* **97B**, 761(1980); 97B, 265(1980); N.Barik and S.N.Jena, *Phys.Rev.* **D26**, 618(1982); N.Barik and M.Das, *Phys.Lett.* **120B**, 403(1983), *Phys.Rev.* **D28**, 2823(1983); ibid. D33, 176(1986a); *Pramana-J.Phys.* **27**, 783(1986b). S.N.Jena, P.Panda and K.P.Sahu, *Int.J.Theo.Phys.Group Th.Non-Linear optics.* **9**, 69(2002); ibid. 8, 351(2002); *J.Phys.G:Nucl.part.Phys.***27**, 1519(2001); S.N.Jena, P.Panda and T.C.Tripathy, *Phys.Rev.* **D63**, 014011(2000).

[15] S.N.Jena and M.R.Behera, *Pramana-J.Phys.***44**, 357(1995); ibid: 47, 233(1996); *Int.J.Mod.Phys.* **E7**, 69(1998); 425(1998); S.N.Jena, M.R.Behera and S.Panda, *J.Phys.G:Nucl.part.Phys.* **24**, 1089(1998).

[16] E.Magyari, *Phys.Lett.* **95B**, 295(1980); C.Quigg and J.L.Rosner *Phys.Lett.* **B71**, 153(1977); S.N.Jena and D.P.Rath, *Pramana-J.Phys.* **27**, 773(1986); N.Barik, S.N.Jena and D.P.Rath, *Phys.Rev.***D41**, 1568(1990); *Int.J.Mod.Phys.* **A7**, 6813(1992); S.N.Jena, K.P.Sahu and P.Panda, *Int.J.Theo.Phys.Group Th.Non-Linear optics,* **8**, 213(2002).

[17] M.Gell-Mann and M.Levy, *Nuovo cimento* **16**, 705(1960).

[18] Y.Nambu and G.Jona-Lasino, *Phys.Rev.* **112**, 345(1961); ibid. 124, 246(1961); R. Brockmann, W.Weise, and E. Werner, *Phys. Lett.* **122B**, 201(1983); S.P.Klevansky, *Rev. Mod. Phys.* **64**, 649(1992).

[19] T.P.Cheng and L.F.Li, *Phys.Rev.Lett.***74**, 2872(1995).

[20] J.Bartelski, A.Szymacha, L.Mankcewicz and S.Tatur, *Phys.Rev.* **D29**, 1035(1984).

[21] E.Eich, D.Rein and R.Rodenberg, PITHA Report.No83/21 (1983); *Z.Phys.* **C28**, 225(1985).

[22] S.N.Jena and S.Panda, *Int.J.Mod.Phys.* **A7**, 2841(1992); *Pramana-J.Phys.* **35**, 21(1990); ibid. 37, 47(1991), *J.Phys.G.* **18**, 273(1992), *J.Phys.G.* **19**, 837(1993); *Int.J.Mod.Phys.* **A8**, 4563(1993).

[23] A.W.Thomas, *Nucl.Phys.* **B13**, 1(1983) and references cited therein.

[24] S.N.Jena, M.R.Behera and S.Panda, *J.Phys.G: Nucl.Part.* **24**, 1089-1103(1998); S.N.Jena, M.R.Behera and S.Panda, *Phys.Rev.* **D55**, 291(1997).

[25] L.Ya.Glozman, hep-ph/0004229 (unpublished).

In: Theoretical Physics and Nonlinear Optics
Editors: Thomas F. George et al

ISBN: 978-1-61122-939-4
© 2012 Nova Science Publishers, Inc.

Chapter 7

BELL'S THEOREM REFUTED WITH A KOLMOGOROVIAN COUNTEREXAMPLE

J.F. Geurdes[*]
C.vd. Lijnstraat 164, 2593 NN Den Haag
Netherlands

Abstract

The statistics behind Bell's inequality is demonstrated to allow a Kolmogorovian model of probabilities that recovers the quantum covariance. It shown that $P(A=+1)=P(A=-1)=1/2$ given $A=1$, or $A=-1$ is the result of measurement with device A. The occurrence of irregular integration prevents the use of Schwarz's inequality, hence, prevents the derivation of Bell's inequality. The obtained result implies that not all local hidden variables theories are ruled out by Aspect's (see Ref 4) experiment.

1. Introduction

In the early beginnings of quantum theory (QM) questions of interpretation arose. In a later stage, an important step in the debate was made by Bell[1].

Based on Einstein's criticism (EPR paradox) of completeness[2], Bell formulated an expression for the relation between distant (spin) measurements such as described by Bohm[3]. In Bell's expression, local hidden variables (LHV's) to restore locality and causality to the theory are introduced through a probability density function and through their influence upon the elementary measurement functions in the two separate wings (denoted by the A-and the B-wing) of the experiment.

Many experiments and theoretical developments arose from Bell's original paper. The most important experiment was performed by Aspect[4]. Aspect's results were interpreted as a confirmation of the completeness of quantum mechanics and the impossibility of local hidden causality. From that point onwards, QM was considered a non-local theory. In a previous

---
[*] E-mail address: geurd030@planet.nl

paper, the present author has discussed this proof[5] arguing that there was insufficient ground for such a far reaching conclusion.

Let us describe briefly a typical Bell experiment. In this kind of experiment, from a single source, two particles with, e.g. opposite spin, are send into opposite directions. We could think of a positron and an electron arising from para-Positronium, drawn apart by dipole radiation. Subsequently, in the respective wings of the experiment, the spin of the individual particle is measured with a Stern-Gerlach magnet. The measurements are found to covary and depend on the, unitary, parameter vectors, (a_1, a_2, a_3) and (b_1, b_2, b_3) of the magnets. Quantum mechanics predicts the correlation to be

$$P(\vec{a}, \vec{b}) = -\sum_{k=1}^{3} a_k b_k \qquad (1)$$

Bell's theorem states that local hidden variables cannot recover P(a,b), because all local hidden distributions run into inequalities similar to the one stated below. Vectors a, b, c, d are unitary.

$$|P_{LHV}(\vec{a}, \vec{b}) - P_{LHV}(\vec{a}, \vec{d})| + |P_{LHV}(\vec{c}, \vec{b}) + P_{LHV}(\vec{c}, \vec{d})| \le 2 \qquad (2)$$

In the present paper, Eq. (1) will be derived from local hidden variables. Obviously this cannot be done with regular integration.

2. Preliminary Remarks

In this section the proposed model for a LHV explanation of the EPR paradox is introduced. The basis of the model is the Gaussian probability density.

When a Gaussian density is normal with a mean zero and a standard deviation of unity we write, N(0,1). The Gaussian density with mean zero, standard deviation, |T|, is

$$\delta_T(x) = \frac{1}{|T|\sqrt{2\pi}} \exp\left[-\frac{x^2}{2|T|^2}\right] \qquad (3)$$

Hence, $\delta_T(x) = N(0,|T|)(x)$. This is equal to the Dirac delta, when $T \to 0$, T and x independent. Associated to the previous function is the 'Heaviside' function which is

$$H_T(x) = \int_{-\infty}^{x} \delta_T(y)\, dy \qquad (4)$$

From the fact that, $H_T(0) = 1/2$, for all standard deviations, T>0, it follows that in the limit, $T \to 0$,

$$\lim_{T \to 0} H_T(x) = \begin{cases} 1, & x>0 \\ 1/2, & x=0 \\ 0, & x<0 \end{cases} \quad (5)$$

In the following the behaviour of the Heaviside function in zero needs to be altered. We therefore define an alternative form of Heaviside function.

$$\theta_T(x) = H_T(x) + 2H_T(x)H_T(-x). \quad (6)$$

In the limit, T→0, we have

$$\lim_{T \to 0} \theta_T(x) = \theta(x) = 1, \quad x \geq 0,$$
$$\lim_{T \to 0} \theta_T(x) = \theta(x) = 0, \quad x < 0. \quad (7)$$

Furthermore, differentiation to, x, and taking the limit, T→0, gives the result

$$\lim_{T \to 0} \frac{d}{dx} \theta_T(x) = \delta(x) - 2\delta(x) sign(x). \quad (8)$$

Because, sign(x)=H_T(x)-H_T(-x), for, T→0, and, δ_T(x)=δ_T(-x), we see that

$$\delta_{alt}(x) = \lim_{T \to 0} \frac{d}{dx} \theta_T(x) \quad (9)$$

is unequal to zero only when x=0, and coincides with, δ(x), because, by definition, sign(x), is zero in x=0. Hence, the alternative Heaviside function in Eq. (6), has the proper behaviour. In view of the definition, δ_T(x)=N(0,|T|)(x), it is remarked that, |T|δ_T(x), lies in the interval [0,1]. Subsequently let us define the sum terms

$$\sigma_A = \sum_{k=1}^{3} a_k sign(y_k)$$
$$\sigma_B = \sum_{k=1}^{3} b_k sign(y_k) \quad (10)$$
$$|\sigma_A| \leq \sqrt{3}, \quad |\sigma_B| \leq \sqrt{3}$$

with $a_1^2+a_2^2+a_3^2=1$, and, $b_1^2+b_2^2+b_3^2=1$, and the variables y_k, are normally distributed, $N(0,1)(y_k)$. The entities, a_k, and, b_k, represent the unitary directional parameters of the measuring devices.

3. Probabilistic Behaviour of the Measurement Devices

In the model, a major role is played by two sets of each three variables carrying the index A or B. The indices refer to the measuring device in the left-hand (A) or the right-hand (B) wing. Here in the definition section the use of the indices can be suppressed because we assume that identical measuring devices interact identically with the to-be-measured entity.

In the first place, let us introduce the probability density for the T variable employed earlier. We have

$$\rho_T(T) = \begin{cases} \delta(T-0^+), & T \in [-\epsilon, \epsilon] \\ 0, & T \notin [-\epsilon, \epsilon] \end{cases} \quad (11)$$

with, ϵ, a small positive real number. Here, the 0^+ means the limit from above to zero which may come arbitrary close to zero, but staying unequal to zero at the same time. Hence, for an arbitrary otherwise well-behaved function, f, we have

$$\int_{-\infty}^{\infty} \rho_T(T) f(T) \, dT = \lim_{0 < T \to 0} f(T) = f(0) \quad (12)$$

Furthermore, let us define a distribution for a variable, β, such that,

$$\rho_B(\beta) = \delta(\beta - |T|), \quad \beta \in \mathbb{R}. \quad (13)$$

Before we introduce the probability density of the x variable used previously, we note beforehand that

$$\lim_{0 < T \to 0} T^T = \lim_{0 < T \to 0} \exp[T \ln T] = \exp[\lim_{0 < T \to 0} T \ln T] =$$
$$\exp[\lim_{0 < T \to 0} \frac{1/T}{-1/T^2}] = \exp(0) = 1 \quad (14)$$

This limit is crucial in the construction and evaluation of the model. The probability density (pdf) for x is then defined by

$$\rho_X(x; T, \beta) = \begin{cases} 1/|T|^{\theta(x)}, & x \in [-\frac{1}{2}|T|^\beta, \frac{1}{2}|T|] \\ 0, & x \notin [-\frac{1}{2}|T|^\beta, \frac{1}{2}|T|] \end{cases} \quad (15)$$

Note that the Heaviside (theta) function, $\theta(x)$, in the previous definition is equal to the one defined above, that is, $\theta(x)=1$, when, $x \geq 0$, while, $\theta(x)=0$, when, $x<0$. Note also that

Given all the previous definitions of this section, we can now come to the evaluation of the associated integral over, T, over, β, and, over, x. This integral will return in the model in relation to the A and B indexed variables. Suppressing those indices for this moment we have, for some auxiliary variable, W_0,

$$W_0 = \int_{-\infty}^{\infty} dT \int_{-\infty}^{\infty} d\beta \int_{-\infty}^{\infty} dx \, \rho_T(T) \, \rho_B(\beta) \, \rho_x(x; T, \beta) \, . \tag{16}$$

Evaluation of the integral over x and its density gives

$$\int_{-\infty}^{\infty} dx \, \rho_x(x; T, \beta) = \int_{-\frac{1}{2}|T|^\beta}^{\frac{1}{2}|T|} dx \, \frac{1}{|T|^{\theta(x)}} \tag{17}$$

This leads to the integral

$$\int_{-\infty}^{\infty} dx \, \rho_x(x; T, \beta) = \int_{-\frac{1}{2}|T|^\beta}^{0} dx + \int_{0}^{\frac{1}{2}|T|} dx \, \frac{1}{|T|} = \frac{1 + |T|^\beta}{2}$$

Hence,

$$W_0 = \int_{-\varepsilon}^{\varepsilon} dT \delta(T - 0^+) \int_{-\infty}^{\infty} d\beta \, \delta(\beta - |T|) \, \frac{1 + |T|^\beta}{2} \tag{19}$$

such that

$$W_0 = \int_{-\varepsilon}^{\varepsilon} dT \delta(T - 0^+) \, \frac{1 + |T|^{|T|}}{2} = \lim_{0 < T \to 0} \frac{1 + T^T}{2} = \frac{1 + 1}{2} = 1 \tag{20}$$

This means that when x, T, and, δ, carry an index, A or B, then in the evaluation of the total density, de pdf's of those variables will result in unity in integration. This is one of the necessities of a proper total pdf. Hence, we may conclude that the densities of, x, T, and, δ, with indices A or B can be introduced in a total density.

Another point of discussion is the probability description necessary for a Kolmogorovian theory. Because, T and δ follow a delta function, which is a Gaussian with infinitesimal small standard deviation, we can focus our attention to the density of x. Note that we already have shown that

$$\int_{-\frac{1}{2}|T|^\beta}^{\frac{1}{2}|T|} \frac{dx}{|T|^{\theta(x)}} = 1 \tag{21}$$

when, $0 < T \to 0$, and, $\beta = |T|$. If we take ζ in the interval $[0, |T|/2]$, probabilities in that interval are between 0 and 1/2 in the limit, $0 < T \to 0$, and, $\beta = |T|$. In this limit, if we then also may take, ζ, in $[-|T|^\beta/2, 0)$. This is a uniform probability between 0 and 1/2 in the interval $[-1/2, 0)$.

Note that the previous is based on

$$\int_{-\frac{1}{2}|T|^{\beta}}^{\xi} \frac{dx}{|T|^{\theta(x)}} = \frac{1}{|T|^{\theta(\xi)}} \{\xi + \frac{1}{2}|T|^{\beta}\} \tag{22}$$

Hence, a probability measure can be associated to the density of, x, when, $0<T\to 0$. For completeness the following is given to illustrate the previous statement.

If, $\xi \geq 0$, then, ξ, can be written as, $\xi = s|T|/2$, with $0 \leq s \leq 1$. The previous integral, then becomes

$$\int_{-\frac{1}{2}|T|^{\beta}}^{\xi} \frac{dx}{|T|^{\theta(x)}} = \frac{1}{|T|} \frac{s|T|}{2} + \frac{|T|^{\beta}}{2} = \frac{1}{2}(1+s) \in [0,1] \tag{23}$$

If, $\xi<0$, then, ξ, can be written as, $\xi = -s|T|^{\beta}/2$, with $0<s\leq 1$. The integral in Eq. 21 then becomes

$$\int_{-\frac{1}{2}|T|^{\beta}}^{\xi} \frac{dx}{|T|^{\theta(x)}} = \frac{-s|T|^{\beta}}{2} + \frac{|T|^{\beta}}{2} = \frac{|T|^{\beta}}{2}(1-s) \in [0,1] \tag{24}$$

Recall that $0<T\to 0$, and, $\beta = |T|$.

4. Definition of the Behaviour of the 'Bridge' between the Local Measurements

After definition of the probabilistic behaviour of the measuring instruments -e.g. Stern-Gerlach magnets-, the connection, or bridge, between the two measurements can be a relatively simple three-dimensional Gaussian density in y_k, k=1,2,3. Hence, in a simple notation,

$$N(0,1)(\vec{y}) = \frac{1}{(2\pi)^{3/2}} \exp[-\frac{1}{2}(y_1^2+y_2^2+y_3^2)] \tag{25}$$

If we, subsequently, inspect the integrations of the product of sign functions and the three-dimensional Gaussian density, we find,

$$\int\int\int_{-\infty}^{\infty} d^3y\, N(0,1)(\vec{y})\, sign(y_k) = 0, \qquad (k=1,2,3) \tag{26}$$

because of the symmetry of the Gaussian. Moreover, the Kronecker delta function ($\delta_{kj}=1$, if k=j, $\delta_{kj}=0$, if k unequal j, and k and j integer) arises in a similar manner from the product of signs and the Gaussian density. We see

$$\int\int\int_{-\infty}^{\infty} d^3y\, N(0,1)(\vec{y})\, sign(y_k)\, sign(y_j) = \delta_{k,j} \qquad (27)$$

The fact that there is a bridge between the two measurements can be demonstrated from the following considerations. If the measurement functions enable the 'extraction' of the sum terms from Eq. (10), then an integration over the Gaussian combined with the sign functions in the sum terms gives

$$\int\int\int_{-\infty}^{\infty} d^3y\, N(0,1)(\vec{y})\, \sigma_A \sigma_B =$$

$$\sum_{k=1}^{3}\sum_{j=1}^{3} a_k b_j \int\int\int_{-\infty}^{\infty} d^3y\, N(0,1)\, sign(y_k)\, sign(y_j) = \qquad (28)$$

$$\sum_{k=1}^{3}\sum_{j=1}^{3} a_k b_j \delta_{k,j} = \sum_{k=1}^{3} a_k b_k.$$

Hence, if the respective measurement functions can be construed such that each sum term enters a Gaussian integration, the quantum mechanical result is recovered from local hidden causality.

Of course there is the difficulty that the, to be defined, A and B measurement functions need to be either, +1, or -1, while the sum terms σ_A, and, σ_B, project in the interval [$-3^{1/2}, 3^{1/2}$]. Hence, we may expect the integrations on the x, T, and, β, in relation to the A or B measurement function, to be irregular. This irregularity, on the other hand, entails that Bell's inequality cannot be derived.

5. The A and B Measurement Functions

Before introducing the measurement functions, A and B, let us recall the σterm from which the above mentioned extraction can occur. Here, we have for the A and B

$$u_{T_A}(x_A; \sigma_A) = \sigma_A |T_A|^{\theta(x_A)} \frac{d}{dx_A} \theta_{T_A}(x_A)$$

$$u_{T_B}(x_B; \sigma_B) = \sigma_B |T_B|^{\theta(x_B)} \frac{d}{dx_B} \theta_{T_B}(x_B) \qquad (29)$$

Note the difference between the, $\theta(x)$, and the, $\theta_T(x)$, in the previous equation. Writing u_T more explicitly, suppressing the indices A, or, B, for the moment, we arrive at

$$u_T(x;\sigma) = \begin{cases} \sigma(-1)^{1-\theta(-x)}|T|\delta_T(x), & x \geq 0 \\ 3\sigma\delta_T(x), & x < 0 \end{cases} \tag{30}$$

As already has been discussed earlier (Eq. (10)), the term, $\sigma|T|\delta_T(x)$, lies in the interval [-1,1]. Ultimately, only x=0, will produce, $\lim_{0<T\to 0}\{|T|\delta_T(x)\} \neq 0$. Hence, the u_T function can be put in a sign function which integration variable runs from -1 to 1. Let us subsequently define two uniform pdfs

$$p_A(\mu_A) = \begin{cases} 1/2, & \mu_A \in [-1,1] \\ 0, & \mu_A \notin [-1,1] \end{cases}$$

$$p_B(\mu_B) = \begin{cases} 1/2, & \mu_B \in [-1,1] \\ 0, & \mu_B \notin [-1,1] \end{cases} \tag{31}$$

Strictly speaking those two uniform densities also belong to the probabilistic behaviour of the measuring instruments, but it makes more sense to introduce them here. Given the two pdfs, we can now define the measurement functions, A and B.

$$A = sign\{u_{T_A}(x_A;\sigma_A) - \mu_A\} \tag{32}$$

$$B = -sign\{u_{T_B}(x_B;\sigma_B) - \mu_B\}$$

The extraction of, σ_A, can now be demonstrated. In the first place, we may write in the evaluation of the statistical expectation of the function A,

$$\frac{1}{2}\int_{-1}^{1} d\mu_A\, sign\{u_{T_A}(x_A;\sigma_A) - \mu_A\} = u_{T_A}(x_A;\sigma_A) \tag{33}$$

Later in the paper a more precise way of writing down the 'extraction plus bridge' process of integration will be given. If however, we accept for the moment that u_T can be obtained this way, then we may be interested in the role of the, x_A, T_A, and β_A hidden variables. Hence, we are interested

$$W_1 = \int_{-\infty}^{\infty} dT_A \int_{-\infty}^{\infty} d\beta_A \int_{-\infty}^{\infty} dx_A\, \rho_{T_A}(T_A)\, \rho_{B_A}(\beta_A)\, \rho_{x_A}(x_A;T_A,\beta_A)\, u_{T_A}(x_A;\sigma_A) \tag{34}$$

For the auxiliary variable, W_1, we firstly integrate over x_A. Hence,

$$\int_{-\infty}^{\infty} dx_A \, \rho_{X_A}(x_A; T_A, \beta_A) \, u_{T_A}(x_A; \sigma_A) =$$

$$\int_{-\frac{1}{2}|T_A|^{\beta_A}}^{\frac{1}{2}|T_A|} dx_A \, \frac{u_{T_A}(x_A; \sigma_A)}{|T_A|^{\theta(x_A)}} \tag{35}$$

Furthermore, we see from the form, $u_T = \sigma |T|^{\theta(x)} \delta_T(x)$, that

$$\int_{-\frac{1}{2}|T_A|^{\beta_A}}^{\frac{1}{2}|T_A|} dx_A \, \frac{u_{T_A}(x_A; \sigma_A)}{|T_A|^{\theta(x_A)}} = \sigma_A \int_{-\frac{1}{2}|T_A|^{\beta_A}}^{\frac{1}{2}|T_A|} dx_A \, \frac{d}{dx_A} \theta_{T_A}(x_A) =$$

$$\sigma_A \{ \theta_{T_A}(\frac{1}{2}|T_A|) - \theta_{T_A}(-\frac{1}{2}|T_A|^{\beta_A}) \} \tag{36}$$

Subsequently, this implies for W_1,

$$W_1 =$$

$$\sigma_A \int_{-\epsilon}^{\epsilon} dT_A \int_{-\infty}^{\infty} d\beta_A \{ \delta(T_A - 0^+) \delta(\beta_A - |T_A|)$$

$$\times \{ \theta_{T_A}(\frac{1}{2}|T_A|) - \theta_{T_A}(-\frac{1}{2}|T_A|^{\beta_A}) \} \} . \tag{37}$$

Further evaluation gives

$$W_1 = \sigma_A \int_{-\epsilon}^{\epsilon} dT_A \, \delta(T_A - 0^+) \{ \theta_{T_A}(\frac{1}{2}|T_A|) - \theta_{T_A}(-\frac{1}{2}|T_A|^{|T_A|}) \} =$$

$$\sigma_A \lim_{0 < T_A \to 0} \{ \theta_{T_A}(\frac{1}{2}|T_A|) - \theta_{T_A}(-\frac{1}{2} T_A^{T_A}) \} = \sigma_A \{ \theta(0) - \theta(-\frac{1}{2}) \} \tag{38}$$

Hence, $W_1 = \sigma_A$. This result shows that σ_A can be extracted from the A measurement function. Similar result can be obtained from the B measurement function. Moreover, we see that the integration is irregular because the u_T lies in [-1,1], while, its result, σ_A, can be larger than 1 and smaller than -1. Hence, irregular integration is the reason that Schwarz's inequality in certain phases of the evaluation of the expectation of A, of B, and/or of AB, fails.

Let us also show in more detail why the Bell inequality cannot be obtained from the model.

Firstly, we note that generally for a model of this type, we have for the covariance of A and B

$$\langle AB \rangle = \int d\lambda\, \rho_1(\lambda) \int d\mu\, \rho_2(\mu) \int d\nu\, \rho_3(\nu)\, A(\lambda,\mu,\vec{a}) B(\lambda,\nu,\vec{b}) \qquad (39)$$

where, λ, μ, ν, denote groups of variables. Secondly, we note that Bell's inequality depends on Schwarz's inequality. Hence, when A=+1, or, -1, and similar for B, we write

$$|\langle AB \rangle| \leq \int d\lambda\, \rho_1(\lambda) \int d\mu\, \rho_2(\mu) \int d\nu\, \rho_3(\nu)\, |A(\lambda,\mu,\vec{a})|\, |B(\lambda,\nu,\vec{b})| \qquad (40)$$

This then, in a regular case, leads to, $|AB|\leq 1$, hence, the absolute covariance between A and B is less than or equal to unity.

If, however, this inequality is valid, we also must have

$$\langle AB \rangle |$$
$$\int d\lambda\, \rho_1(\lambda)\, |\int d\mu\, \rho_2(\mu)\, A(\lambda,\mu,\vec{a})|\, |\int d\nu\, \rho_3(\nu)\, B(\lambda,\nu,\vec{b})| \qquad (41)$$

In our case we can identify

$$W_1| =$$
$$\int_{-\varepsilon}^{\varepsilon} dT\, \delta(T-0^+) \int_{-\infty}^{\infty} d\beta\, \delta(\beta-|T|) \int_{-\frac{1}{2}|T|^\beta}^{\frac{1}{2}|T|} \frac{dx}{2|T|^{\theta(x)}} \int_{-1}^{1} d\mu\, sign(u_T - \mu) |. \qquad (42)$$

Here, $W_1 = W_1(A)$, is an expression in terms of the model, of the first term between absolute lines on the right hand of Eq. (40). A similar remark applies to $W_1(B)$, which is the second term between absolute marks. In a regular case, $|W_1|\leq 1$, however, if we evaluate like it is done in the previous sections, it follows

$$|W_1| =$$
$$\left| \int_{-\varepsilon}^{\varepsilon} dT\, \delta(T-0^+) \int_{-\infty}^{\infty} d\beta\, \delta(\beta-|T|) \int_{-\frac{1}{2}|T|^\beta}^{\frac{1}{2}|T|} \frac{dx}{2|T|^{\theta(x)}} \int_{-1}^{1} d\mu\, sign(u_T - \mu) \right| =$$

$$\left| \int_{-\varepsilon}^{\varepsilon} dT\, \delta(T-0^+) \int_{-\infty}^{\infty} d\beta\, \delta(\beta-|T|) \int_{-\frac{1}{2}|T|^\beta}^{\frac{1}{2}|T|} \frac{dx}{|T|^{\theta(x)}}\, u_T(x;\sigma) \right| = \qquad (43)$$

$$\left| \sigma \int_{-\varepsilon}^{\varepsilon} dT\, \delta(T-0^+) \int_{-\infty}^{\infty} d\beta\, \delta(\beta-|T|) \int_{-\frac{1}{2}|T|^\beta}^{\frac{1}{2}|T|} dx\, \frac{d}{dx}\theta_T(x;\sigma) \right| =$$

$$= |\sigma|.$$

The previous expression and, $|W_i| \leq 1$, are contradictory because, for, $a_1 = a_2 = a_3 = (1/3)^{1/2}$, ($a_1^2 + a_2^2 + a_3^2 = 1$), we can have, $|\sigma| > 1$. Hence, Schwarz's inequality cannot be applied. Hence no Bell inequality exists for this type of AB.

Note that, $\sigma|T|\delta_T(x)$, or u_T, projects in the interval $[-1,1]$. Note also, as a rule of the thumb, that $|T|$ in its role as standard deviation gives, Prob$[-3|T| \leq x \leq 3|T|] \approx 0.99$. Hence, 99% of the surface below the Gaussian lies between $-3|T|$ and $3|T|$. Note in addition also that in Generalized function theory, x and T are independent variables[6], hence, no problem arises when interchanging limit with integration.

6. The Complete Model and the Evaluation of the Expectation

The complete pdf in the model is

$$\rho_{A-wing} = \rho_A(\mu_A) \rho_{T_A}(T_A) \rho_{B_A}(\beta_A) \rho_{X_A}(x_A; T_A, \beta_A)$$

$$\rho_{B-wing} = \rho_B(\mu_B) \rho_{T_B}(T_B) \rho_{B_B}(\beta_B) \rho_{X_B}(x_B; T_B, \beta_B)$$

$$\rho_{tot} = \rho_{A-wing} \rho_{B-wing} N(0,1)(\vec{y})$$
(44)

In the evaluation of the statistical expectations of the model, we are in need of a notation to denote the ongoing integration process. Let us continue to use the previously loosely defined bracket notation, which resembles the one from statistical physics. An already partly worked-out example on the expectation of A can now be used to show the workings of the bracket notation we have.

$$\langle A \rangle = \langle u_{T_A}(x_A; \sigma_A) \rangle$$
(45)

Because, $A = \text{sign}(u_{TA} - \mu_A)$, which after integration over, μ_A, gives the above result.

In the employed notation the angular brackets indicate the ongoing integration. In the previous, the pdf weighted integration over, μ_A, has been performed, such as in Eq. (28). Subsequent pdf weighted integration over the, x_A, T_A, and β_A hidden variables, such as in Eqs.(34)-(38), then gives

$$\langle A \rangle = \langle \sigma_A \rangle =$$

$$\int\int\int_{-\infty}^{\infty} d^3 y N(0,1)(\vec{y}) \sigma_A = 0$$
(46)

The vanishing of $\langle A \rangle$ follows from Eq. (26). Similar result follows for $\langle B \rangle = 0$.

Moreover, with the similar notation, it easily follows that, $\langle 1 \rangle = \langle A^2 \rangle = \langle B^2 \rangle = \langle 1 \rangle$

Conclusively, the result for the covariance of A and B can be written as

$$P(\vec{a},\vec{b}) = \langle AB \rangle = -\langle u_{T_A}(x_A;\vec{\sigma}_A) u_{T_B}(x_B;\vec{\sigma}_B) \rangle = -\langle \sigma_A \sigma_B \rangle \qquad (47)$$

This result arises in a similar manner as, σ_A, from the expectation of A, because the pdf weighted integration over, x_A, T_A, and $\vec{\sigma}_A$, is independent from, weighted integration over, x_B, T_B, and $\vec{\sigma}_B$, and, $AB = -\text{sign}(u_{TA}-\mu_A)\text{sign}(u_{TB}-\mu_B)$. From the previous equation, for completeness, it then follows that

$$P(\vec{a},\vec{b}) = \frac{-1}{(2\pi)^{3/2}} \sum_{k=1}^{3} \sum_{j=1}^{3} a_k b_j \int\int_{-\infty}^{\infty} d^3 y N(0,1)(\vec{y}) \, \text{sign}(y_k) \, \text{sign}(y_j) \qquad (48)$$

This results into the following expression.

$$P(\vec{a},\vec{b}) = -\sum_{k=1}^{3} a_k b_k \qquad (49)$$

7. Conclusion & Discussion

The main conclusion is that the quantum result for the covariance of A and B can be obtained from local hidden causality.

Finally, the derivation of separate probabilities for, $A=+1$, and for, $A=-1$ is discussed. Let us inspect the $P(A=+1)$. The question is, is this probability equal to the probability $P(A=-1)$, that is, is $P(A=+1)=P(A=-1)=1/2$?

In the first place we note that $A=+1$, only when, $u_T > \mu_A$. The associated uniform probability then is equal to

$$P_{A=+1}(u_{T_A} > \mu_A) = \begin{cases} \frac{1}{2}(1 + \frac{\sigma_A}{\sqrt{2\pi}}), & x=0 \\ \frac{1}{2}, & x \neq 0 \end{cases}$$

The other variables, save the, y_k, (k=1,2,3) indirectly via, σ_A, do not influence the sign of A. Hence, we may conclude that the other variables run through the complete intervals as defined previously. We then see

$$P(A=+1) = \int_{A=+1} \rho_{tot} = 1/2$$

$P(A=-1)=1/2$ follows similarly.

It can therefore rightfully be claimed that the Kolmogorovian proposed model refutes the theorem that Bell's inequalities prevent local hidden variables to recover quantum covariance. Furthermore, it refutes the notion that no fruitful, classical type, local hidden variable theory is possible. Perhaps that there are no local hidden causalities, but that is unrelated to Bell's theorem. Moreover, because a working model has been created, arguments against LHV theories without Bell inequalities also fail. In conclusion, it can rightfully be claimed that quantum mechanics is not the exclusive explanation for the experimentally obtained quantum covariance.

References

[1] Bell, J.S. (1964) On the Einstein Podolsky Rosen paradox. *Physics*, **1**, 195-200.
[2] Einstein, A., Rosen, N. and Podolsky, B. (1935) Can quantum-mechanical description of physical reality be considered complete? *Phys. Rev.*, **47**, 777-780.
[3] Bohm, D. The Paradox of Einstein, Rosen and Podolsky. In: Quantum Theory and Measurement, page 354-368. Edited by J.A. Wheeler and W.H. Zurek, Princeton Univ. Press, 1983.
[4] Aspect, A., Dalibard, J. and Roger, G. (1982) Experimental test of Bell's inequalities using time-varying analyzers. *Phys. Rev. Lett.*, **49**, 1804-1806.
[5] Geurdes, J.F. (1998 a) Quantum Remote Sensing, *Physics Essays*, **11**, 367-372. Geurdes, J.F. (1998 b) Wigner's variant of Bell's inequality, *Austr. J. Phys.*, **51**(5), 835-842. Geurdes, J.F. (2001) Bell inequalities and pseudo-functional densities, *Int. J. Theor. Phys., Group. Theor. & Nonl. Opt.*, **7**(3), 51. Geurdes, J.F. (2006) A Counter-Example to Bell's theorem with a softened singularity, *Galilean Electrodynamics,* **17**(1), 16.
[6] Lighthill, M.J. Introduction to Fourier Analysis and Generalised Functions, 1958 Camb. Univ Press.

Chapter 8

Ground States and Excitation Spectra of Light and Strange Baryons in a Relativistic Linear Potential Model

S.N. Jena[a,*], T.C. Tripathy[b,†] and H.H. Muni[c]
[a] Department of Physics, Berhampur University,
Berhampur-760 007, Orissa, India
[b] Department of Physics, Roland Institute of Technology,
Surya Vihar, Berhampur-761008, Orissa, India
[c] Department of Physics, R.N.College, Dura, Ganjam,
Orissa, India

Abstract

Assuming the baryons as an assembly of individual quarks confined in a first approximation by an equally mixed scalar-vector potential in linear form which presumably represents the non-perturbative gluon interactions including gluon self-coupling, the mass spectra of the light and strange baryons are investigated. The contributions of the Goldston boson (π, η and K) exchange interactions between the constituent quarks arising from spontaneous breaking of chiral symmetry are taken into account over and above the center-of-mass correction to provide a unified description of the ground states and excitation spectra of baryons. The present model faces problems in explaining the correct level ordering of excited states; still the results obtained for the ground states of baryons are in reasonable agreement with the experimental values.

1. Introduction

Several papers based on the non-relativistic quark models have appeared [1] in the literature in connection with the study of mass splittings of light and strange baryons. Traditional constituent quark models (CQM's) which adopted OGE [2] as the hyperfine interaction between constituent quarks(Q) have been suggested in the study of light baryon spectroscopy but these models faced some intriguing problems such as (i) the wrong level ordering of

---
*E-mail address: snjena@rediffmail.com
†E-mail address: tarinitripathy@yahoo.com

positive- and negative-parity excitations in the N, Δ, Λ, and Σ spectra, (ii) the missing flavour dependence of the Q-Q interaction necessary for a simultaneous description of the correct level ordering in the N and Λ spectra and (iii) the strong spin-orbit splittings that are produced by the OGE interaction but not found in the empirical spectra. All of these effects have been explained to be due to [3] inadequate symmetry properties inherent in the OGE interaction. Several hybrid models advocating meson-exchange Q-Q interactions in addition to the OGE dynamics of CQM's have been suggested for baryons [4]. In the study of N and Δ spectra, especially π and σ exchanges have been introduced to supplement the interaction between constituent quarks.

A few years back, two groups, viz. Valcarce, Gonzalez, Fernandez and Vento [5] and Dziembowski, Fabre and Miller [6] came up with versions of hybrid constituent quark models. They have presented a reasonable description of N and Δ excitation spectra taking into account a sizeable contribution from the OGE interaction. However, the performance of the hybrid constituent quark models have been studied in details by Glozman et al [3] by using the calculations based on accurate solutions of the three quark systems in both variational Schroedinger and a rigorous Faddeev approach. It has been argued that hybrid Q-Q interactions with a sizeable OGE component encounter difficulties in describing baryon spectra due to the specific contributions from one-gluon- and meson- exchanges together. On the contrary, Glozman et al [7] have shown that a chiral constituent quark model with a Q-Q interaction relying solely on Goldstone-boson-exchange(GBE) is capable of providing a unified description not only of the N and Δ spectra but also of all strange baryons in good agreement with phenomenology. They have also presented a constituent quark model with the confinement potential in linear and harmonic [7] forms for the light and strange baryons providing a unified description of their ground states and excitation spectra. Their model which relies on constituent quarks and Goldstone bosons arising as effective degrees of freedom of low energy QCD from the spontaneous breaking of chiral symmetry ($SBCS$) has been found to be quite proficient in reproducing the spectra of the three quark systems from a precise variational solution of a Schroedinger equation with a semi-relativistic Hamiltonian.

Thus several papers based on non-relativistic quark models have appeared in the literature in connection with the study of the mass spectrum of light and strange baryons. Although the phenomenological picture is reasonable at the non-relativistic level, a relativistic approach is quite indispensable on this account in view of the fact that the baryonic mass splittings are of the same order as the constituent quark masses. Of course, the chiral constituent quark model which has been constructed by Glozman et al [7] in a semi-relativistic framework is a step in this direction and shows an essential improvement over non-relativistic approaches. On the other hand, the MIT bag model [8] has also been found to be relatively successful in this respect. In its improved versions, the Chiral Bag Model (CBM) [9] have included the effect of pion self-energy due to baryon-pion coupling at the vertex to provide a better understanding of the baryon masses. Nevertheless, such models still contain some dubious phenomenological elements which are objectionable. The sharp spherical bag boundary, the zero point energy, the exclusion of pions from within the bag or adhoc inclusion of pions within it, are a few such points to be noted in this context. Furthermore it is somewhat difficult to believe that the static spherical bag remains unperturbed even after the creation of a pion. However, the sharp spherical bag boundary in the bag confinement, which is at the root of all objections and difficulties encountered by the otherwise successful CBM, is nonetheless arbitrary and phenomenological in nature and can therefore be replaced by an alternative and suitable phenomenological average potential for individual quarks, presuming at the same time its good features together with its successful predictions in the study of light baryons in their ground states.

The chiral potential models [10] which are comparatively more straight-forward in the above respects are obviously attempts in this direction. In such models the confining potentials which basically represent the interaction of quarks with the gluon field are usually assumed phenomenologically as Lorentz scalars in harmonic and cubic forms. Potentials of a different type of Lorentz structure with equally mixed scalar and vector parts in harmonic [11], non-Coulombic power-law[12], square-root [13], and logarithmic [14] form are also used in this context.

The term in the Lagrangian density for quarks corresponding to the effective scalar part of the potential in such models being chirally non-invariant through all spaces requires the introduction of an additional pionic component everywhere in order to preserve chiral symmetry. The effective potential of individual quarks in these models, which is basically due to the interaction of quarks with the gluon field, may be thought of as being mediated in a self-consistence manner through Nambu-Jona-Lasino (NJL)-type models [15] by some form of instanton induced effective quark-quark contact interaction with position-dependent coupling strength. The position-dependent coupling strength, supposedly determined by the multi-gluon mechanism, is impossible to calculate from first principle, although it is believed to be small at the origin and increases rapidly towards the hadron surface. Therefore, one needs to introduce the effective potential for individual quarks in a phenomenological manner to seek a posteriori justification in finding its conformity with the supposed qualitative behaviour of the position-dependent coupling strength in the contact interaction.

However, with no theoretical prejudice in favour of any particular mechanism for generating confinement of individual quarks, in the present work we prefer to work in an alternative, but similar scheme based on Dirac equation with a purely phenomenological individual quark potential of the form

$$V_q(r) = \frac{1}{2}\left(1 + \gamma^0\right)(a^2 r + V_0) \qquad (1)$$

with $a > 0$. Such a model takes the Lorentz structure of the potential as an equal admixture of scalar and vector parts because of the fact that both the scalar and vector parts in equal proportions at every point render the solvability of the Dirac equation for independent quarks by reducing it to the form of a *Schrödinger*-like equation. This Lorentz structure of the potential also has an additional advantage of generating no spin-orbit splittings, as observed in the experimental baryon spectrum.

This potential has been used in the past to study successfully the static baryon properties[16], the weak-electric and -magnetic form factors for the semileptonic baryon decays[17], electromagnetic properties of nucleons[18] and the magnetic moments of the baryons in the nucleon octet[19] in reasonable agreement with the experimental data. This model has also been adopted to study reasonably well the mass and decay constant of the ($q\bar{q}$) pion [20], S-state mass splittings of the mesons of $s\bar{s}$, $c\bar{c}$ and $b\bar{b}$ systems, ground state mass splittings of the heavy non-self-conjugate mesons in strange, charm and bottom flavour sector [21] and electromagnetic decays of mesons[21,22]. This model has also been employed to explain reasonably well the mass spectrum of octet baryons taking into account the contributions due to the colour-electric and -magnetic energies arising out of the residual OGE interaction alongwith that due to the residual quark-pion coupling arising out of the requirement of the chiral symmetry and the neccessary centre-of-mass(c.m) motion. But in this work the interactions between the constituent quarks arising out of the Goldstone-boson exchange (GBE) which are considered to play an important role in contributing to the energy of the baryon core were not taken into account. Therefore in the present work we wish to take into account the GBE contributions in a perturbative manner to study both the ground state and excitation spectra of octet baryons without taking the OGE correction. Very recently [23] we have also found that the present model can be used successfully to describe the ground state mass spectra of light and strange baryons taking into account the correction due to the energy arising out of the GBE interactions alongwith those due to the energy associated with the residual OGE interaction and c.m motion. In view of this success, we intend to extend the application of the present model to the study of the excitation spectra of baryons in the present work. As has been pointed out by Glozman et al [3] in their studies on the performance of hybrid constituent quark models, the difficulties encountered in describing baryon spectra may be due to specific contributions from OGE and GBE taken together. On the contrary they have shown that a unified description of the ground states and the excited spectra of baryons is possible if one takes into account only the GBE corrections in a CQM. With this contention in mind we are interested to study the ground states and the excitation spectra of baryons in a unified way in the present work taking into account contributions solely from the GBE interactions between the constituent quarks. Necessary correction due to spurious centre-of-mass motion of the baryon core is also included here. Following the lines of arguments of Glozman et al [3] we do not consider here the quark-gluon coupling due to OGE interaction at short distances.

In the present model, baryons are considered as systems of three constituent quarks with dynamical masses which are confined in a first approximation by an effective linear potential and are subjected to interaction by GBE. For the inclusion of the GBE contributions in this model we have followed the guidelines of the chiral constituent quark models suggested by Glozman et al [7]. However, we use these contributions in a perturbative manner alongwith OGE contributions which have been shown by Glozman et al [3] to have some intriguing problems in CQM's. We are mainly interested to see whether or not our model is able to describe simultaneously the ground states and excitation spectra of light and strange baryons in a unified way taking into account only the GBE contributions in a perturbative manner. In this context we may point out that we consider the constituent quarks of flavours u,d,s with masses considerably larger than the corresponding current quark masses so that the underlying chiral symmetry of QCD is spontaneously broken. As a consequence of $SBCS$, at the same time Goldstone bosons appear, which couple directly to the constituent quarks [7,24]. Hence, beyond the scale of $SBCS$ one is left with constituent quarks with dynamical masses related to $<\bar{q}q>$ condensates and with Goldstone bosons as the effective degrees of freedom. This feature, that in the Nambu-Goldstone mode of chiral symmetry constituent quark and Goldstone boson fields prevail together, is well supported, e.g.by the σ model [25] or the NJL model [26]. In the same framework also with the spin and flavour content of the nucleon are naturally resolved [27].

The work is organized as follows. In section-2 we outline the potential model with the solutions for the relativistic bound state of the individually confined quarks in the ground state of baryons and the energy corrections due to the spurious center-of-mass motion are briefly discussed. Section-3 provides a brief account of the corrections due to Goldostone Boson or (π, η and K-meson) exchange interactions between the constituent quarks in a generalised way. Finally, in section-4 we present the results for the ground states and excited states of light and strange baryon masses, which are in reasonable agreement with the corresponding experimental values.

2. Basic Formalism

Leaving behind for the moment, the interaction of the quarks due to GBE arising from SBCS to be treated perturbatively, we begin with the confinement part of the interaction which is believed to be dominant in baryonic dimensions. This particular part of the interaction which is believed to be determined by the multi-gluon mechanism is impossible to calculate theoretically from first principle. Therefore from a phenomenological point of view we assume that the constituent quarks in a baryon core are independently confined by an average flavour-independent relativistic potential of the form given in equation(1). Hence, to a first approximation, the confining of the interaction represented here by an average flavour-independent potential is believed to provide zeroth order constituent quark dynamics inside such baryons. Here a and V_0 are the potential parameters. We further assume that the constituent independent quarks or antiquarks obey the Dirac equation with potential $V_q(r)$ implying thereby a Lagrangian density of zeroth order as

$$\mathcal{L}_q^0(x) = \overline{\psi}_q(x) \left[\frac{i}{2}\gamma^\mu \overleftrightarrow{\partial}_\mu - m_q - V_q(r) \right] \psi_q(x) \qquad (2)$$

which leads to Dirac equation for individual quark of mass m_q as

$$\left[\gamma^0 E_q - \vec{\gamma}\cdot\vec{p} - m_q - V_q(r) \right] \psi_q(\vec{r}) = 0 \qquad (3)$$

where the normalized quark wave function $\psi_q(\vec{r})$ can be written in two component form as

$$\psi_{nlj}(\vec{r}) = N_{nl} \begin{pmatrix} if_{nlj}(r)/r \\ (\vec{\sigma}\cdot\hat{r})g_{nlj}(r)/r \end{pmatrix} Y_{ljm}(\hat{r}). \qquad (4)$$

Here, the normalised spin angular part

$$Y_{ljm}(\hat{r}) = \sum_{m_l, m_s} \langle l, m_l, \frac{1}{2}, m_s | j, m_j \rangle Y_l^{m_l} \chi_{\frac{1}{2}}^{m_s} \qquad (5)$$

and N_{nl} is the overall normalisation constant. The reduced radial part $f_{nlj}(r)$ of the upper component of Dirac spinor $\psi_{nlj}(\vec{r})$ satisfies the equation

$$f''_{nlj}(r) + \left[\lambda_{nl}\{E_{nl}^q - m_q - V(r)\} - \frac{l(l+1)}{r^2} \right] f_{nlj}(r) = 0 \qquad (6)$$

where

$$\lambda_{nl} = E_{nl}^q + m_q \qquad (7)$$

The present model can in principle provide the quark orbitals $\psi_{nlj}(\vec{r})$ and the zeroth order binding energies of the confined quark for various possible eigen modes through equations(4)-(7). However, the ground state baryons, in which all the constituent quarks are in their lowest eigen states, the corresponding quark orbitals can be expressed as

$$\psi_{1s}(\vec{r}) = N_{nl} \begin{pmatrix} \phi_{1s}(\vec{r}) \\ \frac{\vec{\sigma}\cdot\vec{p}}{\lambda_{nl}}\phi_{1s}(r) \end{pmatrix} \chi \uparrow \qquad (8)$$

where $\phi_{1s}(\vec{r})$ is the radial angular part of the upper component $\psi_{1s}(\vec{r})$ and is given by $\phi_{1s}(\vec{r}) = \frac{i}{\sqrt{4\pi}} f_{1s}(r)/r$. For the ground state equation(6) reduces to

$$f''_{1s}(r) + \left[\lambda_{1s}(E^q_{1s} - m_q - a^2 r - V_0)\right] f_{1s}(r) = 0 \qquad (9)$$

which can be transformed into a convenient dimensionless form

$$f''_{1s}(\rho) + (\epsilon_{1s} - \rho) f_{1s}(\rho) = 0 \qquad (10)$$

where ($\rho = r/r_{0q}$) is a dimensionless variable with $r_{0q} = (\lambda_q a^2)^{-1/3}$ and

$$\epsilon_{1s} = \left(\frac{\lambda_{1s}}{a^4}\right)^{\frac{1}{3}} (E^q_{1s} - m_q - V_0) \qquad (11)$$

The equation (10) is the basic eigen value equation, which can be solved as follows: With $z = \rho - \epsilon_q$, equation(10) reduces to the Airy equation

$$f''_q(z) - z f_q(z) = 0 \qquad (12)$$

The solution $f_q(z)$ of equation (12) is the Airy function $Ai(z)$. Since at $r = 0$ we require $f_q(r) = 0$, we have $Ai(z) = 0$ at $z = -\epsilon_q$. If z_q are the roots of the Airy function such that $Ai(z_n) = 0$, then we have $z = -\epsilon_q = z_n$. For the ground state of quarks or antiquarks, the ϵ_{1s} value is given by the first root z_1 of the Airy function so that

$$\epsilon_q = \epsilon_{1s} = -z_1 \qquad (13)$$

The value of this root z_1= -2.33811 and hence ϵ_{1s}=2.33811.

Now the individual quark binding energy E^q_{1s} of zeroth order in the hadron ground state can be obtained from equation(11) through the relation

$$E^q_{1s} = m_q - V_0 + a x_q \qquad (14)$$

where x_q is the solution of the root equation obtained through substitution from equation (11) in the form

$$x_q^4 + b x_q^3 - \epsilon_q^3 = 0 \qquad (15)$$

with $b = \frac{2E_q + V_0}{a}$. Solution for the quark binding energy E_q in the zeroth order corresponding to the ground state of the baryon immediately leads to the ground state mass of the baryon core in zeroth order as

$$M^0_B = E^0_B = \sum_q E^q_{1s} \qquad (16)$$

Similarly equation (6) can be solved for 2S and 1P states to obtain the individual quark binding energy E_{2S}^q and E_{1P}^q respectively, with the help of the standard numerical method which yields $\epsilon_{2S} = 4.08795$ and $\epsilon_{1P} = 3.3615699$. These values lead to the corresponding masses of the excited states of baryon core in zeroth order in the same way as in case of the ground state.

The overall normalization constant N_{nl} of $\psi_{nlj}(\vec{r})$ appearing in equation(4) is of the form

$$N_{nl}^2 = \left[1 + \frac{(E_{nl}^q - m_q - V_0 - a << r >>_{nl})}{\lambda_{nl}}\right]^{-1} \tag{17}$$

where $<< r >>_{nl}$ is the expectation value of r with respect to $\psi_{nlj}(\vec{r})$.

In this model there would be a sizeable spurious contribution to the energy E_{nl}^q from the motion of the centre-of-mass of the three-quark system. Unless this aspect is duly accounted for, the concept of the independent motion of quarks inside the baryon core will not lead to a physical baryon state of definite momentum. Although there is still some controversy on this subject, we follow the technique adopted by Bartelski et al. and E.Eich et al. [28], which is just one way of accounting for the c.m motion. Following their prescription a ready estimate of the c.m momentum $\vec{P}_B$ of the baryon core can be obtained as

$$< \vec{P}_B^2 >_{nl} = \sum_q < \vec{p}_q^2 >_{nl} \tag{18}$$

where $< \vec{p}_q^2 >_{nl}$ is the average value of the square of the individual quark momentum taken over the single quark states and is given in this model as

$$\begin{aligned}\langle \vec{p}_q^2 \rangle_{nl} =\ & N_{nl}^2 \left[2E_{nl}^q(E_{nl}^q - m_q) - (3E_{nl}^q - m_q - V_0)V_0 \right.\\ & \left. -(3E_{nl}^q - m_q - 2V_0)a^2 \langle\langle r \rangle\rangle_{nl} + a^4 \langle\langle r^2 \rangle\rangle_{nl}\right]\end{aligned} \tag{19}$$

where the double angular brackets represent the expectation values with respect to $\phi_{nl}(\vec{r})$.

Therefore, if E_B^0 is the energy of the baryon core in zeroth order then the centre-of-mass corrected baryon mass is

$$E_B = < E_B^2 >^{\frac{1}{2}} = \left[E_B^{0\,2} - \sum_q < \vec{P}_B^2 >_{nl}\right]^{1/2} \tag{20}$$

3. Energy Correction due to Goldstone Boson Exchange(GBE) Interactions

Looking at the zeroth order Lagrangian density $\mathcal{L}_q^0(x)$ described in section-2, one can note that under a global infinitesimal chiral transformation at least in the (u,d)-flavour sector the axial vector current of quarks is not conserved due to the fact that the scalar terms in $\mathcal{L}_q^0(x)$ which is proportional to $S(r) = [m_q + V(r)/2]$ is chirally odd. Of course the vector part of the potential poses no problem in this respect.

In order to restore the chiral $SU(2) \otimes SU(2)$ symmetry within the partial conservation of axial vector current (PCAC) limit, one can introduce in the usual manner an elementary field $\phi(x)$ of the pion, which is a Goldstone boson of small but finite mass $m_\pi = 140$ MeV through additional terms in the original Lagrangian density $\mathcal{L}_q^0(x)$ so as to write,

$$\mathcal{L}(x) = \mathcal{L}_q^0(x) + \mathcal{L}_\pi^0(x) + \mathcal{L}_I^\pi(x) \tag{21}$$

where $\mathcal{L}_\pi^0(x) = \frac{1}{2}(\partial_\mu \phi)^2 - \frac{1}{2}m_\pi^2(\phi)^2$. The Lagrangian density $\mathcal{L}_I^\pi(x)$ corresponding to the quark-pion interaction is taken to be linear in iso-vector pion field $\phi(x)$ such that

$$\mathcal{L}_I^\pi = \frac{1}{f_\pi} S(r)\overline{\psi_q}(x)\gamma^5 \left(\vec{\tau}\cdot\vec{\phi}\right)\psi_q(x) \tag{22}$$

where $f_\pi = 93$ MeV is the phenomenological pion decay constant. Then the four divergence of the total axial-vector current becomes

$$\partial_\mu A^\mu(x) = -f_\pi m_\pi^2 \phi(x) \tag{23}$$

yielding the usual PCAC relation.

Restoration of chiral symmetry in this manner in the PCAC-limit brings to light the possible quark-pion interaction over and above the dominant confining interactions arising out of the non-perturbative multigluon mechanisms. This additional residual interaction between core-quarks and the surrounding pions may be treated as a small perturbation over the solutions obtained due to the dominant confining part mainly responsible for assembling the quarks in the baryon core. Consequently such pion coupling of the non-strange quarks would give rise to pionic self-energy of the baryons which would ultimately contribute to the physical masses of the baryons. The chiral symmetry, which is almost exact in the light u and d flavour sector is however only approximate in QCD when strangeness is included, because of the large mass of the s-quark.

Following the Hamiltonian technique [29] as has been used in the CBM, we can describe the effect of pion coupling in low-order perturbation theory as follows.

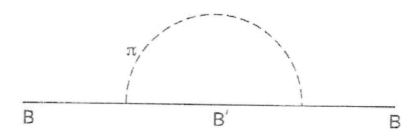

Figure 1. Baryon self-energy due to coupling with pion.

The pionic self-energy of the baryons can be evaluated with the help of the single-loop self-energy diagram (figure 1) as

$$\sum_B (E_B) = \sum_k \sum_{B'} \frac{V^{\dagger BB'} V^{BB'}}{(E_B - \omega_k - M_{B'}^0)} \tag{24}$$

where $\sum_k = \sum_j \int \frac{d^3\vec{k}}{(2\pi)^3}$.

Here j corresponds to the pion-isospin index and B' is the intermediate baryon state. $V^{BB'}(\vec{k})$ is the general baryon-pion absorption vertex function obtained [30] in this model as

$$V_j^{BB'}(\vec{k}) = i\sqrt{4\pi}\frac{f_{BB'\pi}}{m_\pi}\frac{ku(k)}{\sqrt{2\omega_k}}\left(\vec{\sigma}^{BB'}\cdot\hat{k}\right)\tau_j^{BB'} \qquad (25)$$

where $\vec{\sigma}_j^{BB'}$ and $\tau_j^{BB'}$ are spin and isospin matrices and $\omega_k^2 = \vec{k}^2 + m_\pi^2$. The form factor u(k) in this model can be expressed as

$$u(k) = \frac{5N_{nl}^2}{3\lambda_{nl}g_A}[(2m_q + V_0) << j_0(|\vec{k}|)r) >> +a^2 << rj_0(|\vec{k}|r) >>$$
$$+a^2 << j_1(|\vec{k}|r)/k >>] \qquad (26)$$

where $j_0(|\vec{k}|r)$ and $j_1(|\vec{k}|r)$ represent the zeroth and first order spherical Bessel functions, respectively. The double angular brackets stand for the expectation values with respect to $\phi_{nl}(r)$. In this model the axial vector coupling constant g_A for the beta decay of the neutron is given by

$$g_A = \frac{5}{9}\left(\frac{8E_q + 10m_q + V_0}{4E_q + 2m_q - V_0}\right)$$

Now with the vertex function $V_j^{BB'}(\vec{k})$ at hand, it is possible to calculate the pionic self-energy for various baryons with appropriate baryon intermediate states contributing to the process. For degenerate intermediate states on mass shell with $M_B^0 = M_{B'}^0$, the self-energy correction becomes

$$(\delta M_B)_\pi = \sum_B \left(E_B^0 = M_B^0 = M_{B'}^0\right) = -\sum_{k,B'}\frac{V^{\dagger BB'}V^{BB'}}{\omega_k} \qquad (27)$$

Now using equation(25), we find

$$(\delta M_B)_\pi = \frac{-I_\pi}{3}\sum_{B'}C_{BB'}f_{BB'\pi}^2 \qquad (28)$$

where

$$C_{BB'} = \left(\vec{\sigma}^{BB'}\cdot\vec{\sigma}^{B'B}\right)\left(\vec{\tau}^{BB'}\cdot\vec{\tau}^{B'B}\right) \qquad (29)$$

and

$$I_\pi = \frac{1}{\pi m_\pi^2}\int_0^\infty \frac{dk k^4 u^2(k)}{\omega_k^2} \qquad (30)$$

For the intermediate baryon states B', we consider only the octet and decuplet ground states. The self-energy $(\delta M_B)_\pi$ for different baryons can be computed by using the values of $f_{BB'\pi}$ and $C_{BB'}$[1] as has been done in our earlier works[31].

The self-energy $(\delta M_B)_\pi$ calculated here contains both the quark self-energy (fig 2(a)) and the one-pion-exchange contributions (fig 2(b)). It must be noted here that this method ignores to a large extent the short-range part of the pion exchange interaction, which is of

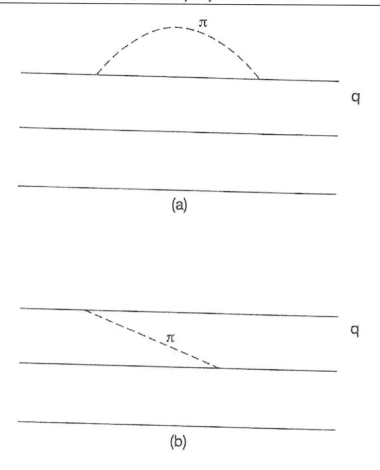

Figure 2. One-pion-exchange contributions to the energy.

crucial importance for splittings. Only when the complete infinte set of all radially excited intermediate state B' is taken into account, the method could be adequate [32].

The $SU(3)_L \times SU(3)_R$ chiral symmetry of QCD Lagrangian is spontaneously broken down to $SU(3)_V$ by the QCD vacuum (in the large N_c limit would be $U(3)_L \times U(3)_R \to U(3)_V$). There are two important generic consequences of the $SBCS$. The first one is an appearance of the octet pseudoscalar mesons of low mass, π, K, η, which represent the associated approximate Goldstone bosons (in the large N_c limit the flavour singlet state η' should be added). The second one is that valence (partially massless) quarks acquire a dynamical mass, which has been called historically constituent mass. Indeed, the non-zero value of quark condensate, $<\bar{q}q> \sim -(250 MeV)^3$, itself implies at the formal level that there must be at low momenta rather big dynamical mass which should be a momentum dependent quantity. Such a dynamical mass is now directly observed on the lattice [33]. Thus the constituent quarks should be considered as quasi-particles whose dynamical mass at low momenta comes from the non-perturbative gluon and quark-antiquark dressing. The flavour-octet axial current conservation in the chiral limit tells that the constituent quarks and Goldstone bosons should be coupled with the strength $g = g_A M / f_\pi$ [24], which is a quark analog of the famous Goldberger-Treiman relation. It has been recently suggested that in the low-energy regime, below the chiral symmetry breaking scale ~ 1 GeV, the low-lying light and strange baryons should be predominantly viewed as systems of three constituent quarks with an effective confining interaction and a chiral interaction mediated by a GBE between the constituent quarks [7].

With this contention in mind, we need to take into account the coupling of Goldstone bosons (π, η and K) to the constituent quarks arising from $SBCS$ in QCD over and above the dominant quark confining interactions arising out of the non-perturbative multigluon interactions. Such GBE interactions between the core quarks can be treated in the perturbative manner like it is done in CBM [9]. Here the fields of the Goldstone bosons may be treated independently without any constraint and their interactions with quarks may be assumed to be linear as it is done in case of pion in equation (30). The corrections due to η-exchange and kaon exchange interactions between the quarks can be calculated following same approach [9] as we have used in case of pion. Thus in general, one writes energy corrections due to GBE interactions as

$$(\delta M_B)_\chi = \frac{-I_\chi}{3} \sum_{B'} C_{BB'} f^2_{BB'\chi} , \qquad (31)$$

where $C_{BB'}$ is given by equation (29) and

$$I_\chi = \frac{1}{\pi m_\chi^2} \int_0^\infty \frac{dk k^4 u^2(k)}{k^2 + m_\chi^2} , \qquad (32)$$

where $\chi = \pi, \eta$, K. Following the discussions given in ref.[1] the baryon-meson coupling constant $f_{BB'\chi}$ can be expressed in terms of the nucleon-meson coupling constant $f_{NN\chi}$. Pion exchange interaction acts only between light quarks whereas η-exchange is allowed in all quark pair states. The kaon exchange interaction takes place in u-s and d-s pair states.

4. Results and Discussion

The zeroth-order mass $M_B^0 = E_B^0$ of ground state of a baryon arising out of the binding energies of the constituent quarks, confined independently by a phenomenological average potential $V_q(r)$ which presumably represents the dominant non-perturbative multigluon interactions is subjected to certain corrections due to the exchange of Goldstone bosons between the constituent quarks, together with that due to the spurious centre-of-mass motion. All of these corrections can be treated independently, as though they are of the same order of magnitude, so that the physical mass of a baryon can be obtained as

$$M_B = E_B + (\delta M_B)_\pi + (\delta M_B)_\eta + (\delta M_B)_K \qquad (33)$$

where $(\delta M_B)_\chi$, with ($\chi = \pi, \eta$, K) is the GBE correction [equation(31)]. Here $E_B = E_B^0 + (\Delta E_B)_{c.m}$ is the c.m corrected baryon mass [equation(20)].

The quantitative evaluation of the c.m corrected baryon mass E_B and the GBE correction $(\Delta M_B)_\chi$ within the frame-work of the model primarily involves the potential parameters (a, V_0), the quark masses m_q and the corresponding binding energy E_{nl}^q alongwith the other relevant model quantities. In the present work where we like to take only the contributions of GBE interaction arising from the spontaneous breaking of chiral symmetry so as to obtain a simultaneous fit to the mass spectra of the ground and the excited states of the light and strange baryons, we choose to fix the potential parameters and quark masses appropriately as discussed below. To take into account the implications of $SBCS$ properly in the present model the constituent quarks of flavours with masses considerably larger than the corresponding current masses are considered here. We choose the constituent quark masses as

$$(m_u = m_d, m_s) = (0.100, 0.270) \quad GeV \tag{34}$$

and the potential parameters as

$$(a, V_0) = (0.256, -0.055) \quad GeV \tag{35}$$

which yield the ground state individual quark binding energies

$$\left(E_{1S}^u = E_{1S}^d, E_{1S}^s\right) = (496.492, 611.395) \quad MeV. \tag{36}$$

from the energy eigen value condition given in equation (11). Now using a standard numerical method with the same set of parameters, the integral I_χ, given by equation (32) is evaluated for 1S, 2S and 1P states as

$$(I_\pi, I_\eta, I_K) = (938.107, 49.002, 62.099) \quad MeV \tag{37}$$

$$(I_\pi, I_\eta, I_K) = (34.363, 0.5088, 0.7315) \quad MeV \tag{38}$$

$$(I_\pi, I_\eta, I_K) = (352.798, 17.018, 21.863) \quad MeV \tag{39}$$

respectively which are used in computation of the GBE corrections. Indeed, in the chiral limit there is only one coupling constant for all Goldstone bosons. Due to explicit chiral symmetry breaking the coupling constant for π, η and K may become different. However, in order to prevent a proliferation of the free parameters we try to keep the number of free parameters as small as possible and assume a single phenomenological pion-nucleon coupling constant $f_{NN\pi}$=0.283 for all mesons (π, η, K) and this value is used here to compute the GBE corrections. The GBE corrections for the ground states as well as the excited states of baryons are presented in Table-1. The energy corrections and the results obtained for the mass spectra of light and strange baryons in their ground states are presented in the Table-2, while those obtained for the excited states are displayed in Table-3. The calculated values of the ground state masses of light and strange baryons are found to agree quite well with the experiment. The model predictions for the mass spectra of the excited states, particularly, for 1P states of N ,Δ and Λ baryons compare well with the experiment. However the results obtained for the mass spectra of 2S states of baryons in the present model are overestimated. Like our other models[34] of this kind the present model also faces problems in explaining the correct level ordering of excited states of baryons. In this model we find that the SU(3) breaking effect due to quark masses $m_u = m_d \neq m_s$ lifts the degeneracy in the baryon masses through the c.m corrected energy term E_B among the groups (N, Δ), $(\Lambda, \Sigma, \Sigma^*)$, (Ξ, Ξ^*) and Ω^-. Then in second step, the GBE corrections $(\delta M_B)_\chi$ arising from $SBCS$ in QCD removes the degeneracy between N and Δ; Λ, Σ and Σ^*; Ξ and Ξ^*.

Table 1. GBE corrections $(\delta M_B)_\chi$ **(where** $\chi = \pi, \eta, K$**) for the ground states and the excited states of baryons (in MeV)**

LS Multiplet	$(\delta M_B)_\pi$	$(\delta M_B)_K$	$(\delta M_B)_\eta^{uu}$	$(\delta M_B)_\eta^{us}$	$(\delta M_B)_\eta^{ss}$	total	$(\delta M_B)_\chi$
N	-490.858	-	50.343	-	-	50.343	-440.515
Δ	-98.172	-	-50.343	-	-	-50.343	-148.515
Λ	-294.515	-135.471	50.343	-	-	50.343	-379.643
Σ	-32.724	-225.785	-16.781	-49.721	-	-66.502	-325.011
Σ*	-32.724	-90.314	-16.781	24.861	-	08.080	-114.958
Ξ	-	-225.785	-	-49.721	-09.208	-58.929	-284.714
Ξ*	-	-90.314	-	24.861	-09.208	15.653	-74.661
Ω^-	-	-	-	-	-27.623	-27.623	-27.623
$\frac{1}{2}^+$, N(1440)	-254.419	-	25.433	-	-	25.433	-228.986
$\frac{1}{2}^-$, N(1520); $\frac{3}{2}^-$, N(1535)	-190.049	-	33.914	-	-	33.914	-156.135
$\frac{3}{2}^+$, Δ(1600)	-50.884	-	-25.433	-	-	-25.433	-76.317
$\frac{1}{2}^-$, Δ(1620); $\frac{3}{2}^-$, Δ(1700)	06.294	-	01.055	-	-	01.055	07.349
$\frac{1}{2}^-$, Λ(1405); $\frac{3}{2}^-$, Λ(1520)	-128.797	-119.573	22.258	-36.104	-	-13.846	-262.216
$\frac{1}{2}^+$, Λ(1600)	-152.651	-68.533	25.433	-	-	25.433	-195.751
$\frac{1}{2}^-$, Λ(1670); $\frac{3}{2}^-$, Λ(1690)	-128.797	-27.990	22.258	-02.374	-	19.884	-136.903

Table 2. The energy corrections $(\delta M_B)_\chi$ **and the physical masses** (M_B) **of ground state baryons (in GeV)**

Baryon	E_B	$(\delta M_B)_\chi$	M_B Calculation	M_B Experiment
N	1.3805	-0.4405	0.940	0.940
Δ	1.3805	-0.1485	1.232	1.232
Λ	1.4956	-0.3796	1.116	1.116
Σ	1.4956	-0.3250	1.1706	1.193
Σ*	1.4956	-0.1149	1.3807	1.385
Ξ	1.6107	-0.2847	1.3260	1.321
Ξ*	1.6107	-0.0747	1.5360	1.533
Ω^-	1.7258	-0.0276	1.6982	1.672

The failure of the present model in explaining the correct level ordering of excited states of baryons may be ascribed to the following facts. In the present model the meson degree of freedom which is taken into account in the same way as in the CBM ignores to a large extent the short-range part of the meson exchange interaction, which is of crucial importance for splittings. Only when the complete infinite set of all radially excited intermediate states B' is taken into account, this method could be adequate[32]. For example, the meson exchange contribution to the N-Δ difference will become much larger. It will also be strongly enhanced when the meson exchange contribution is calculated non-perturbatively. The meson exchange contribution is strongly dependent on the radius of the bare wave function,

Table 3. The energy corrections $(\delta M_B)_\chi$ and the physical masses (M_B) of the excited states of baryons (in GeV)

LS Multiplet	E_B	$(\delta M_B)_\chi$	M_B Calculation	M_B Experiment
$\frac{1}{2}^+$, N(1440)	2.0802	-0.2290	1.8512	1.440
$\frac{1}{2}^-$, N(1520)	1.6860	-0.1561	1.5299	1.527
$\frac{3}{2}^-$, N(1535)				
$\frac{3}{2}^+$, $\Delta(1600)$	2.0802	-0.0763	2.0039	1.600
$\frac{1}{2}^-$, $\Delta(1620)$	1.6860	0.0073	1.6933	1.660
$\frac{3}{2}^-$, $\Delta(1700)$				
$\frac{1}{2}^-$, $\Lambda(1405)$	1.8333	-0.2622	1.5711	1.462
$\frac{3}{2}^-$, $\Lambda(1520)$				
$\frac{1}{2}^+$, $\Lambda(1600)$	2.0423	-0.1958	1.8465	1.600
$\frac{1}{2}^-$, $\Lambda(1670)$	1.8333	-0.1369	1.6964	1.680
$\frac{3}{2}^-$, $\Lambda(1690)$				

i.e. on the type of confinement. This dependence has not been studied in the present work.

Thus we find that the present model which treats the meson degree of freedom perturbatively in the manner like it is done in the CBM [9], can reproduce the ground state mass spectra of light and strange baryons and can provide the simultaneous description of both ground states and radial excitations, but faces difficulties in explaining the correct level ordering of the positive- and negative- parity excitation in the N, Δ, and Λ spectra. Extension of the present work to explain the correct level ordering of excited states in the light and strange baryon spectra would obviously require a phenomenological reparametrization of the potential model and corresponding work in that direction is presently underway.

Acknowledgments

The authors are grateful to Prof.N.Barik, Mayurbhanj Professor of Physics, Utkal University, Bhubaneswar, Orissa, India for his valuable suggestions and useful discussions on this work.

References

[1] N.Isgur and G.Karl, *Phys.Rev.* **D18**, 4187(1978); D20, 1191(1979); Y.Nogami and N.Ohtsuka, *Phys.Rev.* **D26**,261(1982).

[2] A.de.Rujula, H.Georgi and S.L.Glashow, *Phys.Rev.* **D12**, 147(1975).

[3] L.Ya.Glozman, Z.Papp, W.Plessas, K.Varga and R.F.Wagenbrunn, *Phys.Rev.* **C57**, 3406(1998).

[4] I.T.Obukhovsky and A.M.Kusainov, *Phys.Lett.* **B238**, 142(1990).

[5] A.Valcarce, P.Gonzalez, F.Fernandez and V.Vento, *Phys.Lett.* **B367**, 35(1996).

[6] Z.Dziembowski et al *Phys.Rev.* **C53**, R2038(1996).

[7] L.Ya.Glozman, W.Plessas, K.Varga and R.F.Wagenbrunn, *Phys.Rev.* **D58**, 094030(1998); L.Ya.Glozman, D.O.Riska, *Phys.Rep.* **268**, 263-303(1996); L.Ya.Glozman, *Nucl.Phys.***A663**,103(2000)

[8] A.Chodos, R.L.Jaffe, K.Johnson, C.B.Thorn and V.F.Weisskopf, *Phys.Rev.* **D9**, 1471(1974a); A.Chodos, R.L.Jaffe, K.Johnson and C.B.Thorn *Phys. Rev.* **D10**, 2599(1974b); T.De.Grand, R.L.Jaffe, K.Johnson and J.Kiskis, *Phys.Rev.* **D12**, 2060(1975).

[9] A.W.Thomas, *Adv.Nucl.Phys.* **13**, 1(1983).

[10] R.Tegen, R.Brockmann and W.Weise, *Z.Phys.* **A307**, 339(1982); R.Tegen and W.Weise, *Z.Phys.* **A314**, 357(1983); R.Tegen, M.Schedle and W.Weise, *Phys. Lett.* **125B**, 9(1983).

[11] P.Leal.Ferreira and N.Zagury, *Lett.Nuovo-Cimento.* **20**, 511(1977); P.Leal. Ferreira, J.A.Helayel and N.Zagury, *Lett.Nuovo-Cimento.* **A55**, 215(1980); N.Barik, B.K.Dash and M.Das, *Phys.Rev.***D31**,1652(1985); N.Barik and B.K.Dash, *Phys.Rev.***D33**, 1925(1986).

[12] A.Martin, *Phys.Lett.* **93B**, 338(1980); ibid. 100B,5111(1981); N.Barik and S.N.Jena, *Phys.Lett.* **97B**, 761(1980); 97B, 265(1980); N.Barik and S.N.Jena, *Phys.Rev.* **D26**,618(1982); N.Barik and M.Das, *Phys.Lett.* **120B**, 403(1983), *Phys.Rev.* **D28**, 2823(1983); ibid. D33, 176(1986a); *Pramana-J.Phys.* **27**, 783(1986b). S.N.Jena, P.Panda and K.P.Sahu, *Int.J.Theo.Phys.Group Th.Non-Linear optics.* **9**, 69(2002);ibid. **8**, 351(2002); *J.Phys.G:Nucl.part.Phys.***27**, 1519(2001); S.N.Jena, P.Panda and T.C.Tripathy, *Phys.Rev.* **D63**, 014011(2000).

[13] S.N.Jena and M.R.Behera, *Pramana-J.Phys.***44**, 357(1995);ibid: 47, 233(1996); *Int.J.Mod.Phys.* **E7**, 69(1998);425(1998); S.N.Jena, M.R.Behera and S.Panda, *J.Phys.G:Nucl.part.Phys.***24**, 1089(1998).

[14] E.Magyari, *Phys.Lett.* **95B**, 295(1980); C.Quigg and J.L.Rosner *Phys.Lett.* **B71**, 153(1977); S.N.Jena and D.P.Rath, *Pramana-J.Phys.* **27**, 773(1986); N.Barik, S.N.Jena and D.P.Rath,*Phys.Rev.***D41**, 1568(1990); *Int.J.Mod.Phys.* **A7**, 6813(1992); S.N.Jena, K.P.Sahu and P.Panda, *Int.J.Theo.Phys.Group Th.Non-Linear optics,* **8**, 213(2002).

[15] E.Magyari, *Phys.Lett.* **95B**, 295(1980); C.Quigg and J.L.Rosner *Phys.Lett.* **B71**, 153(1977); S.N.Jena and D.P.Rath,*Pramana-J.Phys.* **27**, 773(1986); N.Barik,

S.N.Jena and D.P.Rath,*Phys.Rev.***D41**, 1568(1990); *Int.J.Mod.Phys.* **A7**, 6813(1992); S.N.Jena, K.P.Sahu and P.Panda, *ThInt.J.Theo.Phys.Group .Non-Linear optics,* **8**, 213(2002).

[16] P.Leal.Ferreira, *Lett.Nuovo-Cimento,* **20**, 157(1977); A.P.Kobuskin Report No. ITP-76-58E, Kiev. S.N.Jena and S.Panda, *Pramana-J.Phys.* **35**, 21(1990).

[17] S.N.Jena and S.Panda, *Pramana-J.Phys.* **37**, 47(1990).

[18] S.N.Jena and S.Panda, *Int.J.Mod.Phys.* **A7**,2841(1992) *J.Phys.G:Nucl. Part.Phys.* **18**, 273(1992).

[19] S.N.Jena and S.Panda, *Int.J.Mod.Phys.***A8**, 4563(1993); ibid. A9, 327(1994).

[20] S.N.Jena, M.R.Behera and S.Panda, *Pramana-J.Phys.* **51**, 711(1998).

[21] S.N.Jena, S.Panda and T.C.Tripathy, *Nucl.Phys.* **A658**, 249(1999).

[22] S.N.Jena, S.Panda and J.N.Mohanty, *J.Phys.G:Nucl.Part.Phys.* **24**,1869(1998).

[23] S.N.Jena, T.C.Tripathy and M.K.Muni, *Int.J.Theo.Phys.Group Th.Non-Linear optics.* (communicated in June 2007).

[24] S.Weinberg, *Physica* **A96**, 327(1979); A.Manohar and H.Georgi, *Nucl.Phys.* **B234**, 189(1984); D.I.Diakonov and V.Yu.Petrov, *Nucl.Phys.* **B272**, 457(1986).

[25] M.Gell-Mann and M.Levy, *Nuovo cimento* **16**, 705(1960).

[26] Y.Nambu and G.Jona-Lasino,*Phys.Rev.***112**, 345(1961a); ibid.**124**, 246(1961b); R.Brockmann, W.Weise and E.Werner, *Phys.Lett.***122B**, 201(1983); S.P.Klevansky, *Rev.Mod.Phys.* **64**, 649(1992)

[27] T.P.Cheng and L.F.Li, *Phys.Rev.Lett.***74**, 2872(1995).

[28] J.Bartelski, A.Szymacha, L.Mankcewicz and S.Tatur, *Phys.Rev.* **D29**, 1035(1984); E.Eich, D.Rein and R.Rodenberg, *Z.Phys.* **C28**, 225(1985).

[29] A.W.Thomas, *Nucl.Phys.* **B13**, 1(1983) and references cited therein.

[30] S.N.Jena, M.R.Behera and S.Panda, *J.Phys.G:Nucl.part.Phys.* **24**, 1089(1998); S.N.Jena, M.R.Behera and S.Panda,*Phys.Rev.* **D55**, 291(1997).

[31] S.N.Jena and S.Panda, *Int.J.Mod.Phys.* **A7**, 2841(1992); *Pramana-J.Phys.* **35**, 21(1990); ibid. 37,47(1991), *J.Phys.G.* **18**, 273(1992), *J.Phys.G.* **19**, 837(1993); *Int.J.Mod.Phys.* **A8**, 4563(1993), S.N.Jena, M.R.behera and S.Panda, *Phys.Rev.***D55**, 291(1997).

[32] L.Ya.Glozman, hep-ph/0004229 (unpublished).

[33] S.Aoki et al, *Phys. Rev.Lett.* **82**, 4392(1999).

[34] S.N.Jena, P.Panda and T.C.Tripathy, *Phys.Rev.* **D63**, 014011(2000).

Chapter 9

NEUTRINO MASS IN A SIX DIMENSIONAL E_6 MODEL

Snigdha Mishra and Sarita Mohanty
Department of Physics, Berhampur University,
Berhampur-760 007, Orissa, India

Abstract

We consider a supersymmetric E_6 GUT in a six dimesional $M^4 \otimes T^2$ space. E_6 breaking is achieved by orbifolding the extra two dimensional torus T^2 via a $T^2/(Z_2 \times Z_2' \times Z_2'')$. In effective four dimension we obtain an extended Pati-Salam group with $N = 1$ SUSY, as a result of orbifold compactification. We then discuss the problem of neutrino mass with imposed parity assignment. A light Dirac neutrino is predicted with mass of the order of 10^{-2} eV.

1. Introduction

There are several evidences in favour of supersymmetric grand unified theories (SUSYGUT) [1] to describe physics beyond the standard model. However it suffers serious problem like the problem of doublet-triplet mass splitting and the closely related problem of dimensional-five proton decay operators. This suggests that SUSY GUTs may require some drastic modification to reconcile symmetry breaking with the doublet-triplet mass splitting. Recently it has been shown [2] that in models with one or two extra dimensions, orbifold compactification can break grand unified symmetries in such a manner that it resolves the splitting problem as well as proton decay problem automatically without any fine-tunning. These ideas have been used to build models based on SU(5) and SO(10) [3] in five and six dimensions. The exceptional groups like E_6, E_7 and E_8 have been examined by Haba et al [4] in five dimension with $N = 1$ SUSY. In the present paper we wish to examine the idea in case of E_6 with two extra dimensions. The effective four dimensional theory is obtained by orbifolding the extra dimensional space (which is a torus T^2) with Z_2's. The motivation to study E_6 is that it may be viewed as remnants of superstring theory with E_8 symmetry in ten dimensions.

The paper is organized as follows; In the next section we shall discuss the technical details of the model showing the gauge symmetry breaking explicitly. Section (3) is devoted to the calculation of the four dimensional Yukawa Lagrangian in the neutral sector predicting a light Dirac neutrino. The result and concluding remarks are discussed in the last section.

2. The Model

In the present model, we take an E_6 gauge theory coupled with N=1 SUSY in a six dimensional space as $M^4 \otimes T^2$ where the extra dimensional space is taken as a Torus with radii R_1 and R_2. The breaking of E_6 symmetry with N=1 SUSY is then realised by orbifolding the two dimensional torus T^2 by two discrete groups Z'_2 and Z''_2 leading to two different symmetric subgroups $SU(2) \otimes SU(6)$ (G_{26}) and $SO(10) \otimes U(1)_\chi$ ($G_{10,1}$) in the two orthogonal compact dimensions. The combination $Z'_2 \otimes Z''_2$ breaks E_6 to the maximal common subgroups of G_{26} and $G_{10,1}$ i.e. the extended Pati-Salam group $(SU(2)_L \otimes SU(2)_R \otimes SU(4) \otimes U(1)_\chi)$. Another Z_2 is needed to break the extended supersymmetry (as N=1 SUSY in d=6 is equivalent to N=2 SUSY in four dimensions). The GUT symmetry is broken on the branes which are fixed points of transformations. Here we may note that, we have four fixed points at $Z = (0, 0), \left(\frac{\pi R_1}{2}, 0\right), \left(0, \frac{\pi R_2}{2}\right)$ and $\left(\frac{\pi R_1}{2}, \frac{\pi R_2}{2}\right)$. The corresponding Gauge symmetries are E_6 at (0, 0), $SU(2) \otimes SU(6)$ at $\left(\frac{\pi R_1}{2}, 0\right)$, $SO(10) \otimes U(1)$ at $\left(0, \frac{\pi R_2}{2}\right)$, and the intersection group $SU(2) \otimes SU(2) \otimes SU(4) \otimes U(1)_\chi$ at the $\left(\frac{\pi R_1}{2}, \frac{\pi R_2}{2}\right)$ brane respectively.

Now to visualise the symmetry breaking explicitly we consider N=1 SUSY Yang-Mills theory in six dimensions.

$$L_{6d}^{YM} = Tr\left[-\frac{1}{2} V_{MN} V^{MN} + i \bar{\Lambda} \Gamma^M D_M \Lambda\right] \tag{1}$$

with $V_M = T^A V_M^A$, T^A are the E_6 generators. $M = (\mu, 5, 6)$ for $\mu = 1, ...4$ are the four dimensional indices with coordinates x^μ, and 5, 6 being the two extra dimensional indices with coordinates y and z. Λ is the gaugino consisting of $(\Lambda_1, -i\Lambda_2)$, the chiral fermions with opposite 4d chirality. $D_M \Lambda = \partial_M \Lambda - ig[V_M, \Lambda]$ and $V_{MN} = \frac{1}{ig}[D_M, D_N]$. Γ's are the six dimensional Γ-matrices. The operation Z_2 on T^2 reduces the unwanted N = 2 SUSY to N=1, by suitably grouping the vector and chiral multiplets, i.e. V = (V_μ, Λ_1) and Σ=$(V_{5,6}, \Lambda_2)$ where V and Σ are the matrices in the adjoint representation of E_6. This is realized by the operation of Z_2 on V and Λ such that,

$$P V_\mu(\text{and } \lambda_1)(x,-y,-z)P^{-1} = +V_\mu(\lambda_1)(x,y,z) \quad (2)$$

$$P V_{5,6}(\text{and } \lambda_2)(x,-y,-z)P^{-1} = -V_{5,6}(\lambda_2)(x,y,z) \quad (3)$$

Here P = ± 1 being the eigen value of Z_2. Further the gauge symmetry breaking of E_6 is done by Z'_2 and Z''_2 by operation on the two compact dimensions given as,

$$P' V_\mu(\text{and } \lambda_1)\left(x,-y+\frac{\pi R_1}{2},-z\right)P'^{-1} = +V_\mu(\lambda_1)\left(x,y+\frac{\pi R_1}{2},z\right) \quad (4)$$

$$P'' V_\mu(\text{and } \lambda_1)\left(x,-y,-z+\frac{\pi R_2}{2}.\right)P''^{-1} = V_\mu(\lambda_1)\left(x,y,z+\frac{\pi R_2}{2}\right) \quad (5)$$

Table 1. Parity assignments for the vector multiplets $V_M^A \in 78$ of E_6

G_{2241}	G_{26}	$G_{10,1}$	\multicolumn{3}{c}{V = (V_μ, λ_1)}	\multicolumn{3}{c}{Σ=$(V_{5,6}, \lambda_2)$}
			Z_2	Z'_2	Z''_2	Z_2	Z'_2	Z''_2
(3, 1, 1)$_0$	(3, 1)	45$_0$	+	+	+	-	-	-
(1, 3, 1)$_0$	(1, 35)	45$_0$	+	+	+	-	-	-
(1, 1, 15)$_0$	(1, 35)	45$_0$	+	+	+	-	-	-
(2, 2, 6)$_0$	(2, 20)	45$_0$	+	-	+	-	+	-
(1, 1, 1)$_0$	(1, 35)	1$_0$	+	+	+	-	-	-
(2, 1, 4)$_{-3}$	(2, 20)	16$_{-3}$	+	-	-	-	+	+
(1, 2, $\bar{4}$)$_{-3}$	(1, 35)	16$_{-3}$	+	+	-	-	-	+
(2, 1, $\bar{4}$)$_{-3}$	(2, 20)	$\overline{16}_3$	+	-	-	-	+	+
(1, 2, 4)$_3$	(1, 35)	$\overline{16}_3$	+	+	-	-	-	+

Here P' and P" are the eigen values of Z'_2 and Z''_2 respectively. Here we may note that Z'_2 and Z''_2 operate oppositely for $V_{5,6}$ and λ_2 as shown in equation (3). The component fields belonging to E_6 are again divided according to different parity assignments. Thus the orbifolding to $T^2/(Z_2 \otimes Z'_2 \otimes Z''_2)$ can be viewed through the following parity assignments in the subspaces $E_6 \rightarrow SU(2) \otimes SU(6)$ (Z'_2) and $E_6 \rightarrow SO(10) \otimes U(1)_\chi$ (Z''_2) as noted in table-1.

We observe that the vector bosons belonging to $(3, 1, 1)_0 + (1, 3, 1_0) + (1, 1, 15)_0 + (1,1,1)_0$ in the subspace G_{2241} ($SU(2) \otimes SU(2) \otimes SU(4) \otimes U(1)_\chi$) have all parities positive. These vector bosons remain massless at the compactification scale, which will be obvious from the mode expansion of the fields [5] given by

$$\phi_{+++}(x,y,z) = \frac{1}{\pi\sqrt{R_1 R_2}} \Sigma_{m,n} \frac{1}{2^{\delta m_0 \delta n_0}} \phi_{+++}^{2m,2n}(x) \cos\left(\frac{2m}{R_1} y + \frac{2n}{R_2} z\right) \qquad (6)$$

Here R_1, R_2 are the radii of the Torus T^2. Thus in effective four-dimension, the residual group will be (G_{2241}) ($SU(2) \otimes SU(2) \otimes SU(4) \otimes U(1)_\chi$ i.e. Pati-Salam group extended by $U(1)_\chi$, coupled with N = 1 SUSY.

We shall next consider the matter and the Higgs sector to break G_{2241} model in order to have a viable model consistent with phenomenology. We take two sets of brane localised matter as well as Higgs field in $27_{1,2}$ and $\overline{27}_{1,2}$ representation. The basic idea is to put all matter and Higgs field on a brane such that the structure of original four dimensional E_6 group is left essentially in tact. In other words, wherever matter resides, the Higgs should live on fixed points i.e. in contact with matter fields to give Yukawa couplings.

We shall now choose the parities for the Higgs 27 to project out the zero modes. In the subspaces G_{26}, $G_{10,1}$ the parities Z_2 Z'_2, Z''_2 are assigned corresponding to the following symmetry breaking pattern.

$$E_6 \xrightarrow{M_c} SU(2) \otimes SU(2) \otimes SU(4) \otimes U(1)_\chi \xrightarrow{M_U}$$
$$SU(2) \otimes SU(2) \otimes SU(4) \xrightarrow{M_R} SU(3) \otimes SU(2) \otimes U(1) \xrightarrow{M_W} SU(3) \otimes U(1) \qquad (7)$$

Here the 1st stage of breaking is done by orbifolding at the compactification scale $M_c \sim 10^{19}$ GeV and subsequent stages of breaking are to be achieved at M_U (~10^{15} GeV), M_R (~10^{12} GeV) and M_W (~10^2 GeV) scales respectively. The parities of the multiplet, 27 = L(2, 1, 4)$_1$ ⊕ R(1, 2, $\overline{4}$)$_1$ ⊕ D(1, 1, 6)$_{-2}$ ⊕ H(2, 2, 1)$_{-2}$ ⊕ S(1, 1, 1)$_4$ are assigned in table-2. We shall now take the cubic superpotential on (2, 2, 4, 1) brane, given by:

$$W = 27.27.27 = \lambda_1 HHS + \lambda_2 LRH + \lambda_3 LLD + \lambda_4 RRD + \lambda_5 DDS \qquad (8)$$

Following table 2, we thus have the zero mode sectors as D(1, 1, 6)$_{-2}$ and S(1, 1, 1)$_4$ which have all parities positive. We may have alternate possibilities by choosing Z''_2 of L(2, 1, 4)$_1$ and R(1, 2, $\overline{4}$)$_1$ belonging to (16)$_1$ as +ve instead of (1, 1, 6)$_{-2}$ and H(2, 2, 1)$_{-2}$

belonging to 10_{-2} as + ve. We then have $S(1, 1, 1)_4$ and $L(2, 1, 4)_1$ to be the zero mode sector. In that case, one has to seek for other zero modes for breaking of SU(4) and other SU(2). This will necessiate the enlargement of Higgs sector. Instead we may use the higher dimensional terms more precisely a four fermi interaction term in the Lagrangian. These terms may come as a result of integrating out the heavy modes which attain mass due to orbifold compactification. We then assume the vevs of Higgs scalar (necessary for symmetry breaking at intermediate as well as weak scales) are formed out of heavy fermion (Higgsinos) bilinear condensates, suppressed by the square of compactification scale [6]. We shall show that this proposition will be advantageous to predict light Dirac neutrino [6,7]. We therefore take the parity assignments of table (2) to communicate with low energy phenomenology.

Table 2. Parity assignments for brane-localised Higgs 27

G'_{ps}	G_{26}	$G_{10,1}$	Z_2	Z'_2	Z''_2
$L(2, 1, 4)_1$	$(1, 15)$	16_1	+	+	-
$R(1, 2, \bar{4})_1$	$(2, \bar{6})$	16_1	+	-	-
$D(1, 1, 6)_{-2}$	$(1, 15)$	10_{-2}	+	+	+
$H(2, 2, 1)_{-2}$	$(2, \bar{6})$	10_{-2}	+	-	+
$S(1, 1, 1)_4$	$(1, 15)$	1_4	+	+	+

In order to break the $U(1)_x$ symmetry we assign vev to $S(1,1,1)_4 \sim M_U$ (as shown in equation (7)). But it will automatically give super heavy mass to the exotic colour triplets D, via the term DDS, so that rapid proton decay is nicely avoided. We next consider the intermediate as well as weak scale breaking through the formation of condensates [6] $\langle \bar{\chi}\chi \rangle$ where $\chi \langle \bar{\chi} \rangle$ are the fermion (anti fermion) partner of scalars arising out of higher dimensional terms. The condensates automatically imply the existence of corresponding scalar modes yielding an effective potential of Ginzberg Landau type [8]. We may here note that the presence of condensates could be through some hidden dynamics at the Planck scale as has been considered by Pati [9]. This alternate mechanism is necessary as the doublets H is heavy at the compactification scale. It may be noted that, presence of condensate mechanism, decouples the generation of mass from the generation of vevs [6] allowing a see-saw mechanism to operate, with the condensate scale $\sim \mu \sim \left(M_C^2 M_W\right)^{1/3} \sim 10^{12}$ Gev. We take the following condensates:

$$< \phi^a >_{(2, 2, 1)} \sim \frac{1}{M_C^2} < \bar{\chi}_{\bar{H}_a} X_{\bar{S}_a} > \qquad (9)$$

$$<\Delta_R>_{(1,3,\overline{10})} \sim \frac{1}{M_C^2} <\overline{\chi}_{\overline{R}_1} X_{\overline{R}_2}> \qquad (10)$$

$$<\Delta_L>_{(3,1,10)} \sim \frac{1}{M_C^2} <\overline{\chi}_{\overline{L}_1} X_{\overline{L}_2}> \qquad (11)$$

Here the superscript a = 1, 2 stand for the two sets of condensates corresponding to the two sets 27_1 and 27_2. $\overline{R}_{1,2}, \overline{L}_{1,2}$ corresponds to the $\overline{27}_{1,2}$ respectively. Here we have taken $<\phi> \sim k \sim (0(M_W))$, $<\Delta_R> \sim \vartheta_R \sim O(M_R)$ and $<\Delta_L> \sim \vartheta_L$, such that $\vartheta_R >> k >> \vartheta_L$ [7]. Here we may note that the condensates are formed out of heavy Higgsino fields, similar to the attempt for super symmetry breaking through gaugino condensates.

3. Neutrino Mass Matrix

We can now associate the condensate mechanism to generate a light Dirac neutrino by writing down the Yukawa Lagrangian.

$$L = h_1\left[\bar{f}_L^1 f_R^1 \phi_1 + \bar{f}_L^2 f_R^2 \phi_2\right] + h.c. + h_2\left[\bar{f}_L^{c1} f_L^2 \Delta_L + \bar{f}_R^{c1} f_R^2 \Delta_R\right] + h.c. \qquad (12)$$

Here the brane localized matter sector $F^1\{f_L^1, f_L^{c1}\}$ and $F^2\{f_L^2, f_L^{c2}\}$ correspond to qarks and leptons belonging to 27^1, 27^2 respectively. Obviously in the subspace G_{2241}, $f_L^1 = (2,1,4)_1$ and $f_L^{c1} = (1,2,\overline{4})_1$. Similarly the right handed sectors $\{f_R, f_R^c\}$ belong to the conjugate representation $\overline{27}$ such that $f_R^1 = (2,1,\overline{4})_{-1}$ and $f_R^c = (1,2,4)_{-1}$. We have further assumed that one generation of fermion is heaver than the other set. $\phi_{1,2}, \Delta_L, \Delta_R$ are the condensates as taken in equations (9 – 10). Here ϕ_1 (ϕ_2) is formed out of Higgsinos $\overline{\chi}_1 \chi_1 (\overline{\chi}_2 \chi_2)$ belonging to $F^1(F^2)$. However Δ_L, (Δ_R) is formed by $\overline{\chi}_1 \chi_2$ with $\chi_1 \in F^1$ and $\chi_2 \in F^2$. Here we may note that L_y permits Dirac mass term (1st term) as well as Majorana mass term (2nd term) as a result of an imposed parity. The Lagrangian is invariant under the transformation, $F_1 \to F^1$, $F^2 \to -F^2$, $\Delta_{L,R} \to -\Delta_{L,R}$. The interesting thing to observe that such a transformation can be obtained if one takes parity of F^1 (27^1) as +ve with that of F^2 (27^2) as –ve. As has been mentioned before, the condensates $\Delta_{L,R}$ are formed out of the Higgsinos belonging to (27^1) and (27^2), so that $\Delta_{L,R} \to -\Delta_{L,R}$ can be achieved. In fact this discrete symmetry changes the scenario of conventional GUTs with Majorana mass. On the other hand it introduces a mass mixing between the basic and heavy set of fermions which will yield Dirac neutrino.

We shall now write the mass matrix in the neutral sector $\left(\nu_a, \nu_a^c\right)$ with a = 1, 2 for two sets of neutrinos, taking into account the Yukawa part [10]. Here ν_a and ν_a^c are the left handed neutrino and its conjugate respectively.

	ν_1	ν_2^c	$\nu_1^c \, V_1^c$	ν_2
ν_1		0		A
ν_2^c				
$\nu_1^c \, V_1^c$		A^T		0
ν_2				

where $A = \begin{pmatrix} h_1 k & h_2 \vartheta_L \\ h_2 \vartheta_R & h_1 k' \end{pmatrix}$. Here k and k' are the vevs corresponding to ϕ_1 and ϕ_2 respectively, with the hierarchy of scales $\vartheta_R \gg k \gg k' \gg \vartheta_L$. On diagonalising the mass matrix we get four eigen values each of which consists of degenerate members (mass with opposite sign) yielding two Dirac particles [7]. The masses in the leading order occur as $m_1 \sim h_2 \vartheta_R$ and $m_2 \sim \dfrac{h_2^2 \vartheta_L \vartheta_R - h_1^2 k k'}{h_2 \vartheta_R}$. Phenomenologically we choose $\vartheta_R \sim 10^{12}$ Gev [11], $h_1 k \sim 10^2$ Gev which gives mass to heavier fermion and $h_1 k' \sim 10^{-3}$ Gev which gives masses to ordinary fermions. Further $\vartheta_L \sim \dfrac{h_1^2 k^2}{\vartheta_R} \sim 10^{-8}$ Gev [7] which is obtained by potential minimization. h_2 is taken to be 10^{-3}, so that we have a super heavy neutrino $m_1 \sim 10^9$ Gev and an ultra light Dirac neutrino as $m_2 \sim h_2 \vartheta_L \sim 10^{-2}$ ev.

4. Conclusion

We shall now briefly emphasize the salient features and its relevance to phenomenology. We have studied an E_6 gauge theory on $M^4 \otimes T^2 / (Z_2 \otimes Z'_2 \otimes Z''_2)$ coupled with N = 1 SUSY to obtain an extended Pati-Salam group (G_{2241}) as a result of orbifold compactification. We have taken the E_6 group in six dimension with the point in mind that it may bear the memory of E_8 group in d = 10. Due to this reason, we are also confined to the brane-localised Higgs sector belonging to $27(\overline{27})$ only. The question of doublet-triplet splitting is absent here as the doublet (H) is super heavy at the compactification scale and the triplet (D) attains mass of the order of M_U from Vev of "S". The suppression of rapid proton decay is automatically achieved. We have made use of non-zero of bilinear fermion condensates, which are responsible for symmetry breaking at two different levels. The condensate mechanism is conjectured in such a way that the condensates may be there in an intermediate scale and the corresponding Higgs particle may have a mass in GUT scale and still it has a vev in weak scale. The possibility of fermion condensates may be relevant in case of non observation of Higgs particles. A light Dirac neutrino is predicted in neutral sector with imposed parity assignments. One open question is still to be investigated about the super symmetry breaking which may occur via gaugino condensates.

Acknowledgments

One of the authors (S. Mishra) is thankful to Prof. Asim Ku. Ray for useful discussions on this work.

References

[1] H.P. Nilles, *Phys. Rep.* **110**, 2(1984)
[2] H.E. Haber and G.L. Kane, *Nucl. Phys.* **B232**, 333, (1984)
[3] Y. Kawamura, *Prog. Thor. Phys.* **105**, 999 (2001)
[4] L.J. Hall and Y. Nomura, *Phys. Rev.* **D64**, 055003 (2001)
[5] Hebecker and J. March-Russel, *Nucl. Phys.* **B613**, 3 (2001)
[6] T. Asaka et. al., *Phys. Lett.* **B523**, 199 (2001)
[7] G. Altarelli and F. Feruglio, *Phys. Lett.* **B511**, 257 (2001).
[8] C.H. Albright and S.M. Barr, *Phys. Rev.* **D67**, 013002 (2003)
[9] L. Hall et al., *Phys. Rev.* **D65**, 035008 (2002)
[10] Y. Nomura, et. al., *Nucl. Phys.* **B613**, 147 (2001)
[11] R. Dermisek and A. Mati, *Phys. Rev.* **D65**, 0552002 (2002).
[12] N. Haba and Y. Shimizu, *Phys. Rev.* **D67**, 095001 (2003).
[13] T. Asaka et. Al., *Phys, Lett.* **B523**, 199 (2002).
[14] S. Mahapatra and S.P. Misra, *Phys. Rev.* **D33**, 464 (1986)
[15] S. Mishra and S.P. Misra, *Phys. Lett.* **B186**, 99 (1987)
[16] S. Misra and S.P. Misra, *Phys. Lett.* **B217**, 66 (1989).
[17] S. Mishra et. al., *Phys. Rev.* **D35**, 975 (1986).
[18] R. N. Mohapatra and G.N. Senjanovic, *Phys. Rev.* **D23**, 165 (1981).
[19] J.C. Pati, *Phys. Lett.* **B144**, 375 (1984).
[20] J.C. Pati, *Phys. Rev.* **D30**, 1144 (1984).
[21] S. Mishra and S.P. Misra, *Phys. Lett.* **B217**, 66 (1989).
[22] K. Bhattacharya et. Al., *Phys. Rev.* **D74**, 015003(2006).

In: Theoretical Physics and Nonlinear Optics
Editors: Thomas F. George et al
ISBN: 978-1-61122-939-4
© 2012 Nova Science Publishers, Inc.

Chapter 10

DETERMINATION OF THE PROTON BAG RADIUS BASED ON MIT BAG MODEL

G.R. Boroun[*] and M. Zandi
Department of Physics, Razi University, Kermanshah, 67149 Iran

Abstract

The bag radius of the proton can be determined by MIT bag model based on electric and magnetic form factors of the proton at $0.65 \leq Q^2 \leq 5 (\text{Gev}/c)^2$. The study of our results show that the bag radius decreases as Q^2 increases. Also, the electric and magnetic root-mean-squared radius is determined. In doing so, the static bag radius have been calculated and compared with other results. Comparison of our results with those obtained by others suggests a suitable compatibility.

Introduction

The real proton is not a point particle, so phenomenological methods and experiments can be used to figure out its dimensions. To find out about the structure of the proton, elastic scattering cross section and MIT bag model can be used. In absence of a developed dynamical model, the form factors seem to be still valuable means by which reliable information is acquired in elastic scattering processes [1].

The original MIT bag model was presented about three decades ago [2, 3]. This model is defined by the equation of motion and boundary condition for each field degree of freedom inside the bag and homogeneous boundary condition at the surface of the bag. Hadrons are considered as static extended objects in space. The internal structure of these objects includes quarks and varying gluon field.

In MIT bag model, it is supposed that a region of space called "bag" including hadrons fields are fixed. The pressure of hadron constituents in the surface is constant and the vacuum

---
[*] E-mail address: boroun@razi.ac.ir

around the bag imposes an external pressure on the surface of the bag. As this external pressure increases more than the internal one, the bag shrinks [4]. B is the only parameter of this theory. The characteristic linear dimension of hadron will be scaled by $\left(\frac{1}{B}\right)^{1/4}$ within the spherical cavity of the model. Hadrons constituent fields in the bag can carry any spin or quantum numbers. In this paper we generally supposed that the fields in the bag are massless, that is, free Lagrangian is considered for the fields without any interaction in Lagrangian.

Although the fields in the bag are free in first approximation, at the next level, it is suggested that the fields couple weakly. Weak coupling is considered for the quantum numbers in the hadron. The hadron fields in the bag are colored quarks and gluons [3]. Models such as MIT bag model describe two features of QCD in the quark model [5], asymptotic in short distances and Confinement in large distances.

The goal of this paper is to perform the numerical analysis of the proton bag radius in MIT bag model based on the electric and magnetic form factors of the elastic scattering cross section data and introduce a formula to calculate the its electric and magnetic root-mean-squared radius of the proton. Finally the obtained results are compared to the experimental and previous calculated values.

Calculation Method

Equations of motion and boundary conditions are obtained through the variational principle [3]. Through using massless Dirac equation and boundary conditions considered for the bag, the quark wave function is made as follows [4]:

$$P\psi = 0 \quad (1)$$

$$-e_r \cdot \gamma\psi = \psi\big|_{|x|=R} \quad (2)$$

Or

$$\begin{pmatrix} 0 & -i\sigma_r \\ i\sigma_r & 0 \end{pmatrix}\psi = \psi\big|_{|x|=R} \quad (3)$$

Where R is the bag radius, e_r is the normal unit vector, γ is gamma matrices' and σ_r are Pauli 2×2 matrices. In the static spherical cavity approximation of the bag model the pressure of quarks in the bag must be equal to the external pressure "B".
So that:

$$-\frac{1}{2}\frac{\partial}{\partial r}\sum_\psi \bar{\psi}\psi\Big|_{|x|=R} = B \quad (4)$$

On the other hand, the massless quark wave function in the bag is defined by:

$$\psi = N \begin{pmatrix} j_l(Er)\chi^\mu_k(\theta,\varphi) \\ isgn(k)j_{\bar{l}}(Er)\chi^\mu_{-k}(\theta,\varphi) \end{pmatrix} e^{-iEt} \qquad (5)$$

Where j is spherical Bessel function, $\chi^\mu_{k,-k}$ are Pauli spinors and:

$$N^2_{k=-1} = \frac{ER}{2R^3(ER-1)j_0^2(ER)} \qquad (6)$$

So the eigenvalue of ER can be calculated by MIT linear boundary condition, as:

$$\omega = ER\big|_{k=-1} \approx 2.043, 5.396, 8.578 \qquad (7)$$

Since the surface of the bag is spherical and valence quarks are in the lowest eigenmode, the electric and magnetic form factor for proton can be written as follows [6]:

$$G^P_E(Q^2) = \int_0^R 4\pi r^2 dr j_0(Qr)\left[g^2(r) + f^2(r)\right] \qquad (8)$$

$$G^P_M(Q^2) = 2m_N \int_0^R 4\pi r^2 dr \frac{j_1(Qr)}{Q}\left[2g(r)f(r)\right] \qquad (9)$$

Where m_N is the nucleon mass. The functions $g(r)$ and $f(r)$ are defined by:

$$g(r) = N j_0\left(\frac{\omega r}{R}\right) \qquad (10)$$

$$f(r) = N j_1\left(\frac{\omega r}{R}\right) \qquad (11)$$

Where $\omega = 2.04$ in the lowest mode and $N^2 = \frac{\omega}{8\pi R^3 j_0^2(\omega)(\omega-1)}$.

On the other hand, the cross section can be explained in terms of two form factors, G^P_E and G^P_M. So that:

$$\sigma_{norm} = \tau G^2_M + \varepsilon G^2_E \qquad (12)$$

Where $\tau = \dfrac{Q^2}{4m_p^2}$, m_p is the mass of proton. Since there is no pure coulomb interaction, G_E^P and G_M^P are a function of $Q^2 = -q^2 > 0$ (four- momentum transfer squared). If q=0, then G_E^P and G_M^P are electric charge and magnetic moment of the proton. In other words, to determine form factors, parameters of ε and, σ_{norm}, the normalized cross section can be defined as follows :

$$\varepsilon = \left[1 + 2(1+\tau)\tan^2\left(\frac{\theta}{2}\right)\right]^{-1}, \qquad (13)$$

$$\sigma_{norm} = \varepsilon(1+\tau)\frac{E}{E'}\frac{\sigma_{exp}}{\sigma_{Mott}}. \qquad (14)$$

Now, to extract the electric and magnetic form factors from the measured cross-section published in Ref. [7-9], at each value of q^2, σ_{norm} is plotted versus ε, where from the slope, the value of G_E^2 is obtained. The intersection of the line with σ_{norm} - axis then yields G_M^2 as shown in tables (1-3). According to calculated electric and magnetic form factors, the radius of the bag can be obtained based on the Eqs. (8-9) at each value of q^2. The results are shown in tables (1-3) and figures (1-3). It is seen that the value of R is in reverse proportion to the increase of Q^2, since Q^2 increases cause an external pressure ,B, imposed on the surface of the bag and makes it contract ,hence ,its radius decreases . In these calculations, to obtain the static radius of the bag, R, the limit value of bag radius can be calculated in ($Q^2 \rightarrow 0$) based on the fits of the graphs in figure (1-3) which equals to 1.011 Fermi. According with this value for the static bag radius, we can determination of the magnetic and electric root-mean squared radius, which is defined as follows:

$$G_{E,M}(Q^2) \approx 1 - \frac{Q^2 \langle r^2 \rangle_{E,M}}{6} \qquad (15)$$

As $Q^2 \rightarrow 0$. A direct evolution gives [10]:

$$\langle r_E^2 \rangle_{proton} = 4\pi\pi^2\left(\frac{R}{\omega}\right)^5 \frac{\omega^2(2\omega^3 - 2\omega^2 + 4\omega - 3)}{3(2\omega^2 - 2\omega + 1)} = 0.53R^2 \qquad (16)$$

$$\left\langle r_M^2 \right\rangle_{proton} = \frac{8\pi\pi^2}{5\mu_P} \left(\frac{R}{\omega}\right)^6 \frac{\omega^2(8\omega^3 + 10\omega^2 - 20\omega + 15)}{24(2\omega^2 - 2\omega + 1)} = 0.39R^2. \quad (17)$$

Thus the ratio $\dfrac{\left\langle r_E^2 \right\rangle_p}{\left\langle r_M^2 \right\rangle_p}$ is independent of R. Its value is 1.36 in $\omega = 2.04$. This result shows that the charge root-mean-squared radius is considerably larger than the magnetic root-mean-squared radius. In other words, $G_P^E(Q^2)$ decreases faster than $G_P^M(Q^2)$ at small momentum transfers. According to the calculated static radius, the numerical value of magnetic and electric root-mean-squared radius can be obtained and then compared with the results obtained by others. As it can be seeing in table 4, the charge root-mean-squared radius is larger than the magnetic root-mean-square radius. This analysis leads us towards this means, $\dfrac{\mu_p G_E^P(Q^2)}{G_M^P(Q^2)}$ decreases as Q^2 increases and this is a consequence of the MIT bag model.

Conclusion

The electric and magnetic form factors of the proton are calculated and through using them in the MIT bag model, the bag radius can be calculated and compared with other results. The achieved results show that the radius of the bag decreases as Q^2 increases. In the limit of $Q^2 \to 0$, the static radius of the bag can be calculated and based on this, magnetic and electric square mean root radius can be obtained and compared with the results of others.

Table 1. Comparison of calculated magnetic and electric form factors with Ref. [7] and ratio of these form factors to dipole fit .Additionally, the radius of the bag is shown based on achieved data

$Q^2 (Gev/c)^2$	G_E^P	G_M^P/μ_P	$\dfrac{G_E^P}{G_D}$	$(\dfrac{G_E^P}{G_D})_*$	$\dfrac{G_M^P}{\mu_P G_D}$	$(\dfrac{G_M^P}{\mu_P G_D})_*$	$R(fm)$
1	0.1753	0.1740	1.017	1.001	1.009	1.017	0.851
2.003	0.0784	0.0695	1.145	1.173	1.145	1.014	0.689
2.497	0.0585	0.0498	1.193	1.101	1.016	1.030	0.637
3.007	0.0463	0.0365	1.268	1.231	1.000	1.012	0.593

The values of $(\dfrac{G_M^P}{\mu_P G_D})_*$ and $(\dfrac{G_E^P}{G_D})_*$ has been obtained from Ref. [7] and of G_D is dipole fit.

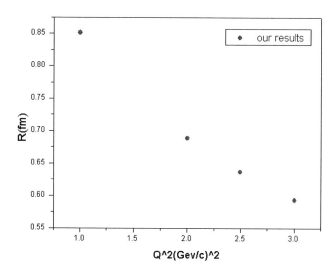

Figure 1. The bag radius according to values of Q^2 Ref. [7].

Table 2. Comparison of calculated magnetic and electric form factors with Ref. [8] and ratio of these form factors to dipole fit. Additionally, the radius of the bag is shown based on achieved data

$Q^2(Gev/c)^2$	G_E^P	G_M^P/μ_P	$\dfrac{G_E^P}{G_D}$	$(\dfrac{G_E^P}{G_D})_*$	$\dfrac{G_M^P}{\mu_P G_D}$	$(\dfrac{G_M^P}{\mu_P G_D})_*$	$R(fm)$
1.75	0.0797	0.0875	0.957	0.956	1.050	1.050	0.725
2.5	0.0432	0.0512	0.833	0.868	1.046	1.054	0.642
3.25	0.0262	0.0338	0.815	0.884	1.051	1.045	0.581
4	0.0201	0.0235	0.885	0.919	1.034	1.031	0.532
5	0.0247	0.0152	1.597	0.942	0.983	1.012	0.479

The values of $(\dfrac{G_M^P}{\mu_P G_D})_*$ and $(\dfrac{G_E^P}{G_D})_*$ has been obtained from Ref.[8] and of G_D is dipole fit.

Table 3. Comparison of calculated magnetic and electric form factors with Ref. [9] and ratio of these form factors to dipole fit. Additionally, the radius of the bag is shown based on achieved data

$Q^2(Gev/c)^2$	G_E^P	G_M^P/μ_P	$\dfrac{G_E^P}{G_D}$	$\dfrac{G_M^P}{\mu_P G_D}$	$R(fm)$
0.65	0.2688	0.2713	0.986	0.995	0.941
0.9	0.1941	0.1971	0.998	1.013	0.873
2.2	0.0557	0.0627	0.936	1.053	0.671
2.75	0.0382	0.0442	0.907	1.050	0.619
3.75	0.0236	0.0262	0.931	1.034	0.546
4.25	0.0202	0.0210	0.986	0.986	0.518
5.25	0.0155	0.0142	1.09	1.000	0.472

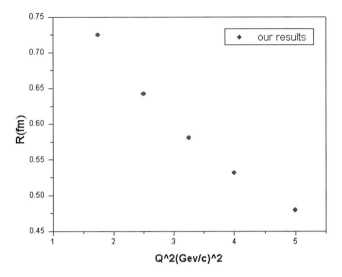

Figure 2. The bag radius according to values of Q² Ref .[8].

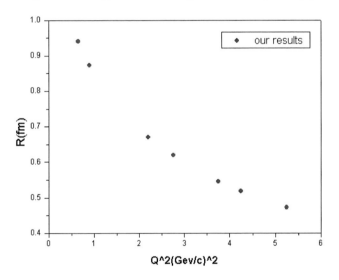

Figure 3. The bag radius according to values of Q² Ref. [9].

Table 4. The results of our calculations and comparison with those obtained by others

	$R_0(fm)$	$\langle r_E^2 \rangle^{1/2}_{\text{Pr}oton}(fm)$	$\langle r_M^2 \rangle^{1/2}_{\text{Pr}oton}(fm)$
Our results	1.011	0.735	0.631
Ref. [4]	1.005	0.73	-
Ref. [11]	0.95	0.84	0.82

References

[1] F. Halzen and , A.D.Martin., 1984, "Quarks and Leptons " , weily publications.
[2] Chodos, R. L. Jaffe, K. Johnson, and C.B. Thron, *Phys. Rev.D,* V. 10, No. 8, 2599 (1974).
[3] A.Chodos et al., *Phys. Rev. D* **9**, 3471 (1974).
[4] Walter Greiner, Andereas Schafer "*Quantum Chromodynamics*", Springer (1994).
[5] V.Šimonis, *Lithuanian Journal of physics*, VOL.37, No.2, 108 (1996).
[6] D.H.Lu, A.W.Thomas, and A.G.Williams, *Phys. Rev. C* **57**, 2628 (1998).
[7] R.C.Walker et al., *Phys. Rev. D* **49**, 5671 (1994).
[8] L.Andivihas et al., *Phys. Rev. D* **50**, 5491 (1994).
[9] M.E.Christy et al., *Phys. Rev. C* **70**, 15206 (2004).
[10] Ding H.Lu, Shin Nan Yang and Anthony W.Thomas *nucl-th/9911065*.
[11] Tong Cheon and Moon Taeg Jeong, *Chinese Journal of Physics* VOL.29, No.5, 451 (1991).

Chapter 11

SEARCHES FOR HIGGS BOSONS IN TWO-HIGGS DOUBLET MODELS: THE FERMIOPHOBIC LIMIT

L. Brücher[a,*] *and R. Santos*[a,b,†]
[a]Centro de Física Teórica e Computacional, Universidade de Lisboa,
Av. Prof. Gama Pinto 2, 1649-003 Lisboa, Portugal
[b]Instituto Superior de Transportes e Comunicações, Campus Universitário
R. D. Afonso Henriques, 2330-519 Entroncamento, Portugal

Abstract

Over the last three decades physicists have tried to discover the famous minimal Higgs boson: the so-called Standard Model (SM) Higgs boson. So far, no scalar boson has been found. This inspired physicists to broaden their view and enlarge the Higgs sector of the SM. The simplest and most natural extension is to include another complex field. Such a model maintains most of the characteristics of the SM, and at the same time has a much richer phenomenology including charged scalars, pseudoscalars and the possibility of including CP-violation. These models are known as Two-Higgs Doublet Models (2HDM). Besides these new features, it comes as a bonus that some 2HDM mimic the Minimal Supersymmetric Model (MSSM) regarding its scalar sector.

Contrary to the SM and MSSM, 2HDM models have an extremely interesting facet: they allow a very weak coupling between scalars and fermions. These types of scalars are called fermiophobic Higgs bosons and the only way to detect them is via vector boson decays, in particular through the two-photon signature which was used by experimentalists at LEP to set a limit on its mass.

In this article the phenomenology of the five Higgs particles of some 2HDM's will be shown and the current status of their searches will be reviewed. Special attention will be paid to the fermiophobic limit, where new searches at the next generation colliders (the Large Hadron Collider and the Linear Collider) are proposed.

---

*E-mail address: bruecher@cii.fc.ul.pt
†E-mail address: rsantos@cii.fc.ul.pt

1. Introduction

LEP II has closed its operation and neither a signal of the minimal Higgs boson nor one of a particle originating from models with a larger Higgs sector was found (see e.g.[1]). Thus, the mechanism of electroweak symmetry breaking still remains unproved. However, in just a few years another opportunity will be made available to physicists. First, with CERN's Large Hadron Collider (LHC) and then with the planned TeV Linear Collider (LC) there is a new chance to understand the mechanism of electroweak symmetry breaking. With the available energy at the LHC several millions of Higgs bosons can be produced every year, depending on its mass and on the model under consideration. A Standard Model (SM) Higgs boson will surely be found in the first year of operation. But if this Higgs boson is not found we have to be ready to a world of possibilities including that of a fermiophobic Higgs boson.

In this paper we will extend our work [2, 3] on the fermiophobic Higgs boson to the possible production and detection at these new colliders. We will emphasize a production mode which to our knowledge has not been evaluated so far. This production mode could be relevant to distinguish between the two possible potentials realized in nature in the fermiophobic limit of the 2HDM. Moreover we will extend our analysis to a wider area of parameters (mainly δ) which in this model is a measure of the ratio of the vacuum expectation values. This approach reflects the higher energy accessible at the forthcoming generations of colliders. To be complete we start by reviewing the 2HDM in the fermiophobic limit in the next two sections. In section 4. we show the theoretical mass limits on the model. In section 5. we discuss the production modes and in section 6. we will present the possible signatures of fermiophobic Higgs bosons. To complete our analysis we will review the current experimental limits on the 2HDM particles in section 7.

2. The Potentials

In this section, we give a review of the general 2HDM extensions of the SM. The most general 2HDM potential has 14 parameters and it explicitly violates CP. If both explicit and spontaneous CP-violation is to be avoided, the potential depends only on 10 real parameters. Defining the complete set of invariants $x_1 = \phi_1^\dagger \phi_1$, $x_2 = \phi_2^\dagger \phi_2$, $x_3 = \Re\{\phi_1^\dagger \phi_2\}$ and $x_4 = \Im\{\phi_1^\dagger \phi_2\}$ we write

$$V_{10} = -\mu_1^2 x_1 - \mu_2^2 x_2 - \mu_3^2 x_3 + \lambda_1 x_1^2 + \lambda_2 x_2^2 + \lambda_3 x_3^2 + \lambda_4 x_4^2 + \lambda_5 x_1 x_2 + \lambda_6 x_1 x_3 + \lambda_7 x_2 x_3 \tag{1}$$

where ϕ_i with $i = 1, 2$ denote two complex scalar doublets with hyper-charge 1. This potential is stable at tree-level provided that we choose a CP-conserving minimum. In fact, we have recently proved [4] that a 10 parameter potential is stable at tree-level regardless of its CP nature. That is, if nature has decided to be a 2HDM, CP-conserving at the potential level, it will remain like this until the end of time unless dramatic changes in the conditions of the Universe take place. When V_{10} chooses one type of minimum to be in, it is always the global one. All others, if they exist, are saddle points. Had nature decided to be in a CP-violating minimum than this would be the global one. If no further symmetry is imposed to the fields the corresponding Yukawa Lagrangian is a source of large tree-level flavor changing neutral currents (FCNC) and the model needs some fine-tuning in that sector.

There is however a way to be absolutely sure that, at tree level, the CP-conserving extremum will never be a minimum [5] regardless of what the conditions of the universe will be. Moreover we have proved that this can be accomplished by using the same symmetries that lead to flavor conservation in the Yukawa sector. If we impose a Z_2 transformation $\phi_1 \to \phi_1$ and $\phi_2 \to -\phi_2$ with appropriate discrete transformations for the quarks in the fermionic sector, we will obtain a CP-conserving, vacuum stable potential with no flavor changing neutral currents. We denote it by V_A and it can be derived from V_{10} by setting $\mu_3^2 = \lambda_6 = \lambda_7 = 0$,

$$V_A = -\mu_1^2 x_1 - \mu_2^2 x_2 + \lambda_1 x_1^2 + \lambda_2 x_2^2 + \lambda_3 x_3^2 + \lambda_4 x_4^2 + \lambda_5 x_1 x_2 \ . \tag{2}$$

However, if instead of applying the Z_2 transformation, we decide to use a global $U(1)$ symmetry $\phi_2 \to e^{i\alpha} \phi_2$, we obtain exactly the same physical particles and structure at the interactions level, but a slightly different potential which corresponds to set $\mu_3^2 = \lambda_6 = \lambda_7 = 0$ and $\lambda_3 = \lambda_4$. Because of the broken $U(1)$ symmetry this potential has a massless pseudo-scalar particle. The way to recover the mass of this particle is to introduce a soft breaking term μ_3^2. This way we obtain the potential used in the Minimal Supersymmetric Model (MSSM) which is

$$V_B = -\mu_1^2 x_1 - \mu_2^2 x_2 - \mu_3^2 x_3 + \lambda_1 x_1^2 + \lambda_2 x_2^2 + \lambda_3 (x_3^2 + x_4^2) + \lambda_5 x_1 x_2 \ . \tag{3}$$

Notice that the addition of a dimension two term does not spoil the renormalizability of the model. Obviously this potential is also invariant under Z_2 except for the soft breaking term $\mu_3^2 x_3$. This term could also be added to V_A leading to a potential that has one degree of freedom more, but quite similar physics.

Choosing the vacuum as

$$\langle \phi_1 \rangle = \begin{pmatrix} 0 \\ \frac{v_1}{\sqrt{2}} \end{pmatrix} \qquad \langle \phi_2 \rangle = \begin{pmatrix} 0 \\ \frac{v_2}{\sqrt{2}} \end{pmatrix} \ , \tag{4}$$

with v_i real, we write

$$V_{AB} = -\mu_1^2 x_1 - \mu_2^2 x_2 - \mu_3^2 x_3 + \lambda_1 x_1^2 + \lambda_2 x_2^2 + \lambda_3 x_3^2 + \lambda_4 x_4^2 + \lambda_5 x_1 x_2 \ , \tag{5}$$

which allow us to study both potentials V_A and V_B at the same time. The minimum conditions are

$$T_1 = v_1 \left[-\mu_1^2 + \lambda_1 v_1^2 + \lambda_+ v_2^2 \right] - \frac{1}{2} \mu_3^2 v_2 = 0 \tag{6a}$$

$$T_2 = v_2 \left[-\mu_2^2 + \lambda_2 v_2^2 + \lambda_+ v_1^2 \right] - \frac{1}{2} \mu_3^2 v_1 = 0 \ , \tag{6b}$$

with $\lambda_+ = (\lambda_3 + \lambda_5)/2$. Each complex doublet ϕ_i can be written as

$$\phi_i = \begin{bmatrix} a_i^+ \\ (v_i + b_i + i c_i)/\sqrt{2} \end{bmatrix} \tag{7}$$

where a_i^+ are complex fields, and b_i and c_i are real fields. This, in turn, enables us to write the mass terms of potential V_{AB} as:

$$V_{AB}^{mass} = \begin{bmatrix} a_1^+ & a_2^+ \end{bmatrix} M_a \begin{bmatrix} a_1^- \\ a_2^- \end{bmatrix} + \frac{1}{2} \begin{bmatrix} c_1 & c_2 \end{bmatrix} M_c \begin{bmatrix} c_1 \\ c_2 \end{bmatrix}$$

$$+ \frac{1}{2} \begin{bmatrix} b_1 & b_2 \end{bmatrix} M_b \begin{bmatrix} b_1 \\ b_2 \end{bmatrix},$$

with the matrices M_a, M_b and M_c defined as

$$M_a = \frac{1}{2} \begin{bmatrix} \frac{v_2}{v_1}\mu_3^2 - v_2^2 \lambda_3 & -\mu_3^2 + v_1 v_2 \lambda_3 \\ -\mu_3^2 + v_1 v_2 \lambda_3 & \frac{v_1}{v_2}\mu_3^2 - v_1^2 \lambda_3 \end{bmatrix} \tag{8a}$$

$$M_b = \begin{bmatrix} 2v_1^2 \lambda_1 + \frac{v_2}{2v_1}\mu_3^2 & v_1 v_2 (\lambda_3 + \lambda_5) - \frac{\mu_3^2}{2} \\ v_1 v_2 (\lambda_3 + \lambda_5) - \frac{\mu_3^2}{2} & 2v_2^2 \lambda_2 + \frac{v_1}{2v_2}\mu_3^2 \end{bmatrix} \tag{8b}$$

$$M_c = \frac{1}{2} \begin{bmatrix} v_2^2(\lambda_4 - \lambda_3) + \frac{v_2}{v_1}\mu_3^2 & -v_1 v_2 (\lambda_4 - \lambda_3) - \mu_3^2 \\ -v_1 v_2 (\lambda_4 - \lambda_3) - \mu_3^2 & v_1^2(\lambda_4 - \lambda_3) + \frac{v_1}{v_2}\mu_3^2 \end{bmatrix}. \tag{8c}$$

Diagonalizing the quadratic terms of V_{AB} one obtains the mass eigenstates: 2 neutral CP-even scalar particles, H^0 and h^0, a neutral CP-odd scalar particle, A^0, the would-be Goldstone boson partner of the Z, G_0, a charged Higgs field, H^+ and the Goldstone associated with the W boson, G^+. The relations between the mass eigenstates and the SU(2)⊗U(1) eigenstates are:

$$\begin{bmatrix} H^0 \\ h^0 \end{bmatrix} = R_\alpha \begin{bmatrix} b_1 \\ b_2 \end{bmatrix} \qquad \begin{bmatrix} H^+ \\ G^+ \end{bmatrix} = R_\beta \begin{bmatrix} a_1^+ \\ a_2^+ \end{bmatrix} \qquad \begin{bmatrix} A^0 \\ G_0 \end{bmatrix} = R_\beta \begin{bmatrix} c_1 \\ c_2 \end{bmatrix} \tag{9a}$$

with

$$R_\alpha = \begin{bmatrix} \cos\alpha & \sin\alpha \\ -\sin\alpha & \cos\alpha \end{bmatrix} \qquad R_\beta = \begin{bmatrix} -\sin\beta & \cos\beta \\ \cos\beta & \sin\beta \end{bmatrix}, \tag{9b}$$

and $0 < \beta < \frac{\pi}{2}$, $-\frac{\pi}{2} < \alpha < \frac{\pi}{2}$. The particle masses and the angles are given by

$$m_{H^+}^2 = -\frac{1}{2}v^2 \left[\lambda_3 - \frac{\mu_3^2}{v_1 v_2} \right] \tag{10a}$$

$$m_{A^0}^2 = \frac{1}{2}\mu_3^2 \frac{v^2}{v_1 v_2} + \frac{1}{2}(\lambda_4 - \lambda_3)v^2 \tag{10b}$$

$$m_{H^0,h^0}^2 = \lambda_1 v_1^2 + \lambda_2 v_2^2 + \frac{1}{4}\mu_3^2 \left(\frac{v_2}{v_1} + \frac{v_1}{v_2} \right) \tag{10c}$$

$$\pm \sqrt{\left(\lambda_1 v_1^2 - \lambda_2 v_2^2 + \frac{1}{4}\mu_3^2 \left(\frac{v_2}{v_1} - \frac{v_1}{v_2} \right) \right)^2 + \left(v_1 v_2 (\lambda_3 + \lambda_5) - \frac{1}{2}\mu_3^2 \right)^2} \tag{10d}$$

$$\tan 2\alpha = \frac{2 v_1 v_2 \lambda_+ - \frac{1}{2}\mu_3^2}{\lambda_1 v_1^2 - \lambda_2 v_2^2 + \frac{1}{4}\mu_3^2 \left(\frac{v_2}{v_1} - \frac{v_1}{v_2} \right)} \tag{10e}$$

$$\tan\beta = \frac{v_2}{v_1}. \tag{10f}$$

To derive the cubic and quartic vertices stemming from the potential it is convenient to rewrite the λ_i in terms of the masses and angles. The results are:

$$\lambda_1 = \frac{1}{2v^2 \cos^2 \beta} \left[m_{H^0}^2 \cos^2 \alpha + m_{h^0}^2 \sin^2 \alpha - \frac{\mu_3^2}{2} \tan \beta \right] \tag{11a}$$

$$\lambda_2 = \frac{1}{2v^2 \sin^2 \beta} \left[m_{H^0}^2 \sin^2 \alpha + m_{h^0}^2 \cos^2 \alpha - \frac{\mu_3^2}{2} \tan^{-1} \beta \right] \tag{11b}$$

$$\lambda_3 = -\frac{2}{v^2} \left[m_{H^+}^2 - \frac{\mu_3^2}{\sin(2\beta)} \right] \tag{11c}$$

$$\lambda_4 = \frac{2}{v^2} \left[m_{A^0}^2 - m_{H^+}^2 \right] \tag{11d}$$

$$\lambda_5 = \frac{\sin(2\alpha)}{v^2 \sin(2\beta)} \left[m_{H^0}^2 - m_{h^0}^2 \right] + \frac{2}{v^2} m_{H^+}^2 - \frac{\mu_3^2}{v^2 \sin(2\beta)} \;. \tag{11e}$$

For the sake of completeness we will use the above results to write the cubic and quartic terms that are different in both potentials in appendix C..

3. The Yukawa Lagrangian

3.1. General Overview

Coupling the scalars to the fermions in all possible ways gives inevitably rise to FCNC at tree-level. The way to avoid it is to force the Yukawa coupling matrices to be linearly independent. The most economical way to avoid FCNC is to couple only one of the fields to each one of three types of particles, up-quarks, down-quarks and leptons. This would give eight possible ways to couple the fermions to the scalars. Because the potential is invariant under $\phi_1 \leftrightarrow \phi_2$ accompanied by a redefinition in the parameters, the number of possibilities can be reduced to four. Hence, choosing to always couple ϕ_2 to the up quarks we have the following options:

- Model I - Only ϕ_2 gives mass to all particles.

- Model II - ϕ_2 gives mass to the quarks and ϕ_1 to the leptons.

- Model III - ϕ_2 gives mass to the up quarks and leptons and ϕ_1 to the down quarks.

- Model IV - ϕ_2 gives mass to the up quarks and ϕ_1 to the down quarks and the leptons.

All these conditions can be imposed to the Lagrangian through discrete symmetries and therefore they are protected at higher orders. To be complete we show in appendix B. all couplings to the fermions in the different models.

3.2. The Fermiophobic Limit

Considering Model I we readily see that the lightest Higgs boson couples to all fermions via $\cos \alpha$. As α approaches $\frac{\pi}{2}$ this coupling tends to zero and in the limit it vanishes, giving rise to a fermiophobic Higgs. Notice that the fermiophobic limit can not be attained in the MSSM because the rules of Supersymmetry disallow other models other than Model II. Examining equation (10e) we see that the fermiophobic limit ($\alpha = \frac{\pi}{2}$) can be obtained in potential A in two ways: either $\lambda_+ = 0$ or $v_1 = 0$. In potential B there is only one possibility $2v_1 v_2 \lambda_+ = \frac{1}{2}\mu_3^2$. In this latter case, equations (10) give immediately:

$$m_{A^0}^2 = 2\lambda_+ \left(v_1^2 + v_2^2\right) \tag{12a}$$

$$m_{H^0}^2 = 2\lambda_2 v_2^2 + 2\lambda_+ v_1^2 = m_{A^0}^2 + 2(\lambda_2 - \lambda_+) v^2 \sin^2\beta \tag{12b}$$

$$m_{h^0}^2 = 2\lambda_1 v_1^2 + 2\lambda_+ v_2^2 = m_{A^0}^2 - 2(\lambda_+ - \lambda_1) v^2 \cos^2\beta \ . \tag{12c}$$

In the former case ($V_{(A)}$), $\lambda_+ = 0$ gives

$$m_{H^0}^2 = 2\lambda_2 v_2^2 \tag{13a}$$

$$m_{h^0}^2 = 2\lambda_1 v_1^2 \tag{13b}$$

while $v_1 = 0$ gives a massless h^0. In this analysis we have assumed that $v_1 < v_2$. The reversed situation leads to similar conclusions since one is then interchanging the role of the two doublets.

The triple couplings involving two gauge bosons and a scalar particle like, for instance $Z_\mu Z^\mu h^0$, are always proportional to the angle $\delta = \alpha - \beta$. In particular, the couplings for h^0 are proportional to $\sin\delta$ whereas the corresponding H^0 couplings are proportional to $\cos\delta$. This general results can be understood if one recalls the argument about the role played by the neutral scalars in restoring the unitarity in the scattering of longitudinal W's, i.e. in $W_L^+ W_L^- \to W_L^+ W_L^-$. The restoration of unitarity requires that the sum of the squares of the $W^+ W^- h^0$ and $W^+ W^- H^0$ couplings adds up to a constant proportional to the $SU(2)$ gauge coupling, g.

3.3. Loop Effects

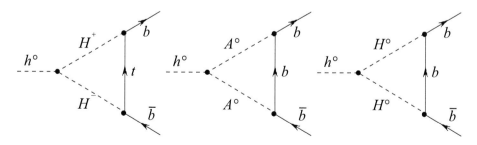

Figure 1. Feynman diagrams containing scalars contribution to $h^0 \to b\bar{b}$.

As pointed out in the previous section, h^0 becomes completely fermiophobic at $\alpha = \pi/2$. However, h^0 can still decay to two fermion pair via $h^0 \to W^*W(Z^*Z) \to 2\bar{f}f$ or $h^0 \to W^*W^*(Z^*Z^*) \to 2\bar{f}f$. We will include these decays in our analysis. It is worthwhile to point out that these processes occur near the $W(Z)$ threshold. Moreover, decays of h^0 to two fermions can be induced by scalar and gauge boson loops (see e.g. fig. 1). In the

2HDM, the angle α has to be renormalized to render $h^0 \to f\bar{f}$ finite. However, at $\alpha = \pi/2$, all one-loop decays $h^0 \to f\bar{f}$ are finite. Thus we can impose the following condition for $\delta\alpha$: the renormalized one-loop decay width for $h^0 \to f\bar{f}$ is equal to the finite unrenormalized decay width. This condition is equivalent to set $[\delta\alpha]_{\alpha=\pi/2} = 0$. We have checked that this condition holds for all fermions. The only relevant one-loop decay is $h^0 \to b\bar{b}$ due to a large contribution of the Feynman diagram shown in fig. 1a to the total decay width.[1]

The couplings of the CP-odd scalar, A^0, and of the charged scalar, $H^\pm$, are proportional to $\cot\beta$. If we want these particles to be fermiophobic as well, β has to approach α ($\beta \to \alpha = \pi/2$). In this limit the coupling of h^0 to the vector bosons, which is proportional to the sine of $\delta \equiv \alpha - \beta$ tends to zero. The differences between potential A and B can be extremely important in this limit since h^0 will have different signatures in each model. In contrast, the heaviest CP-even scalar, H^0, acquires the SM Higgs couplings to the fermions in this limit. We will relax the limit $\beta \approx \pi/2$ and analyze the decays as a function of δ and of the Higgs masses.

4. Theoretical Mass Limits

It is common sense to assume that models in quantum field theory are perturbatively computable. This is important to make reliable predictions on cross sections and decay widths. If one assumes that this is true for the 2HDM, several bounds on the masses and parameters of the Higgs sector can be derived[6, 7]. We will use this bounds, especially the bound from [8] and the extensions in [9], not to leave the solid grounds of perturbation theory. Note that this limits are especially built for potential A, extended with a soft-symmetry breaking term (λ_5) in [9].

They may be extended to potential B, if one adds the soft-breaking dimension two term μ_3. This obviously leads to an additional dependence on m_{A^0} in the limits. We will skip the complete analysis of the equations for potential B. Instead we will show the limits of the physical parameters as functions of the original parameters. These limits are not as strict as the ones derived from the tree-level analysis, but they are more instructive and clearer to the reader.

In potential A we have very strict upper limits on m_h, if δ is small (see fig. 2). In the limit $\delta \to 0$ the light Higgs boson is massless, which can easily be derived from the definition of m_h in the fermiophobic limit.

$$m_{h^0} = \sqrt{2\lambda_1}\, v \sin\delta \qquad (14)$$

If one respects the upper limit on m_h, there are almost no limits on m_{A^0} and m_{H^+} in the small δ region, as can be seen in fig 3. In this figure we have used the equations from [8] to show the m_{A^0}-m_{H^+}-dependence on δ. Note that in [9] more strict overall bounds are derived, which we will respect in our current analysis.

In potential B we do not have a strict overall upper limit on m_h for small δ. In contrast, the mass splitting between m_{A^0} and m_h is restricted. To show this, we have plotted in fig. 4 the splitting of the masses in relation to δ for different values of the original parameter $\Delta\lambda \equiv \frac{1}{2}(\lambda_3 + \lambda_5) - \lambda_1$. The region limited by each value of $|\Delta\lambda|$ is the allowed region

---

[1] The coupling $[H^+ \bar{t} b]$ is proportional to the t-quark mass.

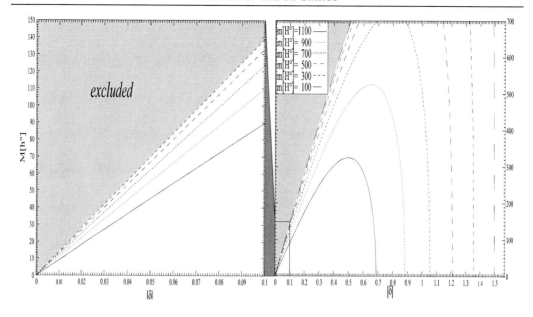

Figure 2. Limit on m_{h^0} as a function of δ for different values of m_{H^0} in Potential A.

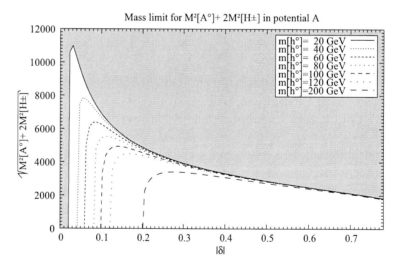

Figure 3. Limit on m_{h^0} as a function of δ for different values of m_{h^0} in Potential A.

for m_{h^0} for a given value of m_A. Although it is most likely that $|\Delta\lambda| < 10$, a value of $|\Delta\lambda| = 100$ cannot strictly be excluded, if one wants to be very conservative. It can clearly be seen that in the tiny δ region the masses of h^0 and A^0 are almost degenerated. As soon as δ increases, the splitting between the masses is relaxed. In the small δ region the restriction on the splitting of the masses vanishes completely.

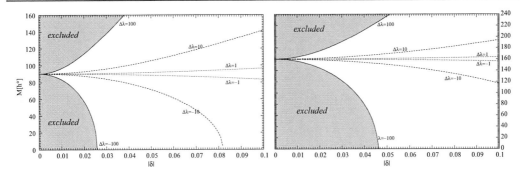

Figure 4. Limit on the splitting of m_{h^0} and m_A as a function of δ for $m_{h^0} = 90$ GeV and $m_{h^0} = 160$ GeV in Potential B.

5. Production Modes

5.1. Hadron Colliders

The production of single fermiophobic Higgs bosons associated either with a gauge boson or with another scalar was already studied at the Tevatron [10] and at the LHC/LC [11]. In this section we review those searches and add a new channel: the production of a pair of fermiophobic Higgs bosons. The most relevant production mode for Higgs production in the SM is gluon fusion. This mode and also $pp \to h^0 t\bar{t}$ are not allowed in the fermiophobic limit. Therefore the production of a scalar associated with a gauge boson together with vector boson fusion become the dominant channels. But besides those channels we have now the production of two scalar Higgs bosons. As stated before most of them were already calculated. However, all channels where scalars couple directly to the incoming fermions have not been considered. It is obvious that for an electron-positron collider the cross sections would be proportional to powers of the electron mass and therefore meaningless. This is also true for a low energy hadron collider where the sea quarks do not play a relevant part. But, when we consider the LHC's energy, two factors become relevant. First, the quarks from the sea start to play an important role. Second, the cross sections are now proportional to the c-quark or the b-quark masses, which are of the order GeV. Therefore they must be taken into account.

The two body tree-level processes available at the LHC with at least one fermiophobic Higgs in the final state are

$$pp \to h^0 h^0 + X \tag{15a}$$
$$pp \to h^0 H^0 + X \tag{15b}$$
$$pp \to h^0 H^\pm + X \tag{15c}$$
$$pp \to h^0 A^0 + X \tag{15d}$$
$$pp \to h^0 Z + X \tag{15e}$$
$$pp \to h^0 W^\pm + X \;. \tag{15f}$$

Notice that for a fermiophobic Higgs the production mode 15a exists only in $V_{(B)}$. This would not be the case in a general 2HDM.[2]

The first two production modes are the new ones. Both modes have the same coupling strength, but because the H^0 mass is larger than the h^0's, the cross section will always be smaller. The production modes for the other two scalars were calculated for the 2HDM and for the MSSM. However, to our knowledge the diagrams corresponding to $pp \to A^{0*}, (G^{0*}) \to Zh^0 + X$ and $pp \to H^{\pm*}, (G^{\pm*}) \to H^\pm h^0 + X$ were not included. But these channels can become dominant in some regions of the parameter space of the model.

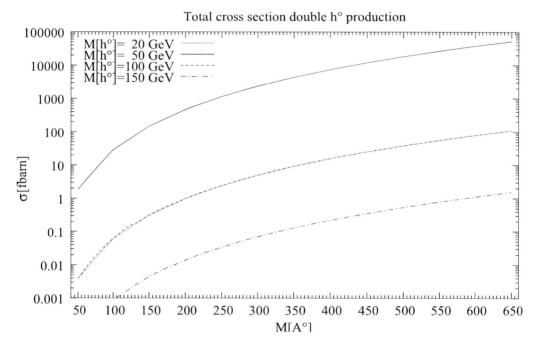

Figure 5. Total cross section $\sigma_H(h^0 h^0 X)$ as function of m_{A^0} for different values of m_{h^0} with $m_{H^0} = 200\text{GeV}$ in Potential B.

5.1.1. Numerical Results

The partonic cross sections for $h^0 Z$ and $h^0 W^\pm$ are the SM ones multiplied by $\sin^2 \delta$ and are shown in appendix A. The partonic cross sections for $h^0 A^0$ and $h^0 H^\pm$ are to large to be shown in here. The limit where all fermions masses are zero can be found in [13]. The partonic cross section for $h^0 h^0$ can be written as:

$$\hat{\sigma}_{\text{LO}}(q\bar{q} \to h^0 h^0) = \frac{G_F^2 m_{A^0}^4}{16\pi} m_q^2 \frac{\left(1 - \frac{4m_q^2}{\hat{s}}\right)^{1/2} \left(1 - \frac{4m_{h^0}^2}{\hat{s}}\right)^{1/2}}{\left(\hat{s} - m_{H^0}^2\right)^2}, \qquad (16)$$

---

[2]We will present the results in this model and in the MSSM soon [12].

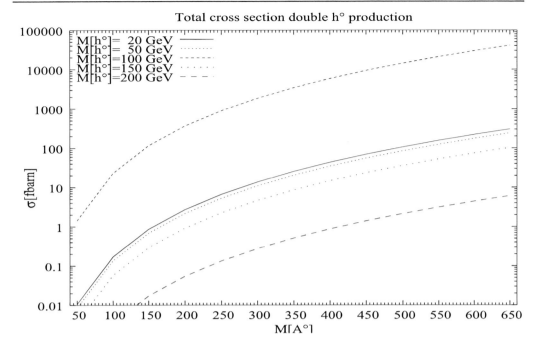

Figure 6. Total cross section $\sigma_H(h^0h^0X)$ as function of m_{A^0} for different values of m_{h^0} with $m_{H^0} = 400$ GeV in Potential B.

where $\hat{s}$ denotes the partonic center of mass energy squared, and m_q is the relevant quark mass. The cross section for h^0H^0 is the same except for the phase space factor.[3] The partonic distributions were taken at the scale of the sum of the masses of the produced particles. We have used CTEQ5 set from [14].

QCD corrections to $pp \to Z(W)h^0 + X$ were calculated in [15]. Corrections to $pp \to h^0A^0 + X$ were calculated in [13] for the MSSM. It was shown in both cases that the tree level cross section is increased by roughly 30 %. We assume that similar correction will affect all production modes. Nevertheless it should be clear that all plots presented correspond to leading order cross sections.

In figs. 5 and 6 we present the cross sections for $pp \to h^0h^0 + X$ as a function of m_{A^0} for several values of the fermiophobic Higgs masses. In fig. 5 we choose $m_{H^0} = 200$ GeV and in fig. 6 we set $m_{H^0} = 400$ GeV. For small values of $m_{A^0} (\leq 100$ GeV$)$ the production rate is meaningless except for very particular values of m_{h^0}. However, as m_{A^0} grows the cross section rises due to its dependence on $m_{A^0}^4$. The effect of a change in m_{H^0} is to shift the maximum of the cross section to a different fermiophobic Higgs mass. In fig. 5 we see that for $m_{H^0} = 200$ GeV the largest cross section value occurs for $m_{h^0} = 50$ GeV and in fig. 6, where $m_{H^0} = 400$ GeV we see that the maximum occurs at around $m_{h^0} = 100$ GeV. Depending on those three parameters m_{A^0}, m_{H^0} and m_{h^0} the numbers of produced Higgs at the LHC can go from zero to several thousands. Notice that, in this limit, the cross section do not depend on δ or on the charged Higgs mass. Hence, if for instances the large

---
[3]The couplings are equal in the fermiophobic limit.

m_{A^0} region is excluded at the LHC it will happen independently of the above mentioned parameters. For the values presented here the $h^0 H^0$ production rate is too small and we have decided not to show it. This channel can be important if $m_{H^0} \approx m_{h^0}$. As a last remark we remind the reader that these conclusions are only valid in potential B in the fermiophobic limit. A study for the a general 2HDM is in progress.

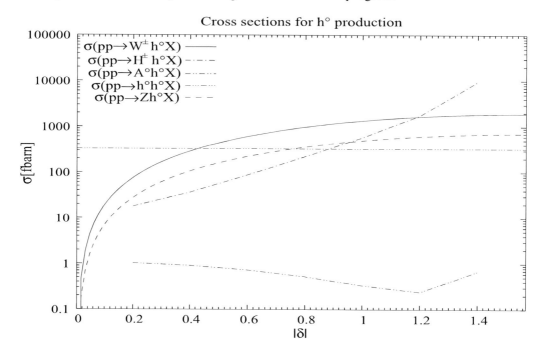

Figure 7. Partial cross sections as function of δ with $m_{H^0} = 600$ GeV, $m_{A^0} = 650$ GeV, $m_{H^+} = 610$ GeV and $m_{h^0} = 100$ GeV in Potential B.

In figs. 7 and 8 we show all five production modes as a function of δ for potential B with all masses close to the largest limit allowed by unitarity constraints, except for m_{h^0} which we take to be 100 GeV and 200 GeV. In both scenarios the $h^0 h^0$ mode shows itself as a straight line because it does not depend on δ. It can be clearly seen from fig. 7 that this is the dominant mode in some of the parameter region. The $h^0 W(Z)$ modes depend on δ through $\sin^2 \delta$ as can be seen from both figures. As $\delta \to \frac{\pi}{2}$ it reaches the SM value. At this limit these production modes depend only on the fermiophobic Higgs mass. Comparing figs. 7 and 8 it is clear that both fall as the mass rises and no change occurs if other Higgs masses are modified. The remaining two modes $h^0 A^0$ and $h^0 H^{\pm}$ are also shown. The production of $h^0 H^{\pm}$ depends on all parameters whereas the production of $h^0 A^0$ does not depend on the mass of $H^{\pm}$. This means that contrary to the other three modes they are strongly affected by the region of the parameter space we choose. They are also model dependent. To show it we plot in fig. 9 the same values we plotted in fig. 8 but for potential A. If it wasn't for the new diagrams we have included for the first time in $h^0 A^0$ and $h^0 H^{\pm}$ production these two plots should be exactly the same as they would not depend on the three scalar vertices, where the models are different but only with couplings to the gauge bosons. For this particular set of values $h^0 H^{\pm}$ grows and $h^0 A^0$ falls when we move from

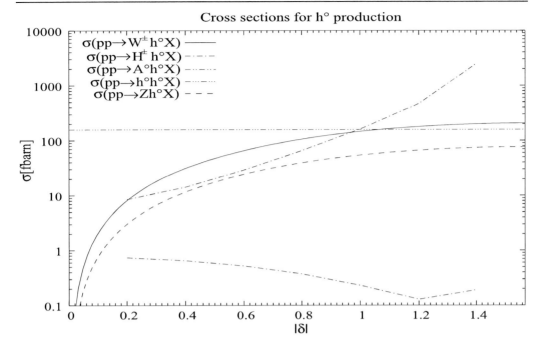

Figure 8. Partial cross sections as function of δ with $m_{H^0} = 600$ GeV, $m_{A^0} = 650$ GeV, $m_{H^+} = 610$ GeV and $m_{h^0} = 200$ GeV in Potential B.

potential B to potential A. To finish this section we show in fig. 10 a plot for potential A with very small values for the Higgs sector masses. This is to show that even for very small masses all production modes can be on the same footing.

5.2. Lepton Collider

A lot of work has been done at LEP II. No fermiophobic Higgs was found and limits were set on its mass [16]. We have used all available experimental results to constrain the model as can be seen in section 7.. The forthcoming LC has probably enough energy to produce a very heavy fermiophobic Higgs. The most important production modes are:

$$e^+e^- \to Z^* \to Zh^0 \quad (17a)$$
$$e^+e^- \to Z^* \to A^0 h^0, \quad (17b)$$

which where already studied in detail.

6. Phenomenology of the Fermiophobic Limit

In this section the signatures of fermiophobic Higgs bosons will be reviewed. The theoretical bounds and the overall picture given by the branching ratios shown in this section, led us to distinguish between four different regions for δ, namely:

- the *tiny* δ region where $|\delta| \leq 0.05$,

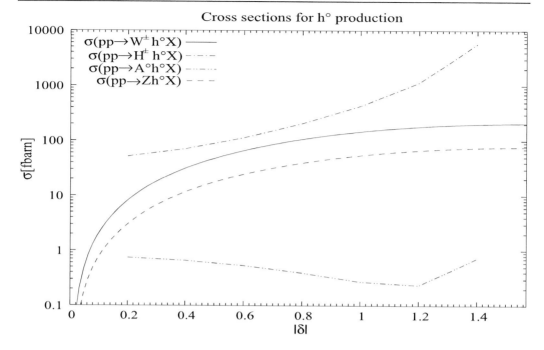

Figure 9. Partial cross sections as function of δ with $m_{H^0} = 600$ GeV, $m_{A^0} = 650$ GeV, $m_{H^+} = 610$ GeV and $m_{h^0} = 200$ GeV in Potential A.

- the *small* δ region with $0.05 < |\delta| \le 0.1$,
- the *medium* δ region when $0.1 < |\delta| \le 1.0$ and
- finally the *large* δ region when $|\delta| > 1.0$.

We will shortly review the tiny and small δ region, which we already worked out in [2] in a region with smaller masses in the Higgs sector. We will extend this analysis in detail to the medium and large δ sector, as this sector is important for detection in hadron colliders[4].

6.1. The Light Scalar Higgs Boson

For the light scalar Higgs boson (h^0) the following tree level couplings have to be considered:

$$h^0 \to W^+W^- \quad ; \quad h^0 \to ZZ \quad ; \quad h^0 \to ZA^0 \quad ;$$
$$h^0 \to W^\pm H^\mp \quad ; \quad h^0 \to A^0 A^0 \quad ; \quad h^0 \to H^+ H^- \quad .$$

As (h^0) has no tree level couplings to the fermions for $\alpha = \pi/2$, the decay to fermions is just present at one-loop. Beside this, the following one-loop decays are important:

$$h^0 \to \gamma\gamma \quad ; \quad h^0 \to Z\gamma \quad ; \quad h^0 \to b\bar{b} \quad .$$

---

[4]Current analysis shows, that gluon fusion is the most important production mode for a standard model Higgs in hadron colliders. Following this analysis we expect a similar scenario for the 2HDM, as has been pointed out in the previous section. In the fermiophobic limit, where h^0 has no coupling to the fermions, this means, that we produce h^0 via other particles of the Higgs sector (i.e. H^0, A^0 and H^+). Their coupling increases proportional to δ, giving a cross section for the production modes increasing with δ.

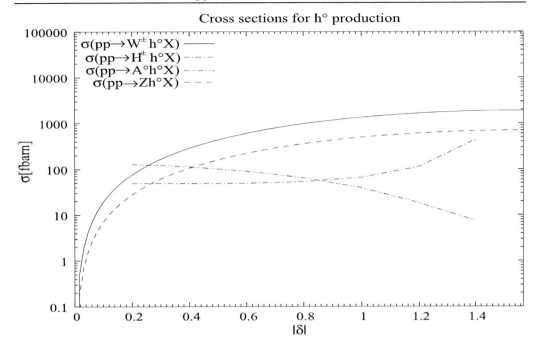

Figure 10. Partial cross sections as function of δ with $m_{H^0} = 180$ GeV, $m_{A^0} = 140$ GeV, $m_{H^+} = 390$ GeV and $m_{h^0} = 50$ GeV in Potential A.

Moreover, we take decays to fermions via virtual vector bosons into account, namely:

$$h^0 \to W^*W^* \to f\bar{f}f\bar{f} \quad ; \quad h^0 \to W^*W \to f\bar{f}W \quad ; \quad h^0 \to W^*H^\pm \to f\bar{f}H^\pm \quad ;$$
$$h^0 \to Z^*Z^* \to f\bar{f}f\bar{f} \quad ; \quad h^0 \to Z^*Z \to f\bar{f}Z \quad ; \quad h^0 \to Z^*A^0 \to f\bar{f}A^0 \quad .$$

The partial tree-level decay widths are listed in appendix A.. The one-loop induced decays have been calculated with *xloops*[17]. For decays via virtual particles we have taken the formulas from [18] and changed them appropriately. For the decays into one vector boson and one scalar the decays via virtual particles (i.e. $h^0 \to W^*H^\pm$ and $h^0 \to Z^*A^0$) have been taken into account. They are only important near the thresholds for the the on-shell decays.

In potential A we have strict upper bounds for m_{h^0} in the tiny and small δ region. Thus we have a clear $\gamma\gamma$-signature below $m_{h^0} \approx m_W$. Then decays via virtual vector bosons play more and more an important role, as can be seen in fig. 11. At $m_{h^0} \approx 110$ GeV we have a clear W^*W signature. The possible decays to other scalars from the Higgs sector are already ruled out by experimental data (see section 7.). We want to emphasize again, that for potential A the upper limit on m_{h^0} has to be respected, as already pointed out in section 4..

In potential B the situation is quite similar in the tiny and small δ region, with two differences. First, instead of having an upper limit on m_{h^0} we have a limit on the splitting between m_{h^0} and m_{A^0}. This allows much higher masses for m_{h^0}. Second, the coupling between the light scalar and the charged Higgs ($[h^0H^+H^-]$) has an extra term. Thus the decay to $\gamma\gamma$ (see fig. 12) may be suppressed or enhanced. This shifts the region where the

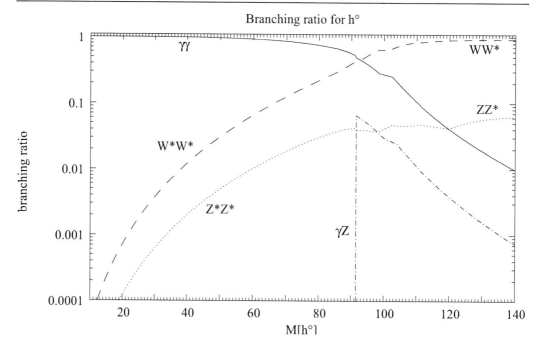

Figure 11. Branching ratio of h^0 at $m_{A^0} = 140$ GeV, $m_{H^+} = 390$ GeV and $m_{H^0} = 180$ GeV and $\delta = 0.1$ in potential A.

signature changes from $\gamma\gamma$ to W^*W^* either to lower or to higher m_{h^0}. Nevertheless we did not find a region of parameters where we have a dominant $\gamma\gamma$-signature above 130 GeV.

In the medium δ region we already have a lower experimental limit on m_{h^0} (see section 7.). Thus it is most likely, that a fermiophobic Higgs will be detected via its decays to vector bosons in this region. Nevertheless we can see two interesting signatures around the lower limit of m_{h^0}. Either we have a $\gamma\gamma$-signature in this region (cf. fig. 13) or, depending on the choice of parameters (i.e. m_{H^+}), we can have a $b\bar{b}$-signature induced via scalar loops. This behavior can still be seen on the lower edge of the large δ region. As δ increases further, we see a dominant $b\bar{b}$-signature in this region for m_{h^0} (fig. 14). If m_{h^0} increases further, we may see decays to scalars or scalars and vector bosons as well (cf. fig. 13 and fig. 14). Hence, if a $b\bar{b}$ pair is detected and identified as originating from the decay of a scalar particle and at the same time the number of $b\bar{b}$ is smaller than what is expected for the SM or the MSSM we could have detected a fermiophobic Higgs through a fermionic signature!

6.2. The Pseudoscalar Higgs Boson

The pseudoscalar Higgs boson has the following tree-level decays:

$$A^0 \to f\bar{f} \quad ; \quad A^0 \to Zh^0 \quad ; \quad A^0 \to W^\pm H^\mp$$

We have also considered the decays via virtual vector bosons near the $W^\pm H^\mp$ and Zh^0 thresholds. Furthermore the following one-loop decays have been taken into account:

$$A^0 \to \gamma\gamma \quad ; \quad A^0 \to Z\gamma \quad ; \quad A^0 \to W^+W^- \quad ; \quad A^0 \to ZZ \quad ; \quad A^0 \to gg \quad ,$$

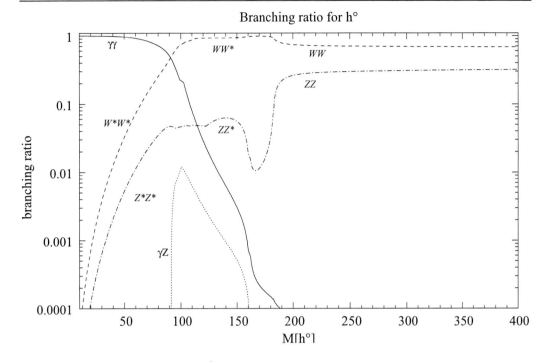

Figure 12. Branching ratio of h^0 at $m_{A^0} = 140$ GeV, $m_{H^+} = 390$ GeV and $m_{H^0} = 180$ GeV and $\delta = 0.05$ in potential B.

where g denotes a gluon.

If only on-shell decays are considered, there are no differences in the decays of A^0 for potential A and potential B. This is a consequence of the fact that couplings of A^0 involving fermions or at least one vector boson are identical in both potentials. Decays to scalars only are kinematically forbidden, as they involve at least one A^0 in the final state due to CP conservation. The only difference in the decays of A^0 in the two potentials originates from the theoretical mass limits (see section 4.). In the tiny δ region the decay $A^0 \to Zh^0$ is kinematically forbidden due to the limited mass splitting between h^0 and A^0 in potential B.

Except for the $W^{\pm}H^{\mp}$ and Zh^0 decay modes, all decays depend on the couplings to the fermions, i.e. $\cot \beta$. Thus, if the $W^{\pm}H^{\mp}$ and Zh^0 channels are not open, the branching ratios for A^0 are independent of δ and just vary with m_{A^0} (see fig. 15). This, of course, is not true for the decay width. Obviously the decay width decreases as $\delta = \pi/2 - \beta$ tends to zero, leading to a stable A^0 in the limit. As soon as the $W^{\pm}H^{\mp}$ and Zh^0 channels are open, A^0 mainly decays into them, if the $t\bar{t}$ channel is not opened. As soon as this channel is open, it competes with the two mixed scalar/vector boson decays in dependence on δ as can be seen in fig. 16. In the small δ region and also in the largest part of the medium δ region A^0 decays into scalars and vector bosons. At the upper edge of the medium δ region and in the large δ region A^0 decays dominantly into a top quark pair.

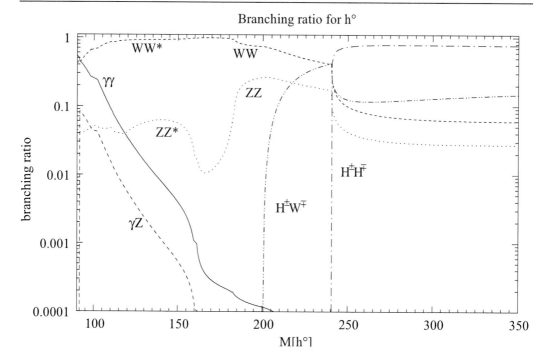

Figure 13. Branching ratio of h^0 at $m_{A^0} = 650$ GeV, $m_{H^+} = 120$ GeV and $m_{H^0} = 600$ GeV and $\delta = 0.5$ in potential A.

6.3. The Charged Scalar Higgs Boson

The charged Higgs boson has the following tree-level decays:

$$H^\pm \to f_{u,d}\bar{f}_{d,u} \quad ; \quad H^\pm \to W^\pm h^0 \quad ; \quad H^\pm \to W^\pm A^0 \quad ; \quad H^\pm \to W^\pm H^0$$

The following one-loop decays have to be considered:

$$H^\pm \to W^\pm \gamma \quad ; \quad H^\pm \to W^\pm Z$$

Again, the 16 (32) graphs for $H^\pm \to W^\pm \gamma$ ($H^\pm \to W^\pm Z$) have been calculated with *xloops* [17]. In general, the branching ratios of the charged Higgs boson show no surprises. Quite similar as in the former case of the pseudoscalar Higgs boson the branching ratios are independent of δ below the W threshold, quite contrary to the MSSM. This is a consequence of the fact that all fermions couple to just one Higgs doublet in the fermiophobic limit (cf. sec. 3.). Above the W threshold the situation stays the same in most δ regions. Just in the tiny δ region the decay $H^\pm \to W^\pm \gamma$ gains importance, as can be seen in fig. 18. As soon as the Wh^0 or WA^0 threshold is passed, $H^\pm$ decays into scalars and vector bosons. The sum of these decays is approximately 100% in both potential for a wide range of values for δ. Only in the large δ region the decay $H^+ \to t\bar{b}$ may become dominant, if $m_{H^+} > m_t + m_b$ (see fig. 17).

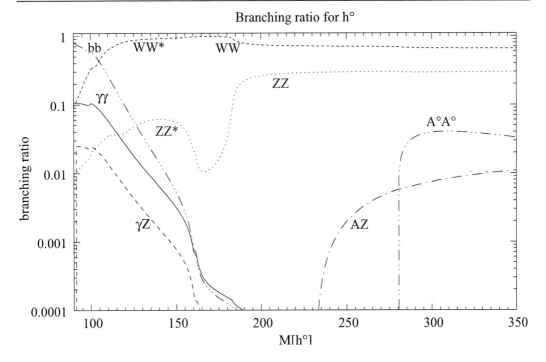

Figure 14. Branching ratio of h^0 at $m_{A^0} = 140$ GeV, $m_{H^+} = 610$ GeV and $m_{H^0} = 600$ GeV and $\delta = 1.4$ in potential A.

6.4. The Heavy Scalar Higgs Boson

To be complete, we show the decay modes of the heavy scalar Higgs boson, although it has large coupling to fermions. Thus one-loop decays may be neglected, as they have no importance in the branching ratio of H^0. We have to consider the following tree-level decays:

$$H^0 \to f\bar{f} \quad ; \quad H^0 \to H^+H^- \quad ; \quad H^0 \to A^0A^0 \quad ; \quad H^0_{(B)} \to h^0h^0 \quad ;$$
$$H^0 \to W^{\pm}H^{\mp} \quad ; \quad H^0 \to ZA^0 \quad ; \quad H^0 \to ZZ \quad ; \quad H^0 \to W^+W^-$$

Note that the decay $H^0 \to h^0 h^0$ vanishes in the fermiophobic limit (i.e. for $\alpha = \pi/2$) in potential A but not in potential B. This fact was already considered, when discussing the production modes in section 5..

A typical plot of the branching ratio as a function of the mass is shown in fig. 19. Obviously, the heavy scalar Higgs boson mainly decays into $b\bar{b}$ below, and into WW above the two vector boson threshold. This behavior is typical for both potentials. The only difference between the potentials can be recognized in the purely scalar decay modes. In potential A their contribution varies from 0% to $\approx 20\%$ depending on the parameters chosen. In potential B the decays to scalars can be the major decay modes for some values of δ, m_{H^+} and m_{A^0}, as can be seen in fig. 19, where the decay $H^0 \to A^0 A^0$ is the major decay mode between $280 < m_{H^0} < 320$ GeV.

In the large δ region, H^0 mainly decays to $t\bar{t}$. Notice that the branching ratios shown in fig. 20 are similar those obtained for the SM Higgs boson, if the scalar decays are ignored.

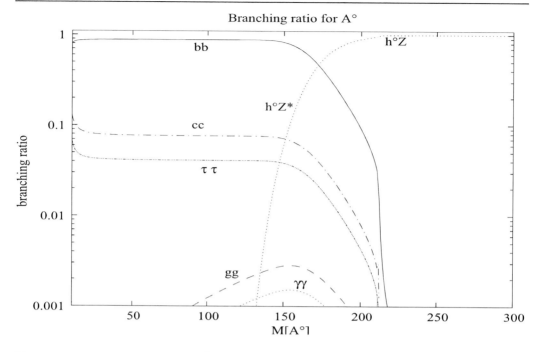

Figure 15. Branching ratio of A^0 at $m_{h^0} = 120$ GeV, $m_{H^+} = 390$ GeV and $m_{H^0} = 600$ GeV and $\delta = 0.1$ in both potentials.

7. Current Limits on Fermiophobic Higgs Bosons

At LEP II there have been extensive searches for a fermiophobic Higgs (see e.g. [16]). We will review these limits in this section.

The LEP collaborations ([16],[19]) have found relative strict combined limits on m_{h^0} and m_{A^0} independent of δ, mainly by exploiting the behavior of the couplings $[ZZh^0]$ and $[Zh^0A]$[5]. These limits cover a sector of an ellipse ($\frac{1}{4}$) in the m_{h^0}-m_{A^0}-plane cutting the axis at $m_{h^0} \approx 100$ GeV and $m_{A^0} \approx 160$ GeV. One can roughly summarize these results as follows:

- for a light neutral Higgs ($m_{h^0} < 40$ GeV) the mass of the pseudoscalar is above $140 - 160$ GeV.

- if 40 GeV $< m_{h^0} < 100$ GeV, then the lower bound for m_{A^0} varies between 40 GeV and 140 GeV.

- if $m_{h^0} > 100$ GeV we have no bounds for the pseudoscalar Higgs boson.

Note that if the decay $Z \to h^0 A^0$ is kinematically forbidden, a light h^0 is still possible, as can be seen in fig. 21 for Potential A and fig. 22 for Potential B. Nevertheless for these values (i.e. $m_{h^0} < 100$ GeV) the upper limit on δ is already very small.

---

[5] $[ZZh^0]$ is proportional to $\sin \delta$, while $[Zh^0A]$ is proportional to $\cos \delta$.

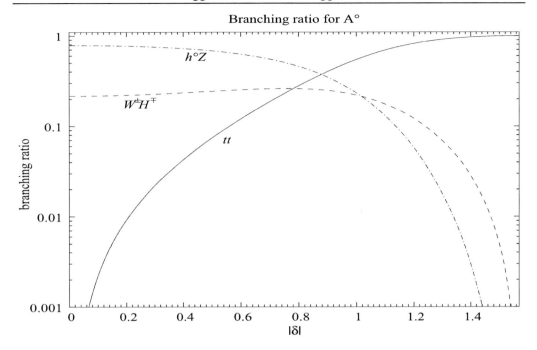

Figure 16. Branching ratio of A^0 at $m_{h^0} = 140$ GeV, $m_{H^+} = 390$ GeV and $m_{H^0} = 180$ GeV and $m_{A^0} = 650$ GeV in both potentials.

For the charged Higgs boson only the general limit for non-observation of charged scalar particles in current colliders can be taken from [20]. It is

$$m_{H^+} > 79.3 \text{GeV} \ . \tag{18}$$

This limit is valid for all δ regions. Note that the improvement of this bound in the MSSM [1] cannot be generally applied in the 2HDM. The MSSM-bounds exploit the fact, that in the MSSM the coupling of the fermions to the Higgs is of type II (cf. sec. 3.). Nevertheless for increasing δ the processes used to derive the MSSM-limits may also be used to derive limits for the 2HDM charged Higgs boson.

8. Conclusion

We have discussed how a fermiophobic Higgs boson can be produced in the forthcoming generation of colliders. A new process for two Higgs production was shown, namely $pp \rightarrow h^0 h^0 (\text{or } H^0) + X$. This process can be dominant in the region of large pseudoscalar mass. We have also included new diagrams in the two scalar channels and have shown that the results can vary significantly. We have presented the possible signatures of these Higgs bosons making it easy for experimentalists to look for them. We have all the tools ready to search in all regions of the parameter space as soon as we have the experimental data. Our analysis of the current experimental limits on this bosons shows that there is still parameter space uncovered. So, one should keep an open mind for surprises.

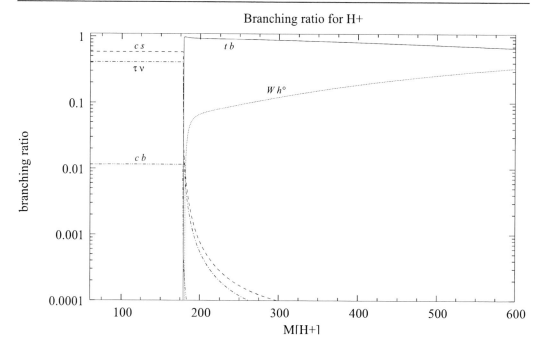

Figure 17. Branching ratio of H^+ at $m_{h^0} = 100$ GeV, $m_{A^0} = 650$ GeV and $m_{H^0} = 600$ GeV and $\delta = 1.0$ in potential A.

Acknowledgments

We thank S.M. Oliveira and Eric Laenen for their support. This work is partially supported by Fundação para a Ciência e Tecnologia under contract No. POCTI/FP/FNU/50155/2003.

A. Formulas for the Cross Sections and Decay Widths

Here we present the most important formulas for the cross sections and decay widths of the Higgs particles. We use the Kallen function $\lambda(x, y, z) = x^2 + y^2 + z^2 - 2xy - 2xz - 2yz$ in the formulas below. For the cross sections with center of mass energy $\hat{s}$ we have the following formulas:

$$\hat{\sigma}_{\rm LO}(q\bar{q} \to h^0 h^0) = \frac{G_F^2 m_{A^0}^4}{16\pi} m_q^2 \frac{\left(1 - \frac{4m_q^2}{\hat{s}}\right)^{\frac{1}{2}} \left(1 - \frac{4m_{h^0}^2}{\hat{s}}\right)^{\frac{1}{2}}}{\left(\hat{s} - m_{H^0}^2\right)^2}$$

$$\hat{\sigma}_{\rm LO}(q\bar{q} \to h^0 Z) = \frac{G_F^2 M_Z^4}{18\pi} (g_v^2 + g_a^2) \, \lambda\left(\hat{s}, m_{h^0}^2, m_Z^2\right)$$
$$\times \frac{m_{h^0}^4 - 2(\hat{s} + m_Z^2)m_{h^0}^2 + m_Z^4 + \hat{s}^2 + 10 m_Z^2 \hat{s}}{\hat{s}(\hat{s} - M_Z^2)^2} \sin^2 \delta$$

$$\hat{\sigma}_{\rm LO}(q\bar{q} \to h^0 W) = \frac{G_F^2 M_W^4}{72\pi} (g_v^2 + g_a^2) \, \lambda\left(\hat{s}, m_{h^0}^2, m_W^2\right)$$

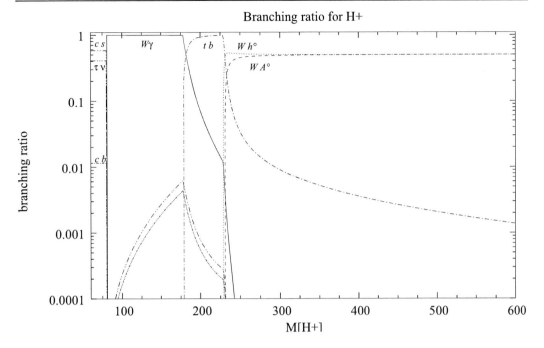

Figure 18. Branching ratio of H^+ at $m_{h^0} = 148$ GeV, $m_{A^0} = 150$ GeV and $m_{H^0} = 600$ GeV and $\delta = 0.01$ in potential B.

$$\times \frac{m_{h^0}^4 - 2(\hat{s} + m_W^2)m_{h^0}^2 + m_W^4 + \hat{s}^2 + 10 m_W^2 \hat{s}}{\hat{s}(\hat{s} - M_W^2)^2} \sin^2 \delta \quad ,$$

where $g_v = (\frac{I_3}{2} - Qsw^2)$ and $g_a = -\frac{I_3}{2}$. Obviously, Q denotes the electric charge, I_3 is the weak isospin component and sw is the sine of the Weinberg angle.

For the lightest scalar the following tree-level decays have been calculated:

$$\Gamma(h^0 \to W^+W^-) = \frac{g^2 m_W^2 \sin^2 \delta}{8\pi m_{h^0}^2} \sqrt{m_{h^0}^2 - 4m_W^2} \left[1 + \frac{(m_{h^0}^2 - 2m_W^2)^2}{8 m_W^4}\right]$$

$$\Gamma(h^0 \to ZZ) = \frac{g^2 m_Z^4 \cos^2 \delta}{16\pi m_W^2 m_{h^0}^2} \sqrt{m_{h^0}^2 - 4m_Z^2} \left[1 + \frac{(m_{h^0}^2 - 2m_Z^2)^2}{8 m_Z^4}\right]$$

$$\Gamma(h^0 \to ZA^0) = \frac{g^2 \cos^2 \delta}{64\pi m_W^2 m_{h^0}^3} \lambda^{\frac{3}{2}}\left(m_{h^0}^2, m_{A^0}^2, m_Z^2\right)$$

$$\Gamma(h^0 \to W^\pm H^\mp) = \frac{g^2 \cos^2 \delta}{64\pi m_{h^0}^3 m_W^2} \lambda^{\frac{3}{2}}(m_{h^0}^2, m_{H^+}^2, m_W^2)$$

$$\Gamma^{(A/B)}(h^0 \to A^0 A^0) = \frac{\sqrt{m_{h^0}^2 - 4m_{A^0}^2}}{32\pi m_{h^0}^2} \left|C_{[h^0 A^0 A^0]}^{(A/B)}\right|^2$$

$$\Gamma^{(A/B)}(h^0 \to H^+H^-) = \frac{\sqrt{m_{h^0}^2 - 4m_{H^+}^2}}{16\pi m_{h^0}^2} \left|C_{[h^0 H^+ H^-]}^{(A/B)}\right|^2 \quad .$$

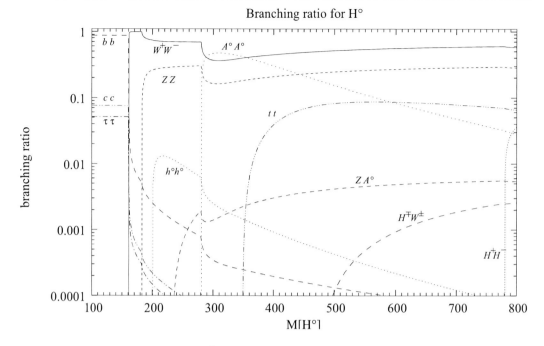

Figure 19. Branching ratio of H^0 at $m_{h^0} = 100$ GeV, $m_{A^0} = 140$ GeV and $m_{H^+} = 390$ GeV and $\delta = 0.1$ in potential B.

The couplings $C^{(A/B)}_{[h^0 H^+ H^-]}$ for either potential A or potential B are listed in appendix B. The one-loop decays have been automatically calculated with *xloops*. Unfortunately the formulas are to large to be shown here. A compact formula for $h^0 \to \gamma\gamma$ in the MSSM can be found in ref. [21].

For the pseudo-scalar Higgs boson one gets:

$$\Gamma(A^0 \to f\bar{f}) = N_c \frac{g^2 m_f^2 \cot^2\beta}{32\pi m_W^2} \sqrt{\frac{m_{A^0}^2}{4} - m_f^2}$$

$$\Gamma(A^0 \to Zh^0) = \frac{g^2 \cos^2\delta}{64\pi \, m_W^2 m_{A^0}^3} \lambda^{\frac{3}{2}}\left(m_{A^0}^2, m_{h^0}^2, m_Z^2\right)$$

$$\Gamma(A^0 \to W^\pm H^\mp) = \frac{g^2}{64\pi \, m_W^2 m_{A^0}^3} \lambda^{\frac{3}{2}}\left(m_{A^0}^2, m_{H^+}^2, m_W^2\right)$$

$$\Gamma(A^0 \to \gamma\gamma) =$$
$$\frac{m_{A^0}^3}{32\pi} \left| \frac{N_c e^3 m_t^2 \cot\beta}{9 \sin\theta_W \, m_W \, \pi^4} \text{Oneloop3Pt}\left(0, 0, 0, m_{A^0}, \tfrac{1}{2}m_{A^0}, \tfrac{1}{2}m_{A^0}, m_t, m_t, m_t\right) \right|^2$$

N_c denotes the number of quark colors. Beside $A^0 \to \gamma\gamma$ all other formulas for one-loop decays have been skipped. The definition of the OneLoop3Pt function can be found in ref. [22]. An $\mathcal{O}(\alpha_s)$ improved formula for $\Gamma(A^0 \to q\bar{q})$ can be found in ref. [23].

The charged Higgs boson has the following partial decay widths:

$$\Gamma(H^+ \to f_t \bar{f}_b) =$$

Searches for Higgs Bosons in Two-Higgs Doublet Models 155

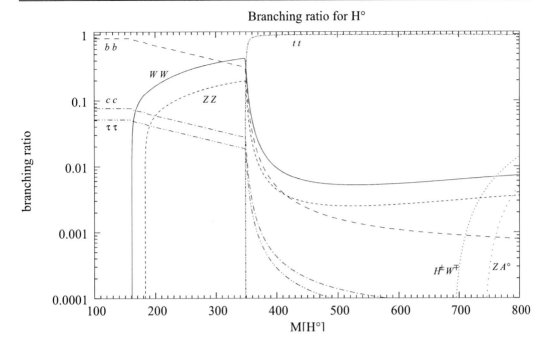

Figure 20. Branching ratio of H^0 at $m_{h^0} = 100$ GeV, $m_{A^0} = 650$ GeV and $m_{H^+} = 610$ GeV and $\delta = 1.4$ in potential A.

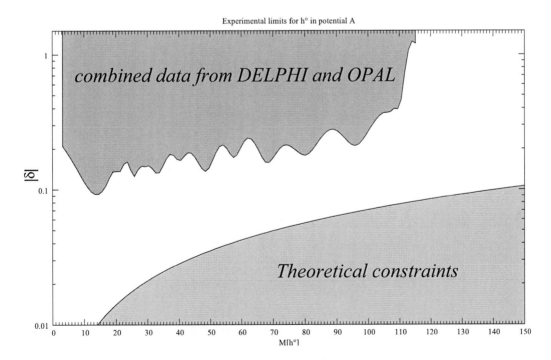

Figure 21. Bounds in the m_{h^0}-δ plane for potential A.

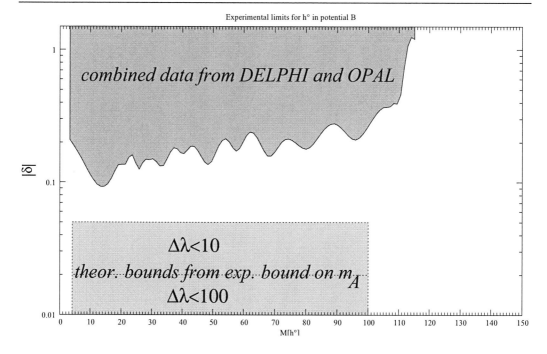

Figure 22. Bounds in the m_{h^0}-δ plane for potential B.

$$\frac{g^2 \cot^2\beta\, N_c\, |V_{tb}|^2}{64\pi\, m_W^2} \frac{\lambda^{\frac{1}{2}}(m_{H^+}^2, m_t^2, m_b^2)}{m_{H^+}^3} \left[(m_{H^+}^2 - m_t^2 - m_b^2)(m_t^2 + m_b^2) + 4\, m_t^2 m_b^2 \right]$$

$$\Gamma(H^+ \to W^+ h^0) = \frac{g^2 \cos^2\delta}{64\pi\, m_{H^+}^3 m_W^2} \lambda^{\frac{3}{2}}(m_{H^+}^2, m_{h^0}^2, m_W^2)$$

$$\Gamma(H^+ \to W^+ A^0) = \frac{g^2}{64\pi\, m_{H^+}^3 m_W^2} \lambda^{\frac{3}{2}}(m_{H^+}^2, m_{A^0}^2, m_W^2)$$

$$\Gamma(H^+ \to W^+ H^0) = \frac{g^2 \sin^2\delta}{64\pi\, m_{H^+}^3 m_W^2} \lambda^{\frac{3}{2}}(m_{H^+}^2, m_{H^0}^2, m_W^2)$$

Note that $\Gamma(H^+ \to f_t \bar{f}_b)$ is also valid for leptons with $m_\nu \equiv m_t = 0$ and $N_c = 1$. Again we skip the formula for $H^+ \to W^+ \gamma$ and $H^+ \to W^+ Z$ due to its length.

For the heavy Higgs boson (H^0) we calculate:

$$\Gamma(H^0 \to f\bar{f}) = \frac{g^2 N_c\, m_f^2 \sin^2\alpha}{64\pi\, m_W^2 \sin^2\beta} \sqrt{m_{H^0}^2 - 4m_f^2} \left[1 - \frac{4m_f^2}{m_{H^0}^2} \right]$$

$$\Gamma(H^0 \to W^+ W^-) = \frac{g^2\, m_W^2 \cos^2\delta}{8\pi\, m_{H^0}^2} \sqrt{m_{H^0}^2 - 4m_W^2} \left[1 + \frac{(m_{H^0}^2 - 2m_W^2)^2}{8\, m_W^4} \right]$$

$$\Gamma(H^0 \to ZZ) = \frac{g^2\, m_Z^4 \cos^2\delta}{16\pi\, m_W^2 m_{H^0}^2} \sqrt{m_{H^0}^2 - 4m_Z^2} \left[1 + \frac{(m_{H^0}^2 - 2m_Z^2)^2}{8\, m_Z^4} \right]$$

$$\Gamma(H^0 \to Z A^0) = \frac{g^2 \sin^2\delta}{64\pi\, m_W^2 m_{H^0}^3} \lambda^{\frac{3}{2}}\left(m_{H^0}^2, m_{A^0}^2, m_Z^2\right)$$

$$\Gamma(H^0 \to W^\pm H^\mp) = \frac{g^2 \sin^2\delta}{64\pi\, m_{H^0}^3 m_W^2} \lambda^{\frac{3}{2}}(m_{H^0}^2, m_{H^+}^2, m_W^2)$$

$$\Gamma^{(A/B)}(H^0 \to A^0 A^0) = \frac{\sqrt{m_{H^0}^2 - 4m_{A^0}^2}}{32\pi\, m_{H^0}^2} \left|C^{(A/B)}_{[H^0 A^0 A^0]}\right|^2$$

$$\Gamma^{(A/B)}(H^0 \to H^+ H^-) = \frac{\sqrt{m_{H^0}^2 - 4m_{H^+}^2}}{16\pi\, m_{H^0}^2} \left|C^{(A/B)}_{[H^0 H^+ H^-]}\right|^2$$

B. Feynman Rules for Fermion Couplings

A more detailed treatment of the Yukawa Lagrangian can be found in [24].

Table 1. Coupling constants for the fermion-scalar interactions

	Mod. I	Mod. II	Mod. III	Mod. IV
α_{eh}	$-\frac{\cos\alpha}{\sin\beta}$	$\frac{\sin\alpha}{\cos\beta}$	$-\frac{\cos\alpha}{\sin\beta}$	$\frac{\sin\alpha}{\cos\beta}$
α_{dh}	$-\frac{\cos\alpha}{\sin\beta}$	$-\frac{\cos\alpha}{\sin\beta}$	$\frac{\sin\alpha}{\cos\beta}$	$\frac{\sin\alpha}{\cos\beta}$
α_{eH}	$\frac{\sin\alpha}{\sin\beta}$	$\frac{\cos\alpha}{\cos\beta}$	$\frac{\sin\alpha}{\sin\beta}$	$\frac{\cos\alpha}{\cos\beta}$
α_{dH}	$\frac{\sin\alpha}{\sin\beta}$	$\frac{\sin\alpha}{\sin\beta}$	$\frac{\cos\alpha}{\cos\beta}$	$\frac{\cos\alpha}{\cos\beta}$
β_e	$-\cot\beta$	$\tan\beta$	$-\cot\beta$	$\tan\beta$
β_d	$-\cot\beta$	$-\cot\beta$	$\tan\beta$	$\tan\beta$

$\bar{e}_i e_i h$	$\frac{ig}{2m_W}\alpha_{eh} m_{e_i}$	$\bar{u}_i u_i h$	$-\frac{ig}{2m_W}\frac{\cos\alpha}{\sin\beta} m_{u_i}$
$\bar{d}_i d_i h$	$\frac{ig}{2m_W}\alpha_{dh} m_{d_i}$	$\bar{e}_i e_i H$	$\frac{ig}{2m_W}\alpha_{eH} m_{e_i}$
$\bar{u}_i u_i H$	$-\frac{ig}{2m_W}\frac{\sin\alpha}{\sin\beta} m_{u_i}$	$\bar{d}_i d_i H$	$\frac{ig}{2m_W}\alpha_{dH} m_{d_i}$
$\bar{e}_i e_i A$	$-\frac{g}{2m_W}\beta_e m_{e_i}\gamma_5$	$\bar{u}_i u_i A$	$-\frac{g}{2m_W}\cot\beta\, m_{u_i}\gamma_5$
$\bar{d}_i d_i A$	$-\frac{g}{2m_W}\beta_d m_{d_i}\gamma_5$	$\bar{e}_i e_i G_0$	$\frac{g}{2m_W} m_{e_i}\gamma_5$
$\bar{u}_i u_i G_0$	$-\frac{g}{2m_W} m_{u_i}\gamma_5$	$\bar{d}_i d_i G_0$	$\frac{g}{2m_W} m_{d_i}\gamma_5$
$\bar{e}_i \nu_i H^+$	$\frac{ig}{\sqrt{2}m_W}\beta_e m_{e_i}\gamma_R$	$\bar{u}_i d_j H^+$	$\frac{ig}{\sqrt{2}m_W} V_{ij}\left[\beta_d m_{d_j}\gamma_R + \cot\beta\, m_{u_i}\gamma_L\right]$
$\bar{\nu}_i e_i H^-$	$\frac{ig}{\sqrt{2}m_W}\beta_e m_{e_i}\gamma_L$	$\bar{d}_i u_j H^-$	$\frac{ig}{\sqrt{2}m_W} V^*_{ij}\left[\beta_d m_{d_i}\gamma_L + \cot\beta\, m_{u_j}\gamma_R\right]$
$\bar{e}_i \nu_i G^+$	$-\frac{ig}{\sqrt{2}m_W} m_{e_i}\gamma_R$	$\bar{u}_i d_j G^+$	$\frac{ig}{\sqrt{2}m_W} V_{ij}\left[-m_{d_j}\gamma_R + m_{u_i}\gamma_L\right]$
$\bar{\nu}_i e_i G^-$	$-\frac{ig}{\sqrt{2}m_W} m_{e_i}\gamma_L$	$\bar{d}_i u_j G^-$	$\frac{ig}{\sqrt{2}m_W} V^*_{ij}\left[-m_{d_i}\gamma_L + m_{u_j}\gamma_R\right]$

C. Feynman Rules for Scalar Couplings

In this section we present the Feynman rules for the triple and quartic interactions of scalar fields which are different in both potentials.

We define the following quantities:

$$A_{\alpha\beta} \equiv \cos^3\beta\sin\alpha + \sin^3\beta\cos\alpha$$
$$B_{\alpha\beta} \equiv \cos^3\beta\cos\alpha - \sin^3\beta\sin\alpha$$
$$C_{\alpha\beta} \equiv \sin^3\alpha\cos\beta + \cos^3\alpha\sin\beta$$
$$D_{\alpha\beta} \equiv \cos^3\alpha\cos\beta - \sin^3\alpha\sin\beta$$

$$E_{\alpha\beta} \equiv (\cos^2\alpha - \sin^2\beta)$$
$$F_{\alpha\beta} \equiv (\cos^2\alpha - \cos^2\beta)$$
$$G_{\alpha\beta} \equiv \cos\beta\sin\beta - 3\cos\alpha\sin\alpha$$
$$H_{\alpha\beta} \equiv \cos 2\beta\cos\delta\cos(\alpha+\beta)$$
$$K_{\alpha\beta} \equiv \cos 2\beta(\cos^2\alpha - \cos^2\beta)$$

C.1. Different Triple Scalar Vertices in V_A

H^+H^-h $\quad -\frac{ig}{m_W}\left(\frac{m_{h^0}^2}{\sin 2\beta}B_{\alpha\beta} - m_{H^+}^2\sin\delta\right)$

H^+H^-H $\quad -\frac{ig}{m_W}\left(\frac{m_{h^0}^2}{\sin 2\beta}A_{\alpha\beta} + m_{H^+}^2\cos\delta\right)$

hHH $\quad -\frac{ig}{2m_W}\frac{\sin 2\alpha \sin\delta}{\sin 2\beta}(2m_{H^0}^2 + m_{h^0}^2)$

hhH $\quad -\frac{ig}{2m_W}\frac{\sin 2\alpha \cos\delta}{\sin 2\beta}(m_{H^0}^2 + 2m_{h^0}^2)$

hhh $\quad -\frac{3ig}{m_W}\frac{m_{h^0}^2}{\sin 2\beta}D_{\alpha\beta}$

HHH $\quad -\frac{3ig}{m_W}\frac{m_{h^0}^2}{\sin 2\beta}C_{\alpha\beta}$

AAh $\quad -\frac{ig}{m_W}\left(\frac{m_{h^0}^2}{\sin 2\beta}B_{\alpha\beta} - M_A^2\sin\delta\right)$

AAH $\quad -\frac{ig}{m_W}\left(\frac{m_{h^0}^2}{\sin 2\beta}A_{\alpha\beta} + M_A^2\cos\delta\right)$

C.2. Different Triple Scalar Vertices in V_B

H^+H^-h $\quad -\frac{ig}{m_W}\left(\frac{m_{h^0}^2}{\sin 2\beta}B_{\alpha\beta} - m_{H^+}^2\sin\delta - \frac{\cos\delta}{\sin 2\beta}m_{A^0}^2\right)$

H^+H^-H $\quad -\frac{ig}{m_W}\left(\frac{m_{H^0}^2}{\sin 2\beta}A_{\alpha\beta} + m_{H^+}^2\cos\delta + \frac{\sin\delta}{\sin 2\beta}m_{A^0}^2\right)$

hHH $\quad -\frac{ig\sin\delta}{2m_W\sin 2\beta}\left(\sin 2\alpha(2m_{H^0}^2 + m_{h^0}^2) + m_{A^0}^2 G_{\alpha\beta}\right)$

hhH $\quad -\frac{ig\cos\delta}{2m_W\sin 2\beta}\left(\sin 2\alpha(m_{H^0}^2 + 2m_{h^0}^2) + m_{A^0}^2 G_{\alpha\beta}\right)$

hhh $\quad -\frac{3ig}{m_W\sin 2\beta}\left(m_{h^0}^2 D_{\alpha\beta} - m_{A^0}^2 E_{\alpha\beta}\cos\delta\right)$

HHH $\quad -\frac{3ig}{m_W\sin 2\beta}\left(m_{H^0}^2 C_{\alpha\beta} + m_{A^0}^2 F_{\alpha\beta}\sin\delta\right)$

AAh $\quad -\frac{ig}{m_W}\left(\frac{m_{h^0}^2}{\sin 2\beta}B_{\alpha\beta} - M_A^2\frac{\cos\delta\cos 2\beta}{\sin 2\beta}\right)$

AAH $\quad -\frac{ig}{m_W}\left(\frac{m_{H^0}^2}{\sin 2\beta}A_{\alpha\beta} - M_A^2\frac{\sin\delta\cos 2\beta}{\sin 2\beta}\right)$

C.3. Different Quartic Scalar Vertices for V_A

$H^+H^-H^+H^-$ $\quad -\frac{2\,ig^2}{\sin^2 2\beta m_W^2}(m_{H^0}^2 A_{\alpha\beta}^2 + m_{h^0}^2 B_{\alpha\beta}^2)$

$AAAA$ $\quad -\frac{3ig^2}{\sin^2 2\beta m_W^2}(m_{H^0}^2 A_{\alpha\beta}^2 + m_{h^0}^2 B_{\alpha\beta}^2)$

AAH^+H^- $\quad -\frac{ig}{\sin^2 2\beta m_W^2}(m_{H^0}^2 A_{\alpha\beta}^2 + m_{h^0}^2 B_{\alpha\beta}^2)$

H^+H^-hh $\quad -\frac{ig^2}{2m_W^2}\left[\frac{1}{\sin^2 2\beta}(m_{H^0}^2 A_{\alpha\beta}\sin 2\alpha\cos\delta + 2m_{h^0}^2 B_{\alpha\beta}D_{\alpha\beta}) + m_{H^+}^2 \sin^2\delta\right]$

H^+H^-HH $\quad -\frac{ig^2}{2m_W^2}\left[\frac{1}{\sin^2 2\beta}(2m_{H^0}^2 A_{\alpha\beta}C_{\alpha\beta} + m_{h^0}^2 B_{\alpha\beta}\sin 2\alpha\sin\delta) + m_{H^+}^2 \cos^2\delta\right]$

$AAhh$ $\quad -\frac{ig^2}{2m_W^2}\left[\frac{1}{\sin^2 2\beta}(m_{H^0}^2 A_{\alpha\beta}\sin 2\alpha\cos\delta + 2m_{h^0}^2 B_{\alpha\beta}D_{\alpha\beta}) + M_A^2 \sin^2\delta\right]$

$AAHH$ $\quad -\frac{ig^2}{2m_W^2}\left[\frac{1}{\sin^2 2\beta}(2m_{H^0}^2 A_{\alpha\beta}C_{\alpha\beta} + m_{h^0}^2 B_{\alpha\beta}\sin 2\alpha\sin\delta) + M_A^2 \cos^2\delta\right]$

H^+H^-Hh $\quad -\frac{ig^2}{2m_W^2}\left[\frac{\sin 2\alpha}{\sin^2 2\beta}(m_{H^0}^2 A_{\alpha\beta}\sin\delta + m_{h^0}^2 B_{\alpha\beta}\cos\delta) - \frac{1}{2}m_{H^+}^2 \sin 2\delta\right]$

$AAHh$ $\quad -\frac{ig^2}{2m_W^2}\left[\frac{\sin 2\alpha}{\sin^2 2\beta}(m_{H^0}^2 A_{\alpha\beta}\sin\delta + m_{h^0}^2 B_{\alpha\beta}\cos\delta) - \frac{1}{2}M_A^2 \sin 2\delta\right]$

$hhhh$ $\quad -\frac{3\,ig^2}{\sin^2 2\beta M_w^2}(4m_{h^0}^2 D_{\alpha\beta}^2 + m_{H^0}^2 \sin^2 2\alpha \cos^2\delta)$

$HHHH$ $\quad -\frac{3\,ig^2}{\sin^2 2\beta m_W^2}(m_{h^0}^2 \sin^2 2\alpha \sin^2\delta + 4m_{H^0}^2 C_{\alpha\beta}^2)$

$hhhH$ $\quad -\frac{3\,ig^2}{2\sin^2 2\beta m_W^2}(4m_{h^0}^2 D_{\alpha\beta}\sin 2\alpha\cos\delta + m_{H^0}^2 \sin^2 2\alpha \sin 2\delta)$

$HHHh$ $\quad -\frac{3ig^2}{2\sin^2 2\beta m_W^2}(m_{h^0}^2 \sin^2 2\alpha \sin 2\delta + 4m_{H^0}^2 C_{\alpha\beta}\sin 2\alpha\sin\delta)$

$hhHH$ $\quad -\frac{ig^2 \sin 2\alpha}{4\sin 2\beta m_W^2}\left[m_{H^0}^2 - m_{h^0}^2 + \frac{3\sin 2\alpha}{\sin 2\beta}(\sin^2\delta m_{H^0}^2 + \cos^2\delta m_{h^0}^2)\right]$

AAG^0G^0 $\quad -\frac{ig^2}{4m_W^2}\left[\frac{\sin 2\alpha}{\sin 2\beta}(m_{H^0}^2 - m_{h^0}^2) + 3(\sin^2\delta m_{H^0}^2 + \cos^2\delta m_{h^0}^2)\right]$

$H^+H^-G^+G^-$ $\quad -\frac{ig^2}{4m_W^2}\left[m_{A^0}^2 + \frac{\sin 2\alpha}{\sin 2\beta}(m_{H^0}^2 - m_{h^0}^2) + 2(\sin^2\delta m_{H^0}^2 + \cos^2\delta m_{h^0}^2)\right]$

G^+G^-AA $\quad -\frac{ig^2}{2m_W^2}\left[m_{H^+}^2 + \frac{1}{\sin 2\beta}(\cos\delta A_{\alpha\beta}m_{H^0}^2 - \sin\delta B_{\alpha\beta}m_{h^0}^2)\right]$

$H^+H^-G^0G^0$ $\quad -\frac{ig^2}{2m_W^2}\left[m_{H^+}^2 + \frac{1}{\sin 2\beta}(\cos\delta A_{\alpha\beta}m_{H^0}^2 - \sin\delta B_{\alpha\beta}m_{h^0}^2)\right]$

$H^+H^-H^\mp G^\pm$ $\quad -\frac{ig^2}{m_W^2 \sin 2\beta}\left[\sin\delta A_{\alpha\beta}m_{H^0}^2 + \cos\delta B_{\alpha\beta}m_{h^0}^2\right]$

$H^+H^-G^0 A$ $\quad -\frac{ig^2}{2m_W^2 \sin 2\beta}\left[\sin\delta A_{\alpha\beta}m_{H^0}^2 + \cos\delta B_{\alpha\beta}m_{h^0}^2\right]$

$AAAG^0$	$-\frac{3ig^2}{2m_W^2 \sin 2\beta} \left[\sin\delta A_{\alpha\beta} m_{H^0}^2 + \cos\delta B_{\alpha\beta} m_{h^0}^2 \right]$
$AAH^\mp G^\pm$	$-\frac{ig^2}{2m_W^2 \sin 2\beta} \left[\sin\delta A_{\alpha\beta} m_{H^0}^2 + \cos\delta B_{\alpha\beta} m_{h^0}^2 \right]$
G^+G^-hh	$-\frac{ig^2}{4m_W^2} \left[\frac{1}{\sin 2\beta} (\sin 2\alpha \cos^2\delta m_{H^0}^2 - 2\sin\delta D_{\alpha\beta} m_{h^0}^2) + 2\cos^2\delta m_{H^+}^2 \right]$
$G^0 G^0 hh$	$-\frac{ig^2}{4m_W^2} \left[\frac{1}{\sin 2\beta} (\sin 2\alpha \cos^2\delta m_{H^0}^2 - 2\sin\delta D_{\alpha\beta} m_{h^0}^2) + 2\cos^2\delta M_A^2 \right]$
G^+G^-HH	$-\frac{ig^2}{4m_W^2} \left[\frac{1}{\sin 2\beta} (2\cos\delta C_{\alpha\beta} m_{H^0}^2 - \sin^2\delta \sin 2\alpha m_{h^0}^2) + 2\sin^2\delta m_{H^+}^2 \right]$
$G^0 G^0 HH$	$-\frac{ig^2}{4m_W^2} \left[\frac{1}{\sin 2\beta} (2\cos\delta C_{\alpha\beta} m_{H^0}^2 - \sin^2\delta \sin 2\alpha m_{h^0}^2) + 2\sin^2\delta M_A^2 \right]$
$H^\mp G^\pm HH$	$-\frac{ig^2}{8m_W^2} \left[\frac{1}{\sin 2\beta} (4\sin\delta C_{\alpha\beta} m_{H^0}^2 + \sin 2\delta \sin 2\alpha m_{h^0}^2) - 2\sin 2\delta m_{H^+}^2 \right]$
$H^\mp G^\pm hh$	$-\frac{ig^2}{8m_W^2} \left[\frac{1}{\sin 2\beta} (\sin 2\delta \sin 2\alpha m_{H^0}^2 + 4\cos\delta D_{\alpha\beta} m_{h^0}^2) + 2\sin 2\delta m_{H^+}^2 \right]$
$AG^0 HH$	$-\frac{ig^2}{8m_W^2} \left[\frac{1}{\sin 2\beta} (4\sin\delta C_{\alpha\beta} m_{H^0}^2 + \sin 2\delta \sin 2\alpha m_{h^0}^2) - 2\sin 2\delta M_A^2 \right]$
$AG^0 hh$	$-\frac{ig^2}{8m_W^2} \left[\frac{1}{\sin 2\beta} (\sin 2\alpha \sin 2\delta m_{H^0}^2 + 4\cos\delta D_{\alpha\beta} m_{h^0}^2) + 2\sin 2\delta M_A^2 \right]$
G^+G^-hH	$-\frac{ig^2 \sin 2\delta}{8m_W^2} \left[\frac{\sin 2\alpha}{\sin 2\beta} (m_{H^0}^2 - m_{h^0}^2) + 2m_{H^+}^2 \right]$
$G^0 G^0 hH$	$-\frac{ig^2 \sin 2\delta}{8m_W^2} \left[\frac{\sin 2\alpha}{\sin 2\beta} (m_{H^0}^2 - m_{h^0}^2) + 2M_A^2 \right]$
$G^\mp H^\pm hH$	$-\frac{ig^2}{4m_W^2} \left[\frac{\sin 2\alpha}{\sin 2\beta} (\sin^2\delta m_{H^0}^2 + \cos^2\delta m_{h^0}^2) - \cos 2\delta m_{H^+}^2 \right]$
$AG^0 hH$	$-\frac{ig^2}{4m_W^2} \left[\frac{\sin 2\alpha}{\sin 2\beta} (\sin^2\delta m_{H^0}^2 + \cos^2\delta m_{h^0}^2) - \cos 2\delta M_A^2 \right]$

C.4. Different Quartic Scalar Vertices for V_B

$H^+H^-H^+H^-$	$-\frac{2ig^2}{\sin^2 2\beta m_W^2} (m_{H^0}^2 A_{\alpha\beta}^2 + m_{h^0}^2 B_{\alpha\beta}^2 - m_{A^0}^2 \cos^2 2\beta)$
$AAAA$	$-\frac{3ig^2}{\sin^2 2\beta m_W^2} (m_{H^0}^2 A_{\alpha\beta}^2 + m_{h^0}^2 B_{\alpha\beta}^2 - m_{A^0}^2 \cos^2 2\beta)$
AAH^+H^-	$-\frac{ig}{\sin^2 2\beta m_W^2} (m_{H^0}^2 A_{\alpha\beta}^2 + m_{h^0}^2 B_{\alpha\beta}^2 - m_{A^0}^2 \cos^2 2\beta)$
H^+H^-hh	$-\frac{ig^2}{2m_W^2} \left[\frac{1}{\sin^2 2\beta} (m_{H^0}^2 A_{\alpha\beta} \sin 2\alpha \cos\delta + 2m_{h^0}^2 B_{\alpha\beta} D_{\alpha\beta} - H_{\alpha\beta} m_{A^0}^2) + (m_{H^+}^2 - M_A^2) \sin^2\delta \right]$

Searches for Higgs Bosons in Two-Higgs Doublet Models

H^+H^-HH	$-\frac{ig^2}{2m_W^2}\left[\frac{1}{\sin^2 2\beta}(2m_{H^0}^2 A_{\alpha\beta}C_{\alpha\beta} + m_{h^0}^2 B_{\alpha\beta}\sin 2\alpha\sin\delta - K_{\alpha\beta}m_{A^0}^2) + \left(m_{H^+}^2 - m_{A^0}^2\right)\cos^2\delta\right]$
$AAhh$	$-\frac{ig^2}{2m_W^2\sin^2 2\beta}\left[m_{H^0}^2 A_{\alpha\beta}\sin 2\alpha\cos\delta + 2m_{h^0}^2 B_{\alpha\beta}D_{\alpha\beta} - H_{\alpha\beta}M_A^2\right]$
$AAHH$	$-\frac{ig^2}{2m_W^2\sin^2 2\beta}\left[2m_{H^0}^2 A_{\alpha\beta}C_{\alpha\beta} + m_{h^0}^2 B_{\alpha\beta}\sin 2\alpha\sin\delta - K_{\alpha\beta}M_A^2\right]$
H^+H^-Hh	$-\frac{ig^2}{2m_W^2}\left[\frac{\sin 2\alpha}{\sin^2 2\beta}(m_{H^0}^2 A_{\alpha\beta}\sin\delta + m_{h^0}^2 B_{\alpha\beta}\cos\delta - m_{A^0}^2\cos 2\beta) - \frac{1}{2}(m_{H^+}^2 - m_{A^0}^2)\sin 2\delta\right]$
$AAHh$	$-\frac{ig^2}{2m_W^2}\frac{\sin 2\alpha}{\sin^2 2\beta}\left[m_{H^0}^2 A_{\alpha\beta}\sin\delta + m_{h^0}^2 B_{\alpha\beta}\cos\delta - M_A^2\cos 2\beta\right]$
$hhhh$	$-\frac{3ig^2}{\sin^2 2\beta M_W^2}\left(4m_{h^0}^2 D_{\alpha\beta}^2 + m_{H^0}^2\sin^2 2\alpha\cos^2\delta - m_{A^0}^2 E_{\alpha\beta}^2\right)$
$HHHH$	$-\frac{3ig^2}{\sin^2 2\beta m_W^2}\left(m_{h^0}^2\sin^2 2\alpha\sin^2\delta + 4m_{H^0}^2 C_{\alpha\beta}^2 - m_{A^0}^2 F_{\alpha\beta}^2\right)$
$hhhH$	$-\frac{3ig^2\sin 2\alpha}{2\sin^2 2\beta m_W^2}\left[4m_{h^0}^2 D_{\alpha\beta}\cos\delta + m_{H^0}^2\sin 2\alpha\sin 2\delta - m_{A^0}^2 E_{\alpha\beta}\right]$
$HHHh$	$-\frac{3ig^2\sin 2\alpha}{2\sin^2 2\beta m_W^2}\left[m_{h^0}^2\sin 2\alpha\sin 2\delta + 4m_{H^0}^2 C_{\alpha\beta}\sin\delta) - m_{A^0}^2(\sin^2\alpha - \sin^2\beta)\right]$
$hhHH$	$-\frac{ig^2\sin 2\alpha}{4\sin 2\beta m_W^2}\left[m_{H^0}^2 - m_{h^0}^2 + m_{A^0}^2\frac{3\sin 2\alpha}{\sin 2\beta}(\sin^2\delta m_{H^0}^2 + \cos^2\delta m_{h^0}^2) - m_{A^0}^2\right]$
AAG^0G^0	$-\frac{ig^2}{4m_W^2}\left[\frac{\sin 2\alpha}{\sin 2\beta}(m_{H^0}^2 - m_{h^0}^2) + 3(\sin^2\delta m_{H^0}^2 + \cos^2\delta m_{h^0}^2) - 2m_{A^0}^2\right]$
$H^+H^-G^+G^-$	$-\frac{ig^2}{4m_W^2}\left[\frac{\sin 2\alpha}{\sin 2\beta}(m_{H^0}^2 - m_{h^0}^2) + 2(\sin^2\delta m_{H^0}^2 + \cos^2\delta m_{h^0}^2)\right]$
G^+G^-AA	$-\frac{ig^2}{2m_W^2}\left[m_{H^+}^2 - m_{A^0}^2 + \frac{1}{\sin 2\beta}(\cos\delta A_{\alpha\beta}m_{H^0}^2 - \sin\delta B_{\alpha\beta}m_{h^0}^2)\right]$
$H^+H^-G^0G^0$	$-\frac{ig^2}{2m_W^2}\left[m_{H^+}^2 - m_{A^0}^2 + \frac{1}{\sin 2\beta}(\cos\delta A_{\alpha\beta}m_{H^0}^2 - \sin\delta B_{\alpha\beta}m_{h^0}^2)\right]$
$H^+H^-H^\mp G^\pm$	$-\frac{ig^2}{m_W^2\sin 2\beta}\left[\sin\delta A_{\alpha\beta}m_{H^0}^2 + \cos\delta B_{\alpha\beta}m_{h^0}^2 - m_{A^0}^2\cos 2\beta\right]$
$H^+H^-G^0A$	$-\frac{ig^2}{2m_W^2\sin 2\beta}\left[\sin\delta A_{\alpha\beta}m_{H^0}^2 + \cos\delta B_{\alpha\beta}m_{h^0}^2 - m_{A^0}^2\cos 2\beta\right]$
$AAAG^0$	$-\frac{3ig^2}{2m_W^2\sin 2\beta}\left[\sin\delta A_{\alpha\beta}m_{H^0}^2 + \cos\delta B_{\alpha\beta}m_{h^0}^2 - m_{A^0}^2\cos 2\beta\right]$
$AAH^\mp G^\pm$	$-\frac{ig^2}{2m_W^2\sin 2\beta}\left[\sin\delta A_{\alpha\beta}m_{H^0}^2 + \cos\delta B_{\alpha\beta}m_{h^0}^2 - m_{A^0}^2\cos 2\beta\right]$
G^+G^-hh	$-\frac{ig^2}{4m_W^2}\left[\frac{1}{\sin 2\beta}(\sin 2\alpha\cos^2\delta m_{H^0}^2 - 2\sin\delta D_{\alpha\beta}m_{h^0}^2) + 2\cos^2\delta(m_{H^+}^2 - m_{A^0}^2)\right]$
G^0G^0hh	$-\frac{ig^2}{4m_W^2}\left[\frac{1}{\sin 2\beta}(\sin 2\alpha\cos^2\delta m_{H^0}^2 - 2\sin\delta D_{\alpha\beta}m_{h^0}^2)\right]$
G^+G^-HH	$-\frac{ig^2}{4m_W^2}\left[\frac{1}{\sin 2\beta}(2\cos\delta C_{\alpha\beta}m_{H^0}^2 - \sin^2\delta\sin 2\alpha m_{h^0}^2) + 2\sin^2\delta(m_{H^+}^2 - m_{A^0}^2)\right]$
G^0G^0HH	$-\frac{ig^2}{4m_W^2}\left[\frac{1}{\sin 2\beta}(2\cos\delta C_{\alpha\beta}m_{H^0}^2 - \sin^2\delta\sin 2\alpha m_{h^0}^2)\right]$
$H^\mp G^\pm HH$	$-\frac{ig^2}{8m_W^2}\left[\frac{1}{\sin 2\beta}(4\sin\delta C_{\alpha\beta}m_{H^0}^2 + \sin 2\delta\sin 2\alpha m_{h^0}^2 + 2m_{A^0}^2 F_{\alpha\beta}) - 2\sin 2\delta(m_{H^+}^2 - m_{A^0}^2)\right]$
$H^\mp G^\pm hh$	$-\frac{ig^2}{8m_W^2}\left[\frac{1}{\sin 2\beta}(\sin 2\delta\sin 2\alpha m_{H^0}^2 + 4\cos\delta D_{\alpha\beta}m_{h^0}^2 + 2m_{A^0}^2 E_{\alpha\beta}) + 2\sin 2\delta(m_{H^+}^2 - m_{A^0}^2)\right]$
AG^0HH	$-\frac{ig^2}{8m_W^2}\left[\frac{1}{\sin 2\beta}(4\sin\delta C_{\alpha\beta}m_{H^0}^2 + \sin 2\delta\sin 2\alpha m_{h^0}^2 + 2m_{A^0}^2 E_{\alpha\beta})\right]$
AG^0hh	$-\frac{ig^2}{8m_W^2}\left[\frac{1}{\sin 2\beta}(\sin 2\alpha\sin 2\delta m_{H^0}^2 + 4\cos\delta D_{\alpha\beta}m_{h^0}^2 + 2m_{A^0}^2 F_{\alpha\beta})\right]$
G^+G^-hH	$-\frac{ig^2\sin 2\delta}{8m_W^2}\left[\frac{\sin 2\alpha}{\sin 2\beta}(m_{H^0}^2 - m_{h^0}^2) + 2(m_{H^+}^2 - m_{A^0}^2)\right]$

$G^0 G^0 h H$	$-\frac{ig^2 \sin 2\delta}{8m_W^2} \left[\frac{\sin 2\alpha}{\sin 2\beta} (m_{H^0}^2 - m_{h^0}^2) \right]$
$G^\mp H^\pm h H$	$-\frac{ig^2}{4m_W^2} \left[\frac{\sin 2\alpha}{\sin 2\beta} (\sin^2 \delta m_{H^0}^2 + \cos^2 \delta m_{h^0}^2 - m_{A^0}^2) - \cos 2\delta (m_{H^+}^2 - m_{A^0}^2) \right]$
$A G^0 h H$	$-\frac{ig^2}{4m_W^2} \left[\frac{\sin 2\alpha}{\sin 2\beta} (\sin^2 \delta m_{H^0}^2 + \cos^2 \delta m_{h^0}^2) \right]$

References

[1] K. A. Assamagan *et al.* [Higgs Working Group Collaboration], *arXiv:hep-ph/0406152*.

[2] L. Brücher and R. Santos, *Eur. Phys. J. C* **12** (2000) 87.

[3] A. Barroso, L. Brücher and R. Santos. *Phys. Rev.* **D 60** (1999) 035005.

[4] P. M. Ferreira, R. Santos and A. Barroso, *Phys. Lett.* **B 603** (2004) 219.

[5] J. Velhinho, R. Santos and A. Barroso. *Phys. Lett.* **B 322** (1994) 213–218.

[6] D. Kominis and R.S. Chivukula. *Phys.Lett.* **B 304** (1993) 152; H. Komatsu. *Prog.Theor.Phys.* **67** (1982) 1177; R.A. Flores and M. Sher. *Ann.Phys.(NY)* **148** (1983) 295; S. Nie and M. Sher. *Phys.Lett.* **B 449** (1999) 89; S. Kanemura, T. Kasai and Y. Okada. *hep-ph* **9903289**.

[7] J. Maalampi, J. Sirkka and I. Vilja. *Phys. Lett.* **B 265** (1991) 371–376.

[8] S. Kanemura, T. Kubota and E. Takasugi. *Phys. Lett.* **B 313** (1993) 155–160.

[9] A.G. Akeroyd, A. Arhrib and E. Naimi. *Phys. Lett.* **B 490** (2000) 119–124.

[10] A. G. Akeroyd and M. A. Diaz, *Phys. Rev. D* **67** (2003) 095007.

[11] A. G. Akeroyd, M. A. Diaz and F. J. Pacheco, *Phys. Rev. D* **70** (2004) 075002.

[12] L. Brücher and R. Santos. *in preparation*.

[13] S. Dawson, S. Dittmaier and M. Spira, *Phys. Rev. D* **58** (1998) 115012.

[14] H. L. Lai *et al.* [CTEQ Collaboration], *Eur. Phys. J. C* **12** (2000) 375.

[15] T. Han and S. Willenbrock, *Phys. Lett. B* **273** (1991) 167. M. Spira, *Fortschr. Phys.* **46** (1998) 203.

[16] J. Abdallah *et al.* [DELPHI Collaboration], *Eur. Phys. J. C* **35** (2004) 313.

[17] L. Brücher, J. Franzkowski and D. Kreimer. *Nucl.Instrum.Meth.* **A 389** (1997) 323–342; L. Brücher, J. Franzkowski and D. Kreimer. *hep-ph* **9710484**; L. Brücher, J. Franzkowski and D. Kreimer. *Comp. Phys. Comm.* **115** (1998) 140–160. L. Brücher, J. Franzkowski and D. Kreimer, Zvenigorod 1993 *Proceedings*, 186-191. L. Brücher, *arXiv:hep-ph/0002028*.

[18] Jorge C. Romão and Sofia Andringa. *Eur. Phys. J.* **C7** (1999) 631.

[19] The OPAL coll. *Search for Neutral Higgs bosons in e^+e^- Collisions at $\sqrt{s} \approx 189$ GeV. OPAL PN382* (1999).

[20] S. Eidelman *et al.* [Particle Data Group Collaboration], *Phys. Lett. B* **592** (2004) 1.

[21] M. Spira, A. Djouadi, D. Graudenz, P. M. Zerwas. *Nuc. Phys.* **B 453** (1995) 17–82.

[22] L. Brücher, J. Franzkowski and D. Kreimer. *Mod. Phys. Lett.* **A9** (1994) 2335–2346; D. Kreimer. *Int. J. Mod. Phys.* **A8** (1993) 1797–1814; L. Brücher and J. Franzkowski. *Mod. Phys. Lett.* **A14** (1999) 881; L. Brücher, J. Franzkowski and D. Kreimer. *Comp. Phys. Comm.* **85** (1995) 153–165; L. Brücher, J. Franzkowski, D. Kreimer. *Comp. Phys. Comm.* **107** (1997) 281–291.

[23] A. Djouadi, J. Kalinowski and P. M. Zerwas. *Z. Phys.* **C 70** (1996) 435–448.

[24] R. Santos and A. Barroso. *Phys. Rev.* **D 56** (1997) 5366–5385.

In: Theoretical Physics and Nonlinear Optics
Editors: Thomas F. George et al

ISBN 978-1-61122-939-4
© 2012 Nova Science Publishers, Inc.

Chapter 12

QUARK-PION COUPLING CONSTANT AND GROUND-STATE BARYON MASSES IN A CHIRAL SYMMETRIC POTENTIAL MODEL

S.N. Jena[1],[*] *M.K. Muni*[2] *and H.R. Pattnaik*[3]
[1]Department of Physics, Berhampur University,
Berhampur-760007, Orissa, India
[2]Department of Mathematics and Science,
SMIT, Ankushpur, Berhampur, Orissa, India
[3]Department of Physics, Temple City
Institute of Technology and Engg., Khurda, Orissa, India

Abstract

Incorporating chiral symmetry into an independent quark model with logarithmic confining potential, the quark-pion coupling constant $G_{qq\pi}$ for quarks in a nucleon are estimated. The value of $\frac{G_{qq\pi}^2}{4\pi}$ obtained with the approximation of a point pion is consistent with that extracted from the experimental vector meson decay width ratio $\Gamma(\rho \to \pi^+\pi^0)/\Gamma(\rho \to \pi^-\nu)$ by Suzuki and Bhaduri. The ground state baryon masses are also calculated in this model taking into account the contributions of the residual quark-pion coupling arising out of the requirement of chiral symmetry and that of the quark-gluon coupling due to one gluon-exchange in a perturbative manner, over and above the necessary center-of-mass correction. The results obtained for the baryon masses agree reasonably well with the corresponding experimental data. The quark-gluon coupling constant $\alpha_c = 0.258$ required here to explain the QCD mass splittings is quite consistent with the idea of considering gluonic correction in lowest- order perturbation theory.

1. Introduction

Several non-relativistic quark models have appeared [1] in the literature in connection with the study of mass spectrum of light and strange baryons. Although the phenomenological

---
[*]E-mail address: snjena@rediffmail.com, Tel: 0680-2282065, Fax: 0680 - 2243322

picture is reasonable at the non-relativistic level, a relativistic approach is quite indispensable on this account in view of the fact that the baryonic mass splitting are of same order as the constituent quark masses. Of course the chiral constituent quark model which has been constructed by Glozman et al.[2] in a semi relativistic framework is a step in this direction and shows an essential improvement over non-relativistic approaches. On the other hand the MIT bag model [3]has also been found to be relatively successful in this respect. In its improved version, the Chiral Bag Model (CBM)[4] have included the effect of pion self-energy due to baryon-pion coupling at the vertex to provide a better understanding of baryon masses. The chiral potential models [5], which are comparatively more straightforward in the above respects, are obviously attempts in this direction. In such models the confining potentials which basically represent the interaction of quarks with the gluon field are usually assumed phenomenologically as Lorentz scalars in harmonic and cubic form. Another attempt with a scalar potential [6] based on the same basic idea has also been made in this direction for the study of the hadronic spectrum. Potentials of different type of Lorentz structure with equally mixed scalar and vector parts in harmonic [7], linear [8], square root [9] and non-coloumbic power law [10] form have also been investigated in this context. The term in the Lagrangian density for quarks corresponding to the effective scalar part of the potential in such models being chirally non-invariant through all space requires the introduction of an additional pionic component everywhere in order to preserve chiral symmetry. The effective potential of individual quarks in these models, which is basically due to interaction of quarks with the gluon field, may be thought of as being mediated in a self consistent manner through Nambu-Jona-Lasino (NJL) type models [11] by some form of instanton induced effective quark-quark contact interaction with position dependent coupling strength. The position dependent coupling strength, supposedly determined by multigluon mechanism, is impossible to calculate from the first principles, although it is believed to be small at the origin and increases rapidly towards the hadron surface. Therefore, one needs to introduce the effective potential for individual quarks in phenomenological manner to seek a posteriori justification in finding its conformity with the supposed qualitative behaviour of the position-dependent coupling strength in the contact interaction.

However, with no theoretical prejudice in favour of any particular mechanism for generating confinement of individual quarks, we prefer to work in an alternative, but similar scheme based on Dirac equation with a purely phenomenological individual quark potential of the form

$$V_q(r) = (1+\gamma^0)[a\ln(r/b)] \qquad (1)$$

with $a, b > 0$. Such a model takes the Lorentz structure of the potential as an equal admixture of scalar and vector parts because of the fact that both the scalar and vector parts in equal proportions at every point render the solvability of the Dirac equation for independent quarks by reducing it to the form of a Schrodinger-like equation. This Lorentz structure of potential also has an additional advantage of generating no spin-orbit splitting, as observed in the experimental baryon spectrum.

In fact the logarithmic potential of the form

$$V(r) = a\log(r/b) \qquad (2)$$

was proposed in the past by Quigg and Rosner [12] for a successful description of the heavy meson spectra in a non-relativistic Schrodinger frame-work. Implications of such a potential in the context of quark confinement and relativistic consistency were also studied by Magyari [13] in reference to heavy meson spectra. It was found that an equally mixed four vector and scalar logarithmic potential of the form given in eqn.(1), when used appropriately in the Dirac equation, can guarantee relativistic quark confinement and also can generate charmonium and upsilon bound state masses in reasonable agreement with experiment. In view of the remarkable success obtained with such a model in mesonic sector, the model has been extended to baryonic sector to study reasonably well the static baryon properties [14], the nucleon electromagnetic form factors[15] and the weak-electric and-magnetic form factors for the semileptonic baryon decays[16]. Then incorporating chiral symmetry in the SU(2)-flavor sector in the usual manner, this model has been used to study the electromagnetic properties of the nucleons [17] and the magnetic moments of the nucleon octet[18] in reasonable agreement with experimental data. In view of the success of such a chiral potential model, we are very much interested to employ it in this work to study the quark-pion coupling constant and ground state baryon masses taking into account the corrections due to : (i) the energy associated with the center-of-mass motion (ii) the pionic self energy of the baryons arising out of the baryon-pion coupling at the vertex (iii) the colour-electric and -magnetic energy arising out of the residual one gluon-exchange interaction. In fact the relativistic logarithmic potential has already been applied successfully in the study of the baryon mass spectra [19] taking into account the contributions of the Goldstone boson exchange (GBE) interactions between the constituent quarks of the baryons over and above the center-of-mass correction. But in these studies the effects of residual one-gluon-exchange (OGE) interactions were not taken into account assuming these to be not so much significant for the electromagnetic properties of baryons. These, however, play an important role in providing color-electrostatic and magneto-static energies to the quark core. Therefore we wish to incorporate the contributions due to the residual OGE interactions between the constituent quarks in this model to study the masses of low-lying baryons in the ground state.

We follow here the usual procedure to incorporate chiral symmetry in the framework of this potential model. As a consequence of preserving the chiral symmetry in the model, there appears a possible residual interaction of the core-quarks with the surrounding elementary pions. This residual interaction must provide additional contributions to the baryonic properties over and above the core contributions. The quark-pion interaction Lagrangian density $\mathcal{L}_I^\pi$ in this model may be found to be proportional to the scalar part of the confining potential which in a way decides the strength of the quark-pion coupling. Therefore, our main objective in this work is to determine the quark-pion coupling constant and subsequently we are also interested to study the effect of quark-pion interaction on physical masses of ground-state baryons together with that of OGE interaction and center-of-mass motion in the present model.

This work is organized as follows. In section 2 we outline the potential model with the solutions for the relativistic bound states of the individually confined quarks in the ground states of baryons. Section 3 provides a brief account of the energy correction due to the spurious center-of-mass motion. We also outline the usual procedure of incorporating the chiral symmetry in the (u-d)-flavour sector only with quark-pion interaction term in the Lagrangian density taken in the linear form. Obtaining the general baryon-pion vertex function in the (u,d,s) sector, we calculate quark-pion coupling constant and the pionic energy for various baryon intermediate states contributing to the physical mass spectrum. This section also provides an account of a further correction to baryon mass due to colour-electric and -magnetic interaction, treated perturbatively. Finally, in section 4 we present the estimation of quark-pion coupling constant which are consistent with that made earlier by others [20]. This section also provides the results for the ground state baryon masses, which are in very good agreement with the corresponding experimental values with a reasonable choice of the quark-gluon coupling constant α_c which is consistent with the idea of treating the one-gluon-exchange in the baryon core in low-order perturbation theory.

2. Basic Formalism

Leaving behind for the moment, the quark-pion interaction in the (u,d)- flavour sector required to preserve chiral symmetry and the quark-gluon interaction originating from one-gluon-exchange at short distances as residual interactions to be treated perturbatively, we begin with the confinement part of the interaction which is believed to be dominant in baryonic dimensions. This particular part of the interaction which is believed to be determined by the multigluon mechanism is impossible to calculate theoretically from the first principles. Therefore, from a phenomenological point of view we assume that the constituent quarks in a baryon core are independently confined by an average flavour-independent potential of the form given in eqn. (1). To a first approximation the confining part of the interaction arising out of the non-perturbative multigluon mechanism including gluon self couplings is believed to provide the zeroth-order quark dynamics inside the baryon core through the quark Lagrangian density in zeroth order.

$$\mathcal{L}_q^0(x) = \overline{\psi}_q(x)\left[\frac{i}{2}\gamma^\mu \overrightarrow{\partial}_\mu - m_q - V_q(r)\right]\psi_q(x) \tag{3}$$

where m_q is the mass of the constituent quark.

Assuming that all quarks in a baryon core are in their ground $1S_{1/2}$ state, the normalized quark wave function $\psi_q(\vec{r})$ satisfying the Dirac equation derivable from $\mathcal{L}_q^0(x)$ as

$$\left[\gamma^0 E_q - \vec{\gamma}\cdot\vec{p} - m_q - V_q(r)\right]\psi_q(\vec{r}) = 0 \tag{4}$$

can be written as a two component form as

$$\psi_q(\vec{r}) = N_q \begin{pmatrix} \Phi_q(\vec{r}) \\ \frac{\vec{\sigma}\cdot\vec{p}}{\lambda_q}\Phi_q(\vec{r}) \end{pmatrix} \chi\uparrow \tag{5}$$

where $\lambda_q = E_q + m_q$, and

$$\Phi_q(\vec{r}) = A_q \frac{f_q(r)}{r} Y_o^0(\theta,\phi) \tag{6}$$

is the normalized radial angular part of $\psi_q(\vec{r})$ with normalization constant A_q. Here N_q is the overall normalization constant of $\psi_q(\vec{r})$ which can be easily obtained as:

$$N_q^2 = \frac{\lambda_q}{2\left[E_q - a <<\ln(r/b)>>_q\right]} \tag{7}$$

where $<<\ln(r/b)>>_q$ is the expectation value with respect to $\Phi_q(\vec{r})$. Finally the reduced radial part $f_q(r)$ can be found to satisfy a Schrodinger-type equation

$$f_q''(r) + \lambda_q \left[E_q - m_q - 2a\ln(r/b) \right] f_q(r) = 0 \tag{8}$$

which can be transformed into a convenient dimensionless form

$$f_q''(\rho) + \left[\varepsilon_q - \ln\rho \right] f_q(\rho) = 0 \tag{9}$$

where $\rho = r/r_{0q}$ is a dimensionless variable with $r_{0q} = (2a\lambda_q)^{-1/2}$ and

$$\varepsilon_q = \left[\frac{E_q' - m_q' + a\ln(2a\lambda_q)}{2a} \right] \tag{10}$$

where $E_q' = E_q - V_0, m_q' = m_q + V_0, \lambda_q = E_q + m_q = E_q' + m_q'$ and $V_0 = -a\log b$.

eqn. (9) provides the basic eigen value equation whose solution by a standard numerical method would give $\varepsilon_q = 1.0443$.

Now the ground state individual quark binding energy E_q is obtainable from energy eigen value condition(10), through the relation

$$E_q' = m_q' - a\ln a + ax_q \tag{11}$$

where x_q is the solution of the root equation

$$x_q + \ln\left[2a\left(x_q + \frac{2m_q' - a\log a}{a} \right) \right] = 2\varepsilon_q \tag{12}$$

In this independent quark model approach mass of the baryon core M_B^0 in zeroth order immediately follows from the ground state individual quark binding energy $E_q = E_q' + V_0$ as

$$M_B^0 = E_B = \sum_q E_q \tag{13}$$

3. Energy Corrections to Baryon Masses

The contribution of quark-binding energy to the mass M_B^0 of the baryon core given in eqn. (13) needs corrections due to center of mass motion, pionic interactions between the constituent quarks and quark gluon interactions due to OGE which need to be calculated separately for obtaining physical masses of the ground state baryons.

3.1. Center-of-Mass Correction

In the present model there would be sizeable spurious contribution to the energy E_q from the motion of the center of mass of the three quark system. If this aspect is not duly accounted for, the concept of the independent motion of the quarks inside the baryon core will not lead to a physical baryon state of definite momentum. Although there is some controversy on this subject, we follow here the prescription followed by Peierls-Yoccoz and other workers[21], which is just one way of accounting for the center of mass motion. Following their prescription, a ready estimate of the center of mass momentum $\vec{P}_B$ can be obtained as

$$<\vec{P}_B^2> = \sum_q <\vec{p}^2>_q \tag{14}$$

where $<\vec{p}^2>_q$ is the average value of the square of the individual quark momentum taken over $1S_{1/2}$ single quark states and is given in this model as

$$<\vec{p}^2>_q = \frac{a(E'_q + m'_q)(E'_q + m'_q + 3a)}{(E'_q + m'_q + a)} \tag{15}$$

In the same way the expression for the physical mass M_B of the bare baryon core can be obtained as

$$<M_B^2/E_B^2> = \left[1 - \sum_q (<\vec{p}^2>_q /E_B^2)\right] \tag{16}$$

which provides the energy correction due to center of mass motion to the baryon mass in eqn. (13) as

$$(\triangle E_B)_{cm} = (M_B - M_B^0) = \left[(E_B^2 - \sum_q <\vec{p}^2>_q)^{\frac{1}{2}} - E_B\right] \tag{17}$$

3.2. Chiral Symmetry Restoration

Looking at the zeroth order Lagrangian density $\mathcal{L}_q^0(x)$ described in section 2, one can note that under a global infinitesimal chiral transformation at least in the (u,d) flavour sector the axial vector current of quarks is not conserved due to the fact that the scalar terms $\mathcal{L}_q^0(x)$, which is proportional to $G(r) = [m_q + V(r)]$, is chirally odd. Of course the vector part of the potential possess no problem in this respect. In order to restore the chiral SU(2) X SU(2) symmetry within the partial conservation of axial vector current (PCAC) limit, one can introduce in the usual manner an elementary pion field $\phi(x)$ of small but finite mass $m_\pi = 140 MeV$ through additional terms in the original Lagrangian density $\mathcal{L}_q^0(x)$ so as to write

$$\mathcal{L}(x) = \mathcal{L}_q^0(x) + \mathcal{L}_\pi^0(x) + \mathcal{L}_I^\pi(x) \tag{18}$$

where $\mathcal{L}_\pi^0(x) = \frac{1}{2}(\partial_\mu \phi)^2 - \frac{1}{2}m_\pi^2(\phi)^2$

The Lagrangian density $\mathcal{L}_I^\pi(x)$ corresponding to the quark-pion interaction is taken to be linear in the iso-vector pion field $\phi(x)$ such that

$$\mathcal{L}_I^\pi(x) = \frac{1}{f_\pi} G(r) \overline{\psi}_q(x) \gamma^5 (\vec{\tau} \cdot \vec{\phi}) \psi_q(x) \tag{19}$$

where $f_\pi = 93 MeV$ is the phenomenological pion decay constant. Then the four divergence of the total axial vector current becomes

$$\partial_\mu A^\mu(x) = -f_\pi m_\pi^2 \phi(x) \tag{20}$$

yielding usual PCAC relation.

Restoration of chiral symmetry in this manner in the PCAC-limit brings to light the possible quark-pion interaction over and above the dominant confining interaction arising out of the non-perturbative multigluon mechanisms. This additional residual interaction between core quarks and the surrounding pions may be treated as a small perturbation over the solution obtained due to the dominant confining part mainly responsible for assembling the quarks in the baryon core. Consequently such pion coupling of the non-strange quarks would give rise to pionic self energy of the baryons which would ultimately contribute to the physical masses of the baryons. Although this consideration can be generalized to include the strange flavour sector for a chiral $SU(3) \otimes SU(3)$ symmetry, we would ignore it because of the large mass of the kaon involved in the process.

3.2.1. Vertex Function

The effects of pion coupling can be studied through the usual perturbative approach [12] with the baryon Hamiltonian constructed in the subspace of the non-exotic colour-singlet baryons. In doing so, one usually divides the space of baryon number one into two subspaces as $(\hat{P} + \hat{Q})$, when

$$\hat{P} = \sum_B |B><B| \tag{21}$$

and $\hat{Q} = (1 - \hat{P})$
where $\sum_B$ is over non-exotic baryons

That is $\hat{P}$ is a projection operator for the non-exotic core states such as N, Δ, R etc. where as $\hat{Q}$ corresponds to the exotic states. The inclusion of corrections arising from the coupling to the $\hat{Q}$-space is formally equivalent to evaluating the lowest order sea quark corrections which are shown numerically to be rather small[24] in bag models. Therefore for the present purposes, we shall neglect the off diagonal terms connecting $\hat{P}$ and $\hat{Q}$. In that case one can construct the core Hamiltonian $\widetilde{H}^0$ in the non-exotic baryon subspace out of H^0 obtained canonically from $\mathcal{L}_q^0(x)$. If one considers the colour singlet non-exotic baryon core state $|B>$ with mass M_B^0 to be eigen states of

$$H^0 = \int d^3\vec{x}\, \bar{\psi}(x) \left[-i\vec{\gamma}.\vec{\nabla} - m_q - V_q(r) \right] \psi(x) \tag{22}$$

then one can write

$$\widetilde{H}^0 = \hat{P} H^0 \hat{P} = \sum_B |B><B| M_B^0 \tag{23}$$

which in terms of conventional second quantized language with $\hat{b}_B^+$ creating a three quark baryon-core state with quantum numbers of N, Δ etc can be

$$\widetilde{H}^0 = \sum_B \hat{b}_B^+ \hat{b}_B M_B^0 \tag{24}$$

Again with the quantized free pion field $\vec{\Phi}(x)$ as

$$\vec{\Phi}_j(x) = \frac{1}{(2\pi)^{3/2}} \int d^3\vec{k}\, (2\omega_k)^{-1/2} \left[\hat{a}_{k,j} e^{i\vec{k}.\vec{x}} + \hat{a}_{k,j}^+ e^{-i\vec{k}.\vec{x}} \right] \tag{25}$$

with $\hat{a}_{k,j}$ and $\hat{a}^{\dagger}_{k,j}$ as the pion destruction and creation operators respectively and $\omega_k = \sqrt{\vec{k}^2 + m_\pi^2}$ as the pion-energy, the Hamiltonian for the free pion field becomes

$$\widetilde{H}_\pi = \sum \int d^3\vec{k}\, \omega_k\, \hat{a}^+_{k,j}\, \hat{a}_{k,j} \tag{26}$$

Finally the interaction Hamiltonian corresponding to $L_I^\pi(x)$ in the already mentioned non-exotic baryon subspace can be written as

$$\widetilde{H}_I^\pi = \hat{P} H_I^\pi \hat{P} = \frac{-i}{f_\pi} \sum_{B,B'} \int d^3\vec{x} <B'| \sum_{q=u,d} \bar{\psi}(x)$$
$$\times\, (\vec{\tau}^q \cdot \vec{\Phi}) \gamma^5 \psi(x) |B> G(r) \hat{b}_B^+ \hat{b}_{B'} \tag{27}$$

Using the expansion (25) for the pion field eqn. (27) becomes

$$\widetilde{H}_I^\pi = \frac{-1}{(2\pi)^{3/2}} \cdot \sum_{B,B',j} \int d^3\vec{k} \left[V_j^{B'B}(\vec{k}) \hat{b}_B^+ \hat{b}_B \hat{a}_{k,j} + h.c \right] \tag{28}$$

where h.c denotes hermitian conjugate and $V_j^{BB'}(\vec{k})$ is the baryon-pion absorption vertex function obtained in the point pion approximation as

$$V_j^{B'B}(\vec{k}) = \frac{-i}{f_\pi}(2\omega_k)^{-1/2} \int d^3\vec{x}\, G(|\vec{x}|) e^{i\vec{k}\cdot\vec{x}} <B'|.\sum_q \bar{\psi}(x)\gamma^5\psi(x)\tau_j^q|B> \tag{29}$$

Here j is the isopin index. Assuming that for the $BB'\pi$- vertex, the spatial orbits of all the quarks in the initial and final baryon states are the same $1S_{1/2}$ ones, one can use expression (5) to write eqn. (29) in a more explicit form as,

$$V_j^{B'B}(\vec{k}) = -\frac{i}{f_\pi}(2\omega_k)^{-1/2} \int d^3(\vec{r}) G(r) e^{i\vec{k}\cdot\vec{r}} <B'|\sum_q \bar{\psi}(\vec{r})\gamma^5\psi(\vec{r})\tau_j^q|B>$$
$$= \frac{-i}{f_\pi}(2\omega_k)^{-1/2}\left(\frac{N_q^2}{\lambda_q}\right) I(k) <B'|\sum_q(\vec{\sigma}_q.\vec{k})\tau_j^q|B> \tag{30}$$

where

$$I(k) = \frac{8\pi}{k} \int dr\, r^2 G(r) j_1(|\vec{k}|r) \Phi_q(r) \Phi'_q(r) \tag{31}$$

with $j_1(|\vec{k}|r)$ = spherical Bessel function of first order. Now with the evaluation of $I(k)$, the expression (30) can be obtained as

$$V_j^{B'B}(k) = \frac{i}{2f_\pi}(2\omega_k)^{-1/2}\left[\frac{3}{5}g_A u(k)\right] <B'|\sum_q(\vec{\sigma}_q.\vec{k})\tau_j^q|B> \tag{32}$$

where g_A is the axial vector coupling constant for nucleon core given as

$$g_A = \frac{5}{9}\left[\frac{3(E'_q + m'_q) - a}{E'_q + m'_q + a}\right] \tag{33}$$

Here $u(k)$ is the vertex form factor which comes out as

$$u(k) = \frac{10N_q^2}{3\lambda_q g_\Lambda}\left[m_q' <<j_0(|\vec{k}|r)>> \right.$$
$$\left. + a <<\ln(\frac{r}{b})j_0(|\vec{k}|r)>> +a<<\frac{j_1(|\vec{k}|r)}{(|\vec{k}|r)}>>\right] \quad (34)$$

where $j_0(|\vec{k}|r)$ and $j_1(|\vec{k}|r)$ are the spherical Bessel functions of order zero and one respectively. The angular brackets appearing in eqn. (34) are the expectation values with respect to $\Phi(\vec{r})$. The vertex form factor $u(k)$ reduces to one for $|\vec{k}| \to 0$ as expected.

Considering in particular the $NN\pi$ vertex with the axial vector coupling constant g_A for the nucleon-core as found in eqn. (33), one can have

$$V_j^{NN}(\vec{k}) = \frac{i}{2f_\pi}(2\omega_k)^{(-1/2)} g_A u(k)\left[(\vec{\sigma}^{NN}.\vec{k})\tau_j^{NN}\right] \quad (35)$$

Using the familiar Goldberger-Treimann relation which establishes a connection between the pseudo-vector nucleon-pion coupling $f_{NN\pi}$ and axial vector coupling g_A through expression,

$$\sqrt{4\pi}\left(\frac{f_{NN\pi}}{m_\pi}\right) = \frac{g_A}{2f_\pi} \quad (36)$$

we can have

$$V_j^{NN}(\vec{k}) = i\sqrt{4\pi}\left(\frac{f_{NN\pi}}{m_\pi}\right)(2\omega_k)^{-1/2}ku(k)\left[(\vec{\sigma}^{NN}.\hat{k})\tau_j^{NN}\right] \quad (37)$$

In the same manner, the general baryon-pion vertex function can be written as

$$V_j^{BB'}(\vec{k}) = i\sqrt{4\pi}\left(\frac{f_{BB'\pi}}{m_\pi}\right)(2\omega_k)^{-1/2}ku(k)\left[(\vec{\sigma}^{BB'}.\hat{k})\tau_j^{BB'}\right] \quad (38)$$

which can enable one to calculate the effects of pion coupling.

3.2.2. Quark-Pion Coupling Constant

We would like to study mainly the coupling of quarks in a nucleon to pions in the present chiral symmetric potential model. Therefore in the view of the fact that chiral $SU(2) \otimes SU(2)$ is experimentally found to be an excellent symmetry of strong interaction having its physical realization in pion with its small mass as the corresponding Goldstone boson. We concentrate our discussion mainly in the (u,d) flavour sector only. Then as a first step in this direction, let us assume that the interaction Lagrangian density in eqn.(19) can be written effectively

$$\mathcal{L}_I^\pi = -iG_{qq\pi}\overline{\psi}(x)\gamma^5\left(\vec{\tau}\cdot\vec{\phi}\right)\psi(x) \quad (39)$$

With $G_{qq\pi}$ is the effective quark-pion coupling strength. Then in a classical field approximation, taking the emitted pion field ϕ_j in the process $q \longrightarrow q+\pi$ as a plane wave with momentum k, we can write the interaction Hamiltonian as,

$$H_{int} \simeq iG_{qq\pi}\int d^3r\overline{\psi}(\vec{r})\gamma^5\psi(\vec{r})e^{(i\vec{k}\cdot\vec{r})}\tau_j \quad (40)$$

Again we can similarly obtain the interaction from $L_I^\pi(x)$ in eqn.(19) as

$$H_{int} \simeq \frac{i}{f_\pi} \int d^3 r \overline{\psi}(\vec{r}) \gamma^5 \psi(\vec{r}) G(r) e^{(i\vec{k}\cdot\vec{r})} \tau_j \qquad (41)$$

where $G(r) = (m_q + V(r))$ essentially decides the strength of quark-pion coupling. Then comparing (40) and (41) we can obtain a much simpler estimate of $G_{qq\pi}$ as

$$G_{qq\pi} = \frac{1}{f_\pi} \frac{\int d^3 r G(r) \overline{\psi}(\vec{r}) \gamma^5 \psi(\vec{r}) e^{(i\vec{k}\cdot\vec{r})}}{\int d^3 r \overline{\psi}(\vec{r}) \gamma^5 \psi(\vec{r}) e^{(i\vec{k}\cdot\vec{r})}} \qquad (42)$$

Now taking the $1S_{1/2}$ spatial wave functions of the quarks as given in eqn. (5), we obtain

$$G_{qq\pi} = \frac{1}{f_\pi}\left[m'_q - V_0 + a <<\log(r/b) j_0(|\vec{k}|r)>> + \frac{<< j_1(|\vec{k}|r)/(|\vec{k}|r) >>}{<< j_0(|\vec{k}|r) >>} \right] \qquad (43)$$

Where $j_0(|\vec{k}|r)$ and $j_1(|\vec{k}|r)$ are the spherical Bessel function of order zero and one respectively. The angular brackets appearing in (43) are the expectation values with respect to $\phi_q(\vec{r})$.

Then with a soft pion approximation we can approximate

$$G_{qq\pi} \simeq \frac{1}{3f_\pi}\left[a + 3(m'_q - V_0) + 3a <<\log(r/b)>> \right] \qquad (44)$$

The expectation value of $<<\log(r/b)>>$ with respect to $\phi_q(\vec{r})$ can be easily obtained as [15]

$$<<\log(r/b)>>_q = \left(\frac{E'_q - m'_q + 2V_0 - a}{2a} \right) \qquad (45)$$

with which the eqn. (44) can be expressed as

$$G_{qq\pi} \simeq \frac{(3E'_q + 3m'_q - a)}{6 f_\pi} \qquad (46)$$

A better estimate of the quark-pion coupling constant can be obtained in a more reasonable way by looking at $NN\pi$-vertex. The $NN\pi$-vertex function, in a point pion approximation, can be obtained from eqn.(32) in the form

$$\begin{aligned} V_j^{N'N}(\vec{k}) &= \frac{i}{2f_\pi}(2\omega_k)^{-\frac{1}{2}}\left[\frac{3}{5}g_A u(k)\right] < N' | \sum_q (\vec{\sigma}_q \cdot \vec{k}) \tau_j^q | N > \\ &= < N' | \sum_q V_j^{qq}(\vec{k}) | N > \end{aligned} \qquad (47)$$

where $V_j^{qq}(\vec{k})$ is the quark-pion vertex operator function given as

$$V_j^{qq}(\vec{k}) = \frac{i}{2f_\pi} \sqrt{\frac{1}{2\omega_k}} \left[\frac{3}{5} g_A u(k)\right] (\vec{\sigma}_q \cdot \vec{k}) \tau_j^q \qquad (48)$$

Now comparing eqn.(48) with the corresponding expression in Chew-Low model [23], which is written in terms of Pseudo-vector $qq\pi$-coupling $f_{qq\pi}$ as

$$V_j^{qq}(\vec{k}) = i\sqrt{\frac{1}{2\omega_k}}\sqrt{4\pi}\left(\frac{f_{qq\pi}}{m_\pi}\right)u(k)(\vec{\sigma}_q \cdot \vec{k})\tau_j^q \qquad (49)$$

We have

$$\sqrt{4\pi}\left(\frac{f_{qq\pi}}{m_\pi}\right) = \frac{1}{2f_\pi}\left[\frac{3}{5}g_A\right] \qquad (50)$$

This is the equivalent Goldberger-Treimann relation which with the familiar equivalence of pseudo-scalar and pseudo-vector coupling constants yields,

$$\left(\frac{G_{qq\pi}}{2\overline{m}_q}\right) = \sqrt{4\pi}\left(\frac{f_{qq\pi}}{m_\pi}\right) = \frac{1}{2f_\pi}\left[\frac{3}{5}g_A\right] \qquad (51)$$

where $\overline{m}_q$ is the effective constituent quark mass taken as one third of the $N-\Delta$ spin-isospin average mass(i.e $\overline{m}_q = 0.391\ GeV$). Then we have

$$\frac{G_{qq\pi}^2}{4\pi} = \frac{i}{4\pi}\left(\frac{\overline{m}_q}{f_\pi}\right)^2\left[\frac{3g_A}{5}\right]^2 \qquad (52)$$

3.2.3. Pionic Self-Energy Contribution

The coupling of the pion field to the non-strange quarks in a minimal way, as given by single loop self-energy diagram shown in fig.1 causes a shift in the energy of the baryon core and thereby gives rise to pionic self-energy of baryons which would ultimately contribute to the physical masses of the baryons. This aspect can be studied in usual perturbative approach. From the second order perturbation theory, the pionic self-energy is usually obtained as

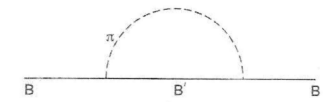

Figure 1. Baryon self-energy due to coupling with pion.

$$\Sigma_B(E_B) = \sum_k \sum_{B'} \frac{V_j^{\dagger BB'} V_j^{BB'}}{(E_B - \omega_k - M_{B'}^0)} \qquad (53)$$

where $\Sigma_k = \Sigma_j \int \frac{d^3\vec{k}}{(2\pi)^3}$. Here, j corresponds to the pion-isospin index and B' is the intermediate baryon state. $V^{BB'}(\vec{k})$ is the general baryon-pion absorption vertex function obtained in this model and is given in eqn. (38)

Treating the possible pionic, gluonic and c.m motion corrections independently at quark-core level $(N,\Delta), (\Lambda, \Sigma, \Sigma^*)$ and (Ξ, Ξ^*) would be separately mass degenerate. Thus for degenerate intermediate states on mass shell with $M_B^0 = M_{B'}^0$, the self-energy correction becomes,

$$(\delta M_B)_\pi = \Sigma_B(E_B^0 = M_B^0 = M_{B'}^0) = -\sum_{k,B'} \frac{V_j^{\dagger BB'} V_j^{BB'}}{\omega_k} \quad (54)$$

Now using eqn. (38) we find

$$(\delta M_B)_\pi = \frac{-I_\pi}{3} \sum_{B'} C_{BB'} f_{BB'\pi}^2 \quad (55)$$

where

$$C_{BB'} = \left(\vec{\sigma}^{BB'} \cdot \vec{\sigma}^{B'B}\right)\left(\vec{\tau}^{BB'} \cdot \vec{\tau}^{B'B}\right) = \left\langle B' \left| \sum_{q,q'} (\vec{\sigma}_q \cdot \vec{\sigma}_{q'})(\vec{\tau}_q \cdot \vec{\tau}_{q'}) \right| B \right\rangle$$

and

$$I_\pi = \frac{1}{\pi m_\pi^2} \int_0^\infty \frac{dk k^4 u^2(k)}{\omega_k^2} \quad (56)$$

where $u(k)$ is given by eqn. (34).

For intermediate baryon states B', we consider only the octet and decuplet ground states. Using the values $f_{BB'\pi}$ and $C_{BB'}$ as summarized in table-1 according to Ref. 1, the pionic self-energy for different baryons can be computed as

$$(\delta M_N)_\pi = \frac{-171}{25} f_{NN\pi}^2 I_\pi$$

$$(\delta M_\Delta)_\pi = \frac{-99}{25} f_{NN\pi}^2 I_\pi$$

$$(\delta M_\Lambda)_\pi = \frac{-108}{25} f_{NN\pi}^2 I_\pi$$

$$(\delta M_\Sigma)_\pi = (\delta M_{\Sigma^*})_\pi = \frac{-12}{5} f_{NN\pi}^2 I_\pi$$

$$(\delta M_\Xi)_\pi = (\delta M_{\Xi^*})_\pi = \frac{-27}{25} f_{NN\pi}^2 I_\pi \quad (57)$$

and of course, $(\delta M_{\Omega^-})_\pi = 0$ since the strange quarks in Ω^- have no interaction with the pion. The self energy $(\delta M_B)_\pi$ calculated here contains both the quark self-energy Fig. 2(a) and 2(b).

Although the Kaon effects arising in the strange flavour sector are not included in this model, we find from eqn.(57) that the mass splitting of the nucleon N differs from that of the strange hyperons due to the coupling of non-strange quarks to pion. For example the

Quark-Pion Coupling Constant and Ground-State Baryon Masses ...

Table 1. Baryon-pion coupling constant, the spin-isospin reduced matrix elements and the pionic self-energy $(\delta M_{BB'\pi})$ (in GeV) for various baryons with appropriate intermediate baryon states together with the total pionic self-energy $(\delta M_B)_\pi$

Baryon B	Baryon intermediate baryon state BB'	$\frac{f_{BB'\pi}}{f_{NN\pi}}$	$\sigma^{BB'} \cdot \sigma^{BB'}$	$\tau^{BB'} \cdot \tau^{BB'}$	$C_{BB'}$	$\delta M_{BB'\pi}$	Total $(\delta M_B)_\pi$
N	NN	1	3	3	9	-0.0689	-0.1571
	NΔ	$(6\sqrt{2})/5$	2	2	4	-0.0882	
Δ	$\Delta\Delta$	1/5	15	15	225	-0.0689	-0.0910
	ΔN	$(6\sqrt{2})/5$	1	1	1	-0.0221	
Λ	$\Lambda\Lambda$	0	3	0	0		
	$\Lambda\Sigma$	$(-2\sqrt{3})/5$	3	3	9	-0.0331	-0.0992
	$\Lambda\Sigma^*$	$-6/5$	2	3	6	-0.0661	
Σ	$\Sigma\Sigma$	4/5	3	2	6	-0.0294	
	$\Sigma\Lambda$	$(-2\sqrt{3})/5$	3	1	3	-0.0110	-0.0551
	$\Sigma\Sigma^*$	$(-2\sqrt{3})/5$	2	2	4	-0.0147	
Σ^*	$\Sigma^*\Sigma^*$	2/5	15	2	30	-0.0368	
	$\Sigma^*\Lambda$	$-6/5$	1	1	1	-0.0110	-0.0551
	$\Sigma^*\Sigma$	$(-2\sqrt{3})/5$	1	2	2	-0.0073	
Ξ	$\Xi\Xi$	$-1/5$	3	3	9	-0.0028	-0.0248
	$\Xi\Xi^*$	$(-2\sqrt{3})/5$	2	3	6	-0.0220	
Ξ^*	$\Xi^*\Xi^*$	1/5	15	3	45	-0.0138	-0.0248
	$\Xi^*\Xi$	$(-2\sqrt{3})/5$	1	3	3	-0.0110	

mass splitting $(\delta M_B)_\pi$ of N and Ξ differ because of the fact that in the case of nucleon N all of the three non-strange quarks constituting it have interaction with the pion whereas in case of Ξ baryon only one non-strange quark in it interacts with the pion, but other two constituent strange quarks don't. Further, on the similar ground the mass splittings $(\delta M_B)_\pi$ of N and Σ also differ.

The bare pseudo-vector nucleon-pion coupling constant $f_{NN\pi}$ can be computed from the usual relation [24]

$$\sqrt{4\pi}\left(\frac{f_{NN\pi}}{m_\pi}\right) = \left(\frac{g_{NN\pi}}{2M_p}\right) \tag{58}$$

where $g_{NN\pi}$ is a pseudoscalar nucleon-pion coupling constant defined as $g_{NN\pi} = G_{NN\pi}(q^2 = -m_\pi^2)$, $G_{NN\pi}$ being the nucleon-pion form factor given by

$$G_{NN\pi}(q^2) = \left(\frac{M_p}{f_\pi}\right) g_A u(q^2) \tag{59}$$

Here M_P is the mass of the proton and g_A given by eqn. (33).

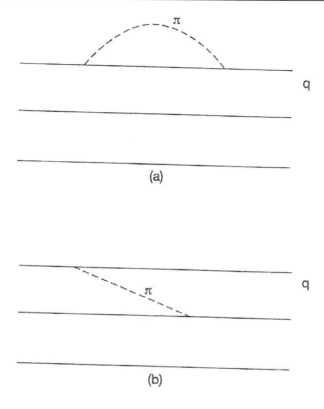

Figure 2. One-pion-exchange contributions due to energy.

The self energy $(\delta M_B)_\pi$ calculated here contains both the quark self energy (fig 2a) and the one pion exchange contributions (fig 2b). It must be noted here that this method ignores to a large extent, the short range part of the pion exchange interaction which is of crucial importance for splittings. Only when the complete infinite set of all radially excited intermediate states B' is taken into account, this method could be adequate [25].

3.3. One Gluon Exchange Correction

The individual quarks in a baryon core are considered so far to be experiencing only the force coming from the average effective potential $V_q(r)$ in eqn. (1), which is assumed to provide a suitable phenomenological description of the non-perturbative gluon interaction including gluon self couplings. All that remains inside the quark core is the hopefully weak one gluon exchange interaction provided by the interaction Lagrangian density

$$\mathcal{L}_I^g(x) = \sum_{\alpha=1}^{8} J_i^{\mu\alpha}(x) A_\mu^\alpha(x) \qquad (60)$$

where $A_\mu^\alpha(x)$ are the 8-vector gluon fields and $J_i^{\mu\alpha}(x)$ is the ith quark colour current. Since at small distances the quarks should be almost free, it is reasonable to calculate the energy shift in the mass spectrum arising out of the quark interaction energy due to their coupling to the coloured gluons, using a first order perturbation theory.

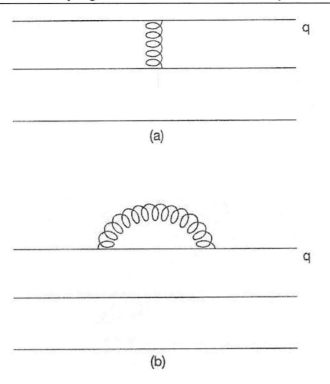

Figure 3. One-gluon-exchange contribution to the energy.

If we only keep the terms of order α_c, the problem reduces to evaluating the diagrams shown in figure 3, where figure 3(a) corresponds to OGE part, while figure 3(b) implies the quark self energy that normally contributes to renormalization of quark masses. If $\vec{E}_i^{\alpha}$ and $\vec{B}_i^{\alpha}$ are the colour-electric and colour-magnetic fields, respectively, generated by the ith quark colour-current

$$J_i^{\mu\alpha}(x) = g_c \bar{\psi}_i(x) \gamma^{\mu} \lambda_i^{\alpha} \psi_i(x) \qquad (61)$$

with λ_i^{α} being the usual Gell-Mann SU(3)- matrices and $\alpha_c = (g_c^2/4\pi)$, then the contribution to the mass due to the relevant diagram can be written as a sum of the colour-electric and colour-magnetic parts as

$$(\triangle E_B)_g = (\triangle E_B)_g^e + (\triangle E_B)_g^m \qquad (62)$$

where

$$(\triangle E_B)_g^e = \frac{1}{8\pi} \sum_{i,j} \sum_{\alpha=1}^{8} \int \int \frac{d^3\vec{r}_i d^3\vec{r}_j}{|\vec{r}_i - \vec{r}_j|} <B|J_i^{o\alpha}(\vec{r}_i) J_j^{o\alpha}(\vec{r}_j)|B> \qquad (63)$$

$$(\triangle E_B)_g^m = \frac{-1}{4\pi} \sum_{i<j} \sum_{\alpha=1}^{8} \int \int \frac{d^3\vec{r}_i d^3\vec{r}_j}{|\vec{r}_i - \vec{r}_j|} <B|\vec{J}_i^{\alpha}(\vec{r}_i) . \vec{J}_j^{\alpha}(\vec{r}_j)|B> \qquad (64)$$

We have not included the self energy diagram in the calculation of the magnetic part of the interaction, which contributes to renormalization of quark masses and can possibly be accounted for in the phenomenological quark masses. However, in principle, the difference between this quark mass renormalization and same effect on a free quark should be included in this model but this cannot be done since there are no free quarks. One of the possibilities for exclusion of this diagram requires that each $\vec{B}_i^\alpha$ should satisfy the boundary condition $\hat{r} \times \vec{B}_i^\alpha = 0$, separately at the edge of the confining region. On the other hand, as the electric field $\vec{E}_i^\alpha$ is in radial direction, the boundary condition is only satisfied if $\hat{r} \times (\sum_i \vec{E}_i^\alpha) = 0$ for a colour singlet state $|B>$ for which $\sum_i \lambda_i^\alpha = 0$. Therefore, in order to preserve boundary conditions we are forced to take into account the self energy diagrams in fig 3 in the calculation of electric part only. Now using eqn. (5) in eqn. (29) we find

$$J_i^{0\alpha}(r_i) = g_c \lambda_i^\alpha N_i^2 \left[\Phi^2(r_i) + \frac{\Phi'^2(r_i)}{\lambda_i^2} \right]$$

$$J_i^\alpha(r_i) = -2g_c \lambda_i^\alpha N_i^2 \left[\frac{(\vec{\sigma}_i \times \hat{r}_i)\Phi(r_i)\Phi'(r_i)}{\lambda_i} \right] \tag{65}$$

Again using eqn. (33) together with the identity

$$\frac{1}{|\vec{r}_i - \vec{r}_j|} = \frac{1}{2\pi^2} \int \frac{d^3\vec{k}}{k^2} \exp\left[-i\vec{k}\cdot(\vec{r}_i - \vec{r}_j) \right] \tag{66}$$

in eqns. (63) and (64) we obtain

$$(\triangle E_B)_g^e = \frac{\alpha_c}{4\pi^2} \sum_{i,j} \left\langle \sum_\alpha \lambda_i^\alpha \lambda_j^\alpha \right\rangle N_i^2 N_j^2 \int \frac{d^3\vec{k}}{k^2} F_i^e(k) F_j^e(k) \tag{67}$$

and

$$(\triangle E_B)_g^m = \frac{-2\alpha_c}{\pi^2} \sum_{i<j} \left\langle \sum_\alpha \lambda_i^\alpha \lambda_j^\alpha \right\rangle N_i^2 N_j^2 \int \frac{d^3\vec{k}}{k^2} \vec{F}_i^m(k)\cdot\vec{F}_j^m(k) \tag{68}$$

where

$$\begin{aligned}F_i^e(k) &= \frac{1}{2\lambda_i}\Big[\{4(E_i' + V_0) - k^2\} << j_0(|\vec{k}|r_i) >> \\ &\quad - 4a << \log(r_i/b) j_0(|\vec{k}|r_i) >> \Big]\end{aligned} \tag{69}$$

and

$$\vec{F}_i^m(k) = \frac{i}{2} << j_0(|\vec{k}|r_i) >> (\vec{\sigma}_i \times \vec{k}) \tag{70}$$

Then eqns. (67) and (68) can be written as

$$(\triangle E_B)_g^e = \frac{\alpha_c}{\pi} \sum_{i,j} \left\langle \sum_\alpha \lambda_i^\alpha \lambda_j^\alpha \right\rangle N_i^2 N_j^2 I_{ij}^e \tag{71}$$

$$(\triangle E_B)_g^m = -\frac{4\alpha_c}{3\pi} \sum_{i<j} <\lambda_i^\alpha \lambda_j^\alpha (\vec{\sigma}_i\cdot\vec{\sigma}_j)> \frac{N_i^2 N_j^2}{\lambda_i \lambda_j} I_{ij}^m \tag{72}$$

where

$$I_{ij}^e = \int_0^\infty dk\, F_i^e(k) F_j^e(k) \tag{73}$$

Table 2. The co-efficients a_{ij} and b_{ij} used in calculation of the colour-electric and colour-magnetic energy contributions due to one-gluon exchange

Baryons	a_{uu}	a_{us}	a_{ss}	b_{uu}	b_{us}	b_{ss}
N	0	0	0	-3	0	0
Δ	0	0	0	+3	0	0
Λ	1	-2	1	-3	0	0
Σ	1	-2	1	1	-4	0
Σ^*	1	-2	1	1	2	0
Ξ	1	-2	1	0	-4	1
Ξ^*	1	-2	1	0	2	1
Ω^-	0	0	0	0	0	+3

$$I_{ij}^m = \int_0^\infty dk k^2 <<j_o(|\vec{k}|r_i)>><<j_o(|\vec{k}|r_j)>> \quad (74)$$

Finally by taking into account the specific quark flavour and spin configuration in various ground state baryons and using the relations $<\sum_\alpha (\lambda_i^\alpha)^2> = \frac{16}{3}$ and $<\sum_\alpha \lambda_i^\alpha \lambda_j^\alpha>_{i \neq j} = -\frac{8}{3}$ for baryons, in general one can write the energy correction due to one gluon exchange as

$$(\triangle E_B)_g^e = \alpha_c(a_{uu}T_{uu}^e + a_{us}T_{us}^e + a_{ss}T_{ss}^e) \quad (75)$$

$$(\triangle E_B)_g^m = \alpha_c(b_{uu}T_{uu}^m + b_{us}T_{us}^m + b_{ss}T_{ss}^m) \quad (76)$$

where a_{ij} and b_{ij} are the numerical co-efficients depending on each baryon listed in table 2 and the terms $T_{ij}^{e,m}$ are

$$T_{ij}^e = \frac{16(E_i' + m_i')(E_j' + m_j')}{3\pi(E_i' + m_i' + a)(E_j' + m_j' + a)} I_{ij}^e \quad (77)$$

$$T_{ij}^m = \frac{32}{9\pi(E_i' + m_i' + a)(E_j' + m_j' + a)} I_{ij}^m \quad (78)$$

From table 3 one can observe that the colour electric contribution for baryon masses vanishes when all constituent quarks masses in a baryon are equal, where as it is non zero otherwise.

4. Results and Discussion

In the preceding section it has been shown that the zeroth order mass $M_B^0 = E_B$ of a ground state baryon arising out of the binding energies of constituent quarks, confined independently by a phenomenological average potential $V_q(r)$ which presumably represents the dominant non-perturbative multigluon interactions, must be subjected to certain corrections due to quark-pion interaction together with OGE interactions and necessary center of mass corrections. All of these corrections can be treated independently as they are of the same

Table 3. Energy Corrections and physical masses of the ground-state baryons (in GeV)

Baryon	E_B^0	$(\triangle E_B)_{cm}$	$(\triangle E_B)_g^e$	$(\triangle E_B)_g^m$	Total	$(\delta M_B)_\pi$	M_B Calculation	Expt.
N	1.3928	-0.1828	0	-0.1129	-0.1129	-0.1571	0.940	0.940
Δ	1.3928	-0.1828	0	0.1129	0.1129	-0.0910	1.232	1.232
Λ	1.5430	-0.1829	-0.0472	-0.1129	-0.1602	-0.0992	1.1007	1.116
Σ	1.5430	-0.1829	-0.0472	-0.0982	-0.1454	-0.0551	1.1595	1.193
Σ^*	1.5430	-0.1829	-0.0472	0.1055	0.0583	-0.0551	1.3632	1.385
Ξ	1.6931	-0.1830	-0.0472	-0.0959	-0.1431	-0.0248	1.3422	1.321
Ξ^*	1.6931	-0.1830	-0.0472	-0.1078	0.0605	-0.0248	1.5459	1.533
Ω	1.8433	-0.1831	0	0.1197	0.1197	0	1.7799	1.672

order of magnitude, so that the physical mass of a low-lying ground state baryon can be obtained as

$$M_B = E_B + (\triangle E_B)_{cm} + (\delta M_B)_\pi \\ + (\triangle E_B)_g^m + (\triangle E_B)_g^e \quad (79)$$

where $(\triangle E_B)_{cm}$ is the energy associated with spurious center of mass motion (eqn. (17)), $(\delta M_B)_\pi$ is the energy arising out of quark-pion interaction (eqn. (57)) and $[(\triangle E_B)_g^e + (\triangle E_B)_g^m]$ is the colour-electric and colour-magnetic interaction energies arising out of the residual one gluon exchange process (eqns. (75) and (76)).

For quantitative evaluation of the terms on the righthand side of eqn. (79), we initially assume the potential parameters a,b to be flavour-independent and take the quark masses as $m_u = m_d \neq m_s$. However, for convenience, we introduce a parameter $V_0 = -a \log b$ which is absorbed appropriately in m_q and E_q of eqn. (8) in order to obtain the solutions leading to individual quark binding energy in terms of $m'_q = m_q + V_0$ and $E'_q = E_q - V_0$ through eqn. (10). Therefore, the computation of the energy correction terms in eqn. (79) and hence the physical mass M_B of the ground state baryon is found to depend on the choice of the effective Lagrangian mass parameters $m'_q (m'_u = m'_d, m'_s)$ and the potential parameter 'a' alone.

In the Lagrangian formalism adopted here, we choose to fix m'_q as

$$(m'_u = m'_d, m'_s) = (0.139, 0.344) \; GeV \quad (80)$$

Then with a suitable choice of the potential parameter

$$a = 0.15 \; GeV \quad (81)$$

the energy eigenvalue condition (10) yields the individual quark effective binding energies

$$(E'_u = E'_d, E'_s) = (0.665, 0.816) \; GeV \quad (82)$$

Using the effective quark masses m'_q, potential parameter 'a' and the energy eigen values E'_q given in eqns. (80) - (82), the quark-pion coupling constant can be computed from eqn. (46) as

$$\frac{G^2_{qq\pi}}{4\pi} = 1.31 \tag{83}$$

which is comparable with the estimate obtained from the observations of Hendry [26] by examining the decay of N and Δ states. Again the eqn. (51) yields

$$\frac{G^2_{qq\pi}}{4\pi} = 0.879 \tag{84}$$

which is comparable with the estimate obtained by Suzuki Bhaduri [27].

The value of I_π in eqn. (56) is then evaluated using a standard numerical method. We find $I_\pi = 0.256\ GeV$, which along with the value of $f_{NN\pi} = 0.299$ enables us to compute energy correction $(\delta M_B)_\pi$ from eqn. (57). The values of $\delta M_{BB'\pi}$ obtained explicitly for various baryons are presented in table 1 along with the total pionic self-energy $(\delta M_B)_\pi$.

We then evaluate the integral expressions for $I_{ij}^{e,m}$ in eqns. (73) and (74) using a standard numerical method and calculate the terms $T_{ij}^{e,m}$ from eqns. (77) and (78), which are necessary for computing $(\triangle E_B)_g^{e,m}$. We find

$$(T_{uu}^e, T_{us}^e, T_{ss}^e) = (0.493, 0.647, 0.619)\ GeV \tag{85}$$

$$(T_{uu}^m, T_{us}^m, T_{ss}^m) = (0.146, 0.132, 0.155)\ GeV \tag{86}$$

Referring to the physical masses of N and Δ which are

$$M_\Delta = \left[E_N^2 - 3 <\vec{p}^2>_u\right]^{1/2} + (\delta M_\Delta)_\pi + 3\alpha_c T_{uu}^m \tag{87}$$

$$M_N = \left[E_N^2 - 3 <\vec{p}^2>_u\right]^{1/2} + (\delta M_N)_\pi - 3\alpha_c T_{uu}^m \tag{88}$$

we find the QCD splitting among the N and Δ masses as

$$6\alpha_c T_{uu}^M = (M_\Delta - M_N) - [(\delta M_\Delta)_\pi - (\delta M_N)_\pi] \tag{89}$$

since $(M_\Delta - M_N) \approx 0.292$ GeV and $[(\delta M_\Delta)_\pi - (\delta M_N)_\pi] \approx 0.066$ GeV, as seen from table 3, we find that $6\alpha_c T_{uu}^M = 0.226$ GeV. This gives $\alpha_c = 0.258$, which is found to be in conformity with the value suggested by QCD [28] and also not so much different from the value 0.3-0.4 obtained in the CBM [29]. However this value was obtained as 0.55 by De Grand et al [24]. It must be pointed out here that we donot need anywhere near as large value of α_c as in the original MIT work, where with out including pionic corrections the QCD splitting was equated with $(M_\Delta - M_N) \simeq 0.292$ GeV. Finally using the combination

$$[M_\Delta - (\delta M_\Delta)_\pi] + [M_N - (\delta M_N)_\pi] = 2\left[E_N^2 - 3<\vec{p}^2>_u\right]^{1/2} \tag{90}$$

we find $E_N = 3(E_u' + V_0)$ which enables us to fix the potential parameter V_0 independent of α_c at a value $V_0 = -0.201\ GeV$. This parameter again fixes the potential parameter 'b' at a value b=3.824 GeV^{-1}. It must be noted here that the value of the $NN\pi$-coupling constant $f_{NN\pi}$ which has been used in the evaluation of pionic corrections, is obtained from eqn. (58) with the help of eqn. (59) and (33). Using the values of m'_u and E'_u from eqns. (80) - (82) respectively in eqn. (33), we find $g_A = 1.317$, which yields from eqn. (58) $f_{NN\pi} = 0.299$ as against the experimental value 0.283. Now, using all the results thus obtained, one can calculate all the individual terms leading to the physical masses of various ground state baryons. The calculated values of the energy corrections and physical masses for the ground-state baryons considered here are presented in table 3. The physical masses of baryons such as $N, \Delta, \Lambda, \Sigma, \Xi$ and Ξ^* are found to be in very good agreement with the experimental values. The quark-gluon coupling constant $\alpha_c = 0.258$ taken in our calculation is quite consistent with the idea of treating one-gluon-exchange effects in lowest order perturbation theory. Thus we draw the following conclusions in the present work

(i) Thus the quark-pion coupling constant determined in the present model is consistent with the estimates made earlier by other workers [20]. (ii) The SU(3)-breaking effect due to the quark-masses $m_u = m_d \neq m_s$ lifts the degeneracy in baryon masses through the energy term $[E_B + (\Delta E_B)_{cm}]$ among the groups $(N,\Delta),(\Lambda,\Sigma,\Sigma^*),(\Xi,\Xi^*)$ and Ω^-. (iii) The constraint of chiral symmetry imposed on the baryon core removes the degeneracy partially through the spin-isospin interaction energy (δM_B) between N and Δ, Λ and Σ, whereas Σ^* still remains degenerate with Σ and Ξ^* with Ξ. (iv) The color-electric and -magnetic interaction energy arising out of the one-gluon exchange with the dominant color-magnetic part giving a spin-spin contribution removes the mass degeneracy completely among these baryons.

In this work we find that a rather strong traditional one-gluon-exchange interaction supplemented by a modest one-meson-exchange interaction arising out of the requirement of chiral symmetry is needed to describe quantitatively the mass spectrum of ground-state baryons with our model. In the past a study with such a contribution, relying on one gluon-exchange supplemented by one-pion and one-sigma exchanges, was performed by Valcarce et al. [30]. Another model with such a combination was investigated in a recent work [31] to reproduce N and Δ splittings. A one-gluon-exchange with a relatively large coupling constant of $\alpha_c = 0.7$ was assumed there in addition to π and η exchanges. On the contrary, some other studies [32] have shown that the combination of stranger one-gluon-exchange and weaker one-meson-exchange residual interactions gives wrong ordering of excited states and one needs a weaker contribution of the gluonic and stronger contribution of the mesonic residual interactions to explain the correct level ordering of excited states as well as in the baryon spectra. However, in order to obtain accurate quantitative results for baryon spectra these studies take non-perturbative gluon-exchange and meson-exchange residual interactions and use three-body calculations which are rigorous and complicated. Certainly this study where we solve Dirac equation for each quark in a single-particle confining potential and treat the residual interactions as low-order perturbation is simple and straightforward.

Acknowledgments

The authors are grateful to Dr. N. Barik, Mayurbhanj, Professor of Physics, Utkal University, Bhubaneswar, India for his valuable suggestions and useful discussions on this work.

References

[1] N. Isgur and G. Karl, *Phys.Rev.* **D 18**,4187 (1978).**D20**, 1191(1979)
C. de Tar, *Phys.Rev.* **D24**, 752(1980)
Y. Nogami and N. Ohtsuka, *ibid.* **26**, 261(1982).

[2] L. Ya. Glozman, Z. Papp, W. Plessas, K. Varga and R. F. Wagenbrunn, *Phys.Rev.* **C57**, 3406 (1998).

[3] A. Chodos, R. L. Jaffe, K. Johnson, C. B. Thorn and V. F. Weisskopf, *Phys.Rev.* **D9**,1471 (1974),**D10**,2599 (1974)
T. De. Grand, R. L. Jaffe, K. Johnson and J. Kiskis, *ibid.* **D12**, 2060(1975).

[4] A. W. Thomas, S. Theberge and G. A. Miller, *Phys.Rev.* **D24**, 216(1981);
A. W. Thomas, *Adv.Nucl.Phys.* **13**,1(1983);
G. A. Miller, *International Review of Nucl. Phys. Vol. 1, edited by W. Weise, (World Scientific)* (1984).

[5] R. Tegen, R. Brockmann and W. Weise, *Z.Phys.* **A307**,339(1982);
R. Tegen, and W. Weise, *ibid*, **A 314**,357(1983);
R. Tegen, M. Schedle and W. Weise, *Phys.Lett.* **B123**, 9(1983).

[6] T. Goldman et.al, *Modern Physics Letter* **A13**, 59(1998).

[7] P. Leal. Ferreira,*Lett.Nuovo – ClimentoSoc.Ital.Fis.* **20**,511(1977).;
P. Leal. Ferreira, J. A. Halayal and N.Zagury, *ibid.* **A 55**,215(1980);
N. Barik and B. K. Dash, *Phys.Rev.* **D 31**,1652(1985); *Pramana – J.Phys.* **24**,707(1985);*PhysRev.* **D 33**, 1925(1986).

[8] P. Leal. Ferreira, *Lett.Nuovo – ClimentoSoc.Ital.Fis* **20**,157(1977)
S. N. Jena and S. Panda, *Pramana – J.Phys.* **35**, 21(1990);*Int – J.Mod.Phys.* **A7**, 2841(1992) ;*J.Phys.* **G 17**, 273(1992);
S. N. Jena, M. R. Behera and S. Panda,*Phys.Rev.* **D55**,291(1997);
S. N. Jena, S. Panda and T. C. Tripathy, *Nucl. Phys.*, **A658**,249(1999);
S. N. Jena, P. Panda and T. C. Tripathy, *Nucl. Phys.*,**A699**,649(2002).

[9] S. N. Jena, M. R. Behera and S. Panda, *J.Phys.* **G 24**,1089(1998);
S. N. Jena and M. R. Behera, *Int – J.Mod.Phys.* **E 7**, 69(1998) ;*Int – J.Mod.Phys.* **E 7**,425(1998) ; *Pramana – J.Phys.* **44**, 357(1995) ;*Pramana – J.Phys.* **47**, 233(1996).

[10] A. Martin, *Phys.Lett.* **B 93**, 338(1980);
N. Barik and S. N. Jena, *ibid.* **B 97**,261(1980) ;*ibid* **B 97**, 265(1980) ; *Phys.Rev.* **D 28**, 612(1982);
N. Barik and M. Das, *Phys.Lett.* **B 120**, 403(1983) ;*Phys.Rev.* **D 28**, 2823(1983) ;*Phys.Rev.* **D 33**,176(1986) ;*Pramana – J.Phys*, **D 27**, 727 (1986);
S. N. Jena, P. Panda and T. C. Tripathy *Phys.Rev.* **D63**,014011(2000); *J.Phys.G.* **27**, 227(2001); ibid 1519(2001).

[11] Y. Nambu and G. Jona- Lasino, *Phys.Rev.* **112**, 345(1961) ;
R. Brockmann, W. Weise and E. Werner, *Phys.Lett.* **B 122**, 201(1983) ;
S. P. Klevansky, *Rev.Mod.Phys.* **64**,649(1992).

[12] C. Quigg and J. L. Rosner, *Phys.Lett.* **B 71**, 153(1977).
C. Quigg and J. L. Rosner, *Phys.Rep.***56**,167 (1979).

[13] E. Magyari,*Phys.Lett.* **B95**, 295 (1980).

[14] S. N. Jena and D. P. Rath, *Phys.Rev.* **D34**,196 (1986) .
S. N. Jena and D. P. Rath, *Pramana – J.Phys.* **27**, 773(1986).

[15] N. Barik, S. N. Jena and D. P. Rath, *Phys.Rev.* **D41**,1568 (1990).

[16] S. N. Jena and D. P. Rath, *Pramana – J.Phys.* **32**, 753(1989).

[17] N. Barik, S. N. Jena and D. P. Rath, *Int.J.Mod.Phys.* **47**,6813 (1992)

[18] N. Barik, S. N. Jena and D. P. Rath, *Int.J.Mod.Phys.* **49**,327 (1994).

[19] S. N. Jena, K. P. Sahu and P. Panda, *Int.J.th.Phys.* **243**,161 (2002) .
S. N. Jena, K. P. Sahu and T. C. Tripathy, *Int.J.th.Phys.* **11**, 69(2004).

[20] A. Suzuki and R. K. Bhaduri, *Phys.Lett.* **B 125**, 347(1983);
D. Faimen and R. W. Hendry, *Phys.Rev.* **173**, 1720(1983);
A. W. Hendry, *Ann.Phys(NY* **140**, 65(1982).

[21] R. E. Peierls and J. Yocoz, *Proc.Phys.Soc.* **A70**, 381(1957);
J. Bartelski, A. Szymacha, L. Mankiewicz and S. Tatur, *Phys.Rev.* **D 29**, 1035(1984) ;
E. Eich, D. Rein and R. Rodenberg, *Z.Phys.* **C 28**, 225(1985);
C. W. Wong, *Phys.Rev.* **D24**, 1416(1981);
I. Duck, *Phys.Lett.* **B77**, 203(1978).

[22] J. F. Donoghue and E. Golowich, *Phys.Rev.* **D15**,3421(1977);
F. E. Close and R. R. Horgan *Nucl.Phys.* **B185**,333(1981).

[23] G. F. Chew, *Phys.Rev.* **94**, 1748-1755(1954);
G. F. Chew and F. E. Low, *Phys.Rev.* **101**, 1570(1955);
Gc. Wick *Rev.Mod.Phys.* **27**, 339(1955).

[24] T. De. Grand, R. L. Jaffe, K. Johnson and J. Kiskis, *Phys.Rev.* **D12**, 2060(1975).

[25] L. Ya. Glozman, hep- ph / 0004229(2000).

[26] A. W. Hendry, *Ann.Phys*(NY) **140**, 65(1982);
D. Faimen and R. W. Hendry, *Phys.Rev.* **173**,1720(1983)

[27] A. Suzuki and R. K. Bhaduri, *Phys.Lett.* **B 125**, 347(1983)

[28] E. Eichten, K. Gottfried, T. Kinshita, J. Kogut, K. D. Lane and T. M. Yan, *Phys.Rev.Lett.* **34**, 369(1975).

[29] S. Theberge, G. A. Miller and A. W. Thomas, *Can.J.Phys.* **60**, 59(1982).

[30] A. Valcarce, P. Gonzalez, F. Fernandez and V. Vento *Phys.Lett.* **B 367**, 35(1996).

[31] Z. Dziembowski et al. *Phys.Rev.* **C35**, R2038(1996).

[32] L. Ya. Glozman and D. O. Riska *Phys.Rep.* **268** (1996) 263;
L. Ya. Glozman, Z. Papp, W. Plessas *Phys.Lett.* **B 381**,1996;
K. Varga and Y. Suzuki *Phys.Rev.* **C52** 1995.

In: Theoretical Physics and Nonlinear Optics
Editors: Thomas F. George et al
ISBN 978-1-61122-939-4
© 2012 Nova Science Publishers, Inc.

Chapter 13

GOLDSTONE-BOSON-EXCHANGE AND ONE-GLUON EXCHANGE CONTRIBUTIONS TO BARYON SPECTRA IN A RELATIVISTIC QUARK MODEL

S.N. Jena,[*] *M.K. Muni*[1,] *and H.R. Pattnaik*[2]
Department of Physics, Berhampur University,
Berhampur-760007, Orissa, India
[1] Department of Mathematics and Science,
SMIT, Ankushpur, Berhampur, Orissa, India
[2] Department of Physics,Temple City Institute of Technology
and Engg.,Khurda, Orissa, India

Abstract

The energy contributions from the Goldstone-boson (π, η, K) exchange and that from one-gluon-exchange interactions between the constituent quarks to the baryon spectra are studied together with that from center-of-mass motion and the mass spectra of light-and strange-baryons are computed in a relativistic quark model. The baryons are assumed here as an assembly of independent quarks confined in a first approximation by an effective logarithmic potential which presumably represents the nonperturbative multigluon interactions including the gluon self-couplings. The results obtained for the ground states and excitation spectra of baryons agree reasonably well with the experimental values. The model yields correct level ordering of the excited states in $N, \triangle$ and Λ spectra.

1. Introduction

A purely phenomenological logarithmic potential of the form

$$V(r) = a\log(r/b) \tag{1}$$

with $a, b > 0$ was proposed in the past by Quigg and Rosner [1,2] for a successful description of the heavy meson spectra in a non-relativistic Schrodinger frame-work. Implications

---
[*]E-mail address: snjena@rediffmail.com, Tel.: 0680-2282065, Fax.: 0680-2243322

of such a potential in the context of quark confinement and relativistic consistency have been studied by Magyari [3] in reference to heavy meson spectra. In his study it has been found that a logarithmic potential with a Lorentz structure in the form of an equal admixture of scalar and vector parts not only can guarantee relativistic quark confinement but also can generate charmonium and upsilon bound state masses in reasonable agreement with experiment. In view of the remarkable success obtained with such a model in mesonic sector, the model has been extended to baryonic sector to study reasonably well the static baryon properties [4,5], the nucleon electromagnetic form factors [6] and the weak-electric and-magnetic form factors for the semileptonic baryon decays [7]. Then incorporating chiral symmetry in the SU[2]-flavor sector in the usual manner, this model has been used to study the electromagnetic properties of the nucleons [8] and the magnetic moments of the nucleon octet [9] in reasonable agreement with experimental data. This model has also been successfully applied in the study of the baryonic mass spectra [10,11] taking into account the contribution of the Goldstone boson exchange (GBE) interaction between the constituent quarks of the baryons over and above the center-of-mass correction. But the studies in such a model have not included the effects of residual one-gluon-exchange (OGE) interactions at short distances assuming these to be not so much significant for the electro-magnetic properties of baryons. These, however, play an important role in providing color-electrostatic and-magnetostatic energies to the quark core in the study of baryon mass spectra. Therefore we wish to incorporate the contribution from the residual OGE interaction between the constituent quarks into this model to study the light-and-strange baryon spectra. In the present work we are mainly interested to study the energy contributions from the GBE and OGE interactions between the constituent quarks to the mass spectra of baryons and we also intend to calculate the mass spectra of baryons in this model taking into account these energy contributions.

In fact several non-relativistic quark models have been used [12-14] in the past for the study of light and strange baryons. Traditional constituent quark models (CQM) adopting one-gluon exchange (OGE) [15] as the hyperfine interaction between constituent quarks(Q) have been suggested in the study of light baryon spectroscopy. Over the years it has become evident that these models face some intriguing problems in explaining the mass spectra of light baryons. Most severe problems faced by CQM are (i) the wrong level ordering of positive and negative-parity excitation in the $N, \triangle, \Lambda$ and Σ spectra (ii) the missing flavor dependence of the Q-Q interaction necessary, e.g. for a simultaneous description of the correct level ordering in the N and Λ spectra and (iii) the strong spin-orbit splittings that are produced by the OGE interaction but not found in the impirical spectra. All of these effects have been explained to be due to [16] inadequate symmetry properties inherent in the OGE interaction. Several hybrid models advocating meson-exchange Q-Q interactions in addition to the OGE dynamics of CQM's have been suggested for baryons [17]. In the study of N and $\triangle$ spectra, especially π and σ exchanges have been introduced to supplement the interaction between constituent quarks.

A few years back two groups, viz. Valcarce, Gonzalez, Fernandez and Vento [18] and Dziembowski, Fabre and Miller [19] came up with Versions of hybrid constituent quark models. They have presented a reasonable description of N and $\triangle$ excitation spectra taking into account a sizeable contribution from the OGE interaction. However, the performance of the hybrid constituent quark models has been studied in detail by Glozman et.al. [16] by using the calculations based on accurate solutions of the three quark systems in both variational Schroedinger and a rigorous faddeev approach. It has been argued that hybrid Q-Q interactions with a sizeable OGE component encounter difficulties in describing baryon spectra due to the specific contributions from one-gluon and meson exchanges together. On the contrary, Glozman et.al. [20-22] have shown that a chiral constituent quark model with a Q-Q interaction relying solely on Goldstone boson exchange (GBE) is capable of providing a unified description not only of the N and Δ spectra but also of all strange baryons in good agreement with phenomenology. They have also presented a constituent quark model with the confinement potential in linear and harmonic [20-22] forms for the light and strange baryons providing a unified description of their ground-states and excitation spectra. Their model which relies on constituent quarks and Goldstone bosons arising as effective degrees of freedom of low energy QCD from the spontaneous breaking of chiral symmetry (SBCS) has been found to be quite proficient in reproducing the spectra of the three quark systems from a precise variational solution of a Schroedinger equation with a semi relativistic Hamiltonian.

Although several non-relativistic quark models have appeared in the literature in connection with the study of mass spectrum of light and strange baryons and the phenomenological picture is reasonable at the non-relativistic level, a relativistic approach is quite indispensable on this account in view of the fact that the baryonic mass splitting are of same order as the constituent quark masses. Of course the chiral constituent quark model which has been constructed by Glozman et al. [20-22] in a semi relativistic framework is a step in this direction and shows an essential improvement over non-relativistic approaches. On the other hand the MIT bag model [23-25] has also been found to be relatively successful in this respect. In its improved version, the chiral bag Model (CBM)[26] have included the effect of pion self-energy due to baryon-pion coupling at the vertex to provide a better understanding of baryon masses. The present model which is used here as an alternative approach to CBM, can also be tried in this context of studying the physical masses of the baryons in the nucleon octet. The chiral potential models [27-29], which are comparatively more straightforward in the above respects, are obviously attempts in this direction. In such models the confining potentials which basically represent the interaction of quarks with the gluon field are usually assumed phenomonologically as Lorentz scalars in harmonic and cubic form. Potentials having different type of Lorentz structure with equally mixed scalar and vector parts in harmonic [30-34], linear [35-39], square root [40-44] and non-coloumbic power law [45-56] form have also been investigated in this context. The term in the Lagrangian density for quarks corresponding to the effective scalar part of the potential in such models being chirally non-invariant through all space requires the introduction of an additional pionic component everywhere in order to preserve chiral symmetry. The effective potential of individual quarks in these models, which is basically due to interaction of quarks with the gluon field, may be thought of as being mediated in a self consistent manner through Nambu-Jona-Lasinio (NJL) type models [57-60] by some form of instanton induced effective quark-quark contact interaction with position dependent coupling strength. The position dependent coupling strength, supposedly determined by multigluon mechanism, is impossible to calculate from the first principles, although it is believed to be small at the origin and increases rapidly towards the hadron surface. Therefore, one needs to introduce the effective potential for individual quarks in phenomenological manner to seek a posteriori justification in finding its conformity with the supposed qualitative behaviour of the position-dependent coupling strength in the contact interaction.

However, with no theoretical prejudice in favour of any particular mechanism for generating confinement of individual quarks, we prefer to work in an alternative, but similar scheme based on Dirac equations with a purely phenomonological individual quark potential of the form

$$V_q(r) = (1+\gamma^0)[a\ln(r/b)] \qquad (2)$$

with $a,b > 0$. Such a model takes the Lorentz structure of the potential as an equal admixture of scalar and vector parts because of the fact that both the scalar and vector parts in-equal proportions at every point render the solvability of the Dirac equation for independent quarks by reducing it to the form of a Schrodinger like equation. This Lorentz structure of potential also has an additional advantage of generating no spin-orbit splitting, as observed in the experimental baryon spectrum.

In this model we assume the baryons as an assembly of three constituent quarks with dynamical masses confined in a first approximation by an effective logarithmic potential of the form (2) which presumably represents the non-perturbative multiglucon interactions including the gluon self-couplings. The present model considers perturbatively the contribution of OGE along with that Q-Q interactions due to GBE over and above the centre-of-mass correction. For the inclusion of GBE correction in this model we have followed the guidelines of the chiral constituent quark models suggested by Glozman et al. [20-22]. In this context we may point out that in the present model we consider the constituent quarks of flavours u,d,s with masses considerably larger than the corresponding current quark masses so that the underlying chiral symmetry of QCD is spontaneously broken. As a consequence of SBCS, at the same time Goldstone bosons appear, which couple directly to the constituent quarks [61-63]. Hence beyond the scale of SBCS one is left with the constituent quarks with dynamical masses related to $<q\bar{q}>$ condensates and with Goldstone bosons as the effective degrees of freedom. This feature, that in the Nambu-Goldstone mode of chiral symmetry constituent quarks and Goldstone boson fields prevail together, is well supported, e.g. by σ model [64] or the NJL model [57-60].

This paper is organized as follows. In sect. 2, we outline the potential model with the solutions for the relativistic bound states of the individually confined quarks in baryons. Section 3 provides a brief account of the energy contributions due to the spurious centre of mass motion and also that of the contributions due to the GBE interaction between the constituent quarks in a generalized way. This section also provides an account of a further contribution to the baryon mass due to color-electric and color-magnetic interaction energies arising from the residual OGE interaction. Finally, in sect. 4, we present the computation of the energy contributions due to GBE and OGE interactions along with those due to c.m. motion. This sect. also provides the results for the ground states and excitation spectra of light-and-strange baryons, which are in reasonable agreement with the corresponding experimental values.

2. Potential Model

Leaving behind for the moment the quark-gluon interaction originating from OGE at short distances and the interaction of constituent quarks due to GBE arising from SBCS to be treated perturbatively, we begin with the confinement part of the interaction which is believed to be dominant in baryonic dimensions. This particular part of the interaction which is believed to be determined by multigluon mechanism is impossible to calculate theoretically from first principle. Therefore from a phenomenological point of view we assume that the constituent quarks in a baryon core are independently confined by an average flavour-independent potential of the form given in eq. (2). To a first approximation the confining part of the interaction arising out of the non-perturbative multigluon mechanism including gluon self couplings is believed to provide the zeroth-order quark dynamics inside the baryon core through the quark Lagrangian density in zeroth order.

$$\mathcal{L}_q^0(x) = \bar{\psi}_q(x)\left[\frac{i}{2}\gamma^\mu \overrightarrow{\partial}_\mu - m_q - V_q(r)\right]\psi_q(x) \qquad (3)$$

where m_q is the mass of the constituent quark.

The normalized quark wave function $\psi_q(\mathbf{r})$ satisfying the Dirac equations derivable from $\mathcal{L}_q^0(x)$ as

$$\left[\gamma^0 E_q - \overrightarrow{\gamma}\cdot\mathbf{p} - m_q - V_q(r)\right]\psi_q(\mathbf{r}) = 0 \qquad (4)$$

can be written as a two component form as

$$\psi_{nlj}(\mathbf{r}) = N_{nl}\begin{pmatrix} if_{nlj}(r)/r \\ \overrightarrow{\sigma}\cdot\hat{r}g_{nlj}(r)/r \end{pmatrix} y_{ljm}(\mathbf{r}) \qquad (5)$$

where the normalized spin angular part

$$y_{ljm}(\mathbf{r}) = \sum_{m_l,m_s} <l,m_l,1/2,m_s|j,m_j> Y_l^{m_l}\chi_{1/2}^{m_s} \qquad (6)$$

and N_{nl} is the overall normalization constant. The reduced radial part $f_{nlj}(r)$ of the upper component of Dirac spinor $\psi_{nlj}(\mathbf{r})$ satisfies the equation

$$f''_{nlj}(r) + \lambda_{nl}[E_{nl}^q - m_q - 2aln(r/b) - l(l+1)/r^2]f_{nlj}(r) = 0 \qquad (7)$$

where

$$\lambda_{nl} = E_{nl}^q + m_q \qquad (8)$$

The present model can in principle provide the quark orbitals $\psi_{nlj}(\vec{r})$ and the zeroth order binding energies of the confined quark for various possible eigen modes through eqs. (5) - (8). However, for the ground state baryons, in which all the constituent quarks are in their lowest eigen states, the corresponding quark orbitals can be expressed as

$$\psi_{1s}(\mathbf{r}) = N_{1s}\begin{pmatrix} \phi_{1s}(\mathbf{r}) \\ \frac{\overrightarrow{\sigma}\cdot\mathbf{p}}{\lambda_{nl}}\phi_{1s}(\mathbf{r}) \end{pmatrix}\chi\uparrow \qquad (9)$$

where $\phi_{1s}(\mathbf{r})$ is the radial angular part of the upper component $\psi_{1s}(\mathbf{r})$ and is given by $\phi_{1s}(\mathbf{r}) = (\frac{i}{\sqrt{4\pi}}) f_{1s}(r)/r$. For the ground state, eqn.(7) reduces to

$$f''_{1s}(r) + \lambda_{1s}\left[E^q_{1s} - m_q - 2a\ln(r/b)\right] f_{1s}(r) = 0 \tag{10}$$

which can be transformed into a convenient dimensionless form

$$f''_{1s}(\rho) + \left[\varepsilon_{1s} - \ln\rho\right] f_{1s}(\rho) = 0 \tag{11}$$

where $\rho = \frac{r}{r_0}$ is a dimensionless variable with $r_0 = (2a\lambda_{1s})^{-1/2}$ and

$$\varepsilon_{1s} = \left[\frac{E^q_{1s} - m_q + a\ln(2ab^2\lambda_{1s})}{2a}\right] \tag{12}$$

Eq. (11) provides the basic eigen value equation whose solution by a standard numerical method would give $\varepsilon_q = 1.0443$.

Now the ground state individual quark binding energy E_q is obtainable from energy eigen value condition(12), through the relation

$$E_q = m_q - a\ln C + ax_q \tag{13}$$

where $C = 2a^2 b^2$ and x_q is the solution of the root equation

$$x_q + \ln\left[x_q + (\frac{2m_q - a\ln C}{a})\right] = 2\varepsilon_q \tag{14}$$

In this independent quark model approach mass of the baryon core M^0_B in zeroth order immediately follows from the ground state individual quark binding energy E^q_{1s} as

$$M^0_B = E_B = \sum_q E^q_{1s} \tag{15}$$

Similarly eq. (7) can be solved for 2S and 1P states to obtain the individual quark binding energy E^q_{2s} and E^q_{1p} respectively with the help of the standard numerical method which yields $\varepsilon_{2s} = 1.8474$ and $\varepsilon_{1p} = 1.643$ [2]. This leads to the corresponding masses of the baryon core in zeroth order in the same way as in case of the ground state. The overall normalization constant N_{nl} of $\psi_{nlj}(\mathbf{r})$ appearing in eq. (5) is of the form

$$N^2_{nl} = \frac{\lambda_{nl}}{2\left[E^q_{nl} - a << \ln(r/b) >>_{nl}\right]} \tag{16}$$

where $<< \ln(r/b) >>_{nl}$ is the expectation value with respect to $\Phi_{nl}(\mathbf{r})$.

3. Energy Contributions to Baryon Masses

The contribution of quark-binding energy to mass M^o_B of baryon core given in eq. (15) needs corrections due to contributions from center-of-mass motion, GBE interaction between the constituent quarks and quark gluon interaction due to OGE which need to be calculated separately for obtaining physical masses of light-and-strange baryons.

3.1. Contribution from Motion of the Center of Mass

In the present model there would be sizeable spurious contribution to the energy E_{nl}^q from the motion of the center of mass of the three quark system. If this aspect is duly accounted for, the concept of the independent motion of the quarks inside the baryon core will lead to a physical baryon state of definite momentum. Although there is some controversy on this subject, we follow the approach of Bartelski et al. and Eich et al. [65,66], which is just one way of accounting for the center of mass motion. Following their prescription, a ready estimate of the center of mass momentum $\mathbf{P}_B$ can be obtained as

$$<\mathbf{P}_B^2> = \sum_q <\mathbf{p}_q^2>_{nl} \tag{17}$$

where $<\mathbf{p}_q^2>_{nl}$ is the average value of the square of the individual quark momentum taken over single quark states and is given in this model as

$$\begin{aligned}
<\mathbf{p}_q^2>_{nl} &= 2N_{nl}^2 \Big[E_{nl}^q (E_{nl}^q - m_q) \\
&\quad - (3E_{nl}^q - m_q) \times a << \ln(r/b) >>_{nl} \\
&\quad + 2a^2 << (\ln(r/b))^2 >>_{nl} \Big]
\end{aligned} \tag{18}$$

with

$$<< \ln(r/b) >>_{nl} = \left[\frac{E_{nl}^q - m_q - a}{2a} \right],$$

$$<< \left(\ln(r/b)\right)^2 >>_{nl} = \left(\frac{E_{nl}^q - m_q}{2a} \right) \left(\frac{E_{nl}^q - m_q - 2a}{2a} \right) + \frac{3}{4}$$

In the same way the expression for the physical mass M_B of the bare baryon core can be obtained as

$$<M_B^2/E_B^2> = \left[1 - \sum_q (<\mathbf{p}_q^2>_{nl} /E_B^2) \right] \tag{19}$$

which provides the energy correction to the baryon mass due to contribution from center-of-mass motion in eq. (15) as

$$(\triangle E_B)_{cm} = (M_B - M_B^0) = \left[(E_B^2 - \sum_q <\mathbf{p}_q^2>_{nl})^{\frac{1}{2}} - E_B \right] \tag{20}$$

3.2. Contribution from Goldstone Boson Exchange Interaction

The $SU(3)_L \times SU(3)_R$ chiral symmetry of QCD Lagrangian is spontaneously broken down to $SU(3)_V$ by QCD vacuum [in the large N_c limit would be $U(3)_L \times U(3)_R \to U(3)_V$]. There are two important generic consequences of the SBCS. The first one is an appearance of the octet of pseudoscalar mesons of low mass, π, K, η which represent the associated approximate Goldstone bosons (in the large N_c limit the flavor singlet state η' should be added). The second one is that valence (partially massless) quarks acquire a dynamical mass, which has been called a historically constituent mass. Indeed, the nonzero value of quark condensate, $<\bar{q}q> \sim -(250\ MeV)^3$, itself implies at the formal level that there must be at low momenta rather big dynamical mass, which should be a momentum dependent quantity. Such a dynamical mass is now directly observed on the lattice [67]. Thus the constituent quarks should be considered as quasi-particles whose dynamical mass at low momenta comes from the non perturbative gluon and quark antiquark dressing. The flavor-octet axial current conservation in the chiral limit tells that the constituent quarks and Goldstone bosons should be coupled with the strength $g = g_A M/f_\pi$,[61-63] which is a quark analog of the famous Goldberger-Treiman relation. It has been recently suggested that in the low-energy regime, below the chiral symmetry breaking scale ~ 1 GeV, the low lying light and strange baryons should be predominantly viewed as systems of three constituent quarks with an effective confining interaction and a chiral interaction mediated by a GBE between the constituent quarks [20-22].

The coupling of Goldstone bosons (π, η and K mesons) to the constituent quarks arising from SBCS in QCD can be taken into account in a perturbative manner in the same way as it has been done in the study of the effect of quark-pion coupling in the CBM [26]. Here the fields of the Goldstone bosons may be treated independently without any constraint and their interactions with the quarks can be assumed to be linear as it is done in case of pion[26].

Following the Hamiltonian technique [26] as has been used in the CBM, we can describe the effect of coupling of mesons ($\chi = \pi, \eta, K$) in the low order perturbation theory as follows. The mesonic self energy of the baryons can be evaluated with the help of single-loop self energy diagram (fig. 1) as

$$\Sigma_B(E_B) = \sum_k \sum_{B'} \frac{V^{\dagger BB'} V^{BB'}}{(E_B - \omega_k - M_{B'}^0)} \qquad (21)$$

where $\sum_k = \sum_j \int \frac{d^3\mathbf{k}}{(2\pi)^3}$.

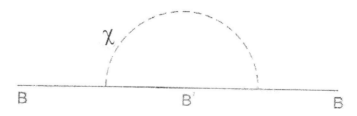

Figure 1. Baryon self-energy due to coupling with meson.

Here, j corresponds to the meson-isospin index and B' is the intermediate baryons state. $V^{BB'}(\mathbf{k})$ is the general baryon-meson absorption vertex function obtained [68,69] in this model as

$$V_j^{BB'}(\mathbf{k}) = i\sqrt{4\pi}\frac{f_{BB'\chi}}{m_\chi}\frac{ku(k)}{\sqrt{2w_k}}(\vec{\sigma}^{BB'}.\hat{\mathbf{k}})\tau_j^{BB'} \quad (22)$$

where $\vec{\sigma}_j^{BB'}$ and $\vec{\tau}_j^{BB'}$ are spin and isospin matrices and $\omega_k^2 = \mathbf{k}^2 + m_\chi^2$. The form factor $u(k)$ in this model can be expressed as

$$\begin{aligned} u(k) &= \frac{10N_q^2}{3\lambda_q g_A^0}\Bigg[m_q << j_0\left(|\mathbf{k}|r\right) >> \\ &+ a << \ln(r/b)j_0\left(|\mathbf{k}|r\right) >> + a << \frac{j_1(|\mathbf{k}|r)}{(|\mathbf{k}|r)} >>\Bigg] \end{aligned} \quad (23)$$

where $j_0(|\mathbf{k}|r)$ and $j_1(|\mathbf{k}|r)$ represents the zeroth order and first order spherical Bessel functions respectively. The double angular brackets stand for the expectation values with respect to $\Phi_q(r)$. In this model the axial vector coupling constant g_A^0 for the beta decay of neutron is given by

$$g_A^0 = \frac{5}{9}\left[\frac{3(E_{nl}^q + m_q) - a}{E_{nl}^q + m_q + a}\right] \quad (24)$$

Now with the vertex function $V_j^{BB'}(\mathbf{k})$ at hand it is possible to calculate the mesonic self energy for various baryons with appropriate baryon intermediate states contributing to the process. For degenerate intermediate states on mass shell with $M_B^0 = M_{B'}^0$, the contribution to the baryon mass due to self energy correction becomes

$$(\delta M_B)_\chi = \sum_B\left(E_B^0 = M_B^0 = M_{B'}^0\right) = -\sum_{k,B'}\frac{V^{\dagger BB'}V^{BB'}}{\omega_k} \quad (25)$$

Now using eq. (22) we find

$$(\delta M_B)_\chi = \frac{-I_\chi}{3}\sum_{B^.}C_{BB'}f_{BB'\chi}^2 \quad (26)$$

where

$$C_{BB'} = (\vec{\sigma}^{BB'}.\vec{\sigma}^{B'B})(\vec{\tau}^{BB'}.\vec{\tau}^{B'B}) \quad (27)$$

and

$$I_\chi = \frac{1}{\pi m_\chi^2}\int_0^\infty\frac{dk k^4 u^2(k)}{\omega_k^2} \quad (28)$$

The self energy $(\delta M_B)_\chi$ for different baryons can be computed by using the values of $f_{BB'\chi}$ and $C_{BB'}$ [12,13] as has been done in our earlier work [70,71]. The self energy $(\delta M_B)_\chi$ calculated here contains both the quark self energy (fig 2a) and the one meson exchange contributions (fig 2b).

In the present work the meson degree of freedom which is taken into account in the same way as in the CBM ignores to a large extent the short-range part of the meson exchange interaction, which is of crucial importance for splttings. Only when the complete infinite set of all radially excited intermediate states B' is taken into account, this method could be

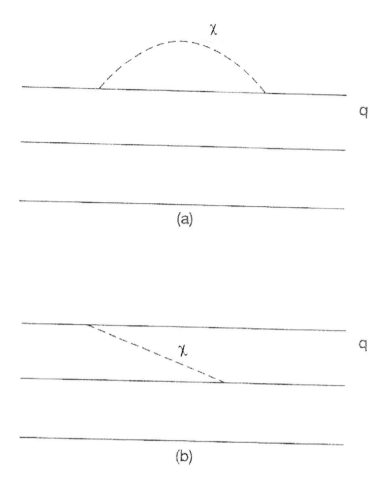

Figure 2. (a) Mesonic self-energy contribution. (b) One-meson-exchange contribution.

adequate [72]. For example, the meson exchange contribution to the $N-\triangle$ difference will become much larger. It will also be strongly enhanced when meson exchange contribution is strongly dependent on the radius of the bare wave function i.e. on the type of confinement. This dependence has not been studied in the present work.

Following the discussions and the table given in refs. [12,13] the baryon-meson coupling constant $f_{BB'\chi}$ can be expressed in terms of the nucleon-meson coupling constant $f_{NN\chi}$. The pion exchange interaction acts only between light quarks where as η-exchange is allowed in all quark pair states. The kaon exchange interaction takes place in $u-s$ and $d-s$ pair states.

3.3. Contribution from One gluon Exchange

The individual quarks in a baryon core are considered so far to be experiencing only the force coming from the average effective potential $V_q(r)$ in eq. (2), which is assumed to provide a suitable phenomenological description of the non-perturbative gluon interaction including gluon self couplings. All that remains inside the quark core is the hopefully weak one gluon exchange interaction provided by the interaction Lagrangian density

$$L_I^g(x) = \sum_{\alpha=1}^{8} J_i^{\mu\alpha}(x) A_\mu^\alpha(x) \tag{29}$$

where $A_\mu^\alpha(x)$ are the 8-vector gluon fields and $J_i^{\mu\alpha}(x)$ is the ith quark colour current. Since at small distances the quarks should be almost free, it is reasonable to calculate the energy shift in the mass spectrum arising out of the quark interaction energy due to their coupling to the coloured gluons, using a first order perturbation theory.

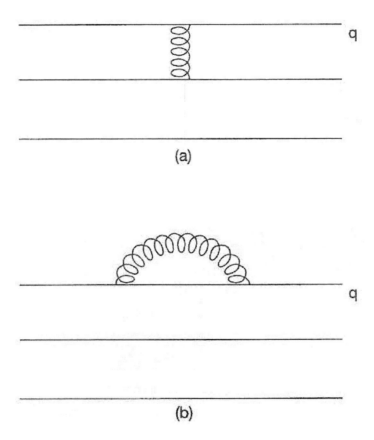

Figure 3. (a) One-gluon-exchange contribution. (b) Gluonic self energy contribution.

If we only keep the terms of order α_c, the problem reduces to evaluating the diagrams shown in fig. 3, where fig. 3(a) corresponds to OGE part, while fig. 3(b) implies the quark self energy that normally contributes to renormalization of quark masses. If $\mathbf{E}_i^\alpha$ and $\mathbf{B}_i^\alpha$ are the colour-electric and colour-magnetic fields, respectively, generated by the ith quark colour-current

$$J_i^{\mu\alpha}(x) = g_c \bar{\psi}_i(x) \gamma^\mu \lambda_i^\alpha \psi_i(x) \tag{30}$$

with λ_i^α being the usual Gell-Mann SU(3)- matrices and $\alpha_c = (g_c^2/4\pi)$, then the contribution to the mass due to the relevant diagram can be written as a sum of the colour-electric and colour-magnetic parts as

$$(\triangle E_B)_g = (\triangle E_B)_g^e + (\triangle E_B)_g^m \tag{31}$$

where

$$(\triangle E_B)_g^e = \frac{1}{8\pi} \sum_{i,j} \sum_{\alpha=1}^{8} \int\int \frac{d^3r_i d^3r_j}{|\mathbf{r}_i - \mathbf{r}_j|} <B|J_i^{o\alpha}(\mathbf{r}_i)J_j^{o\alpha}(\mathbf{r}_j)|B> \qquad (32)$$

$$(\triangle E_B)_g^m = \frac{-1}{4\pi} \sum_{i<j} \sum_{\alpha=1}^{8} \int\int \frac{d^3r_i d^3r_j}{|\mathbf{r}_i - \mathbf{r}_j|} <B|\mathbf{J}_i^{\alpha}(\mathbf{r}_i).\mathbf{J}_j^{\alpha}(\mathbf{r}_j)|B> \qquad (33)$$

We have not included the self energy diagram in the calculation of the magnetic part of the interaction, which contributes to renormalization of quark masses and can possibly be accounted for in the phenomenological quark masses. The exclusion of this diagram, however, requires that each $\mathbf{B}_i^{\alpha}$ should satisfy the boundary condition $\hat{\mathbf{r}} \times \mathbf{B}_i^{\alpha} = 0$, separately at the edge of the confining region, which is a possible case. On the other hand, as the electric field $\mathbf{E}_i^{\alpha}$ is necessarily in the radial direction, it is only possible to satisfy the boundary condition $\hat{\mathbf{r}} \times (\sum_i \mathbf{E}_i^{\alpha}) = 0$ for a colour singlet state $|B>$ for which $\sum_i \lambda_i^{\alpha} = 0$. Therefore, in order to preserve boundary conditions we are forced to take into account the self energy diagrams in fig. 3(b) in the calculation of electric part only. Now using eq. (5) in eq. (30) we find

$$J_i^{0\alpha}(r_i) = g_c \lambda_i^{\alpha} N_i^2 \left[\Phi^2(r_i) + \frac{\Phi'^2(r_i)}{\lambda_i^2} \right]$$

$$J_i^{\alpha}(r_i) = -2g_c \lambda_i^{\alpha} N_i^2 \left[\left(\frac{\vec{\sigma}_i \times \hat{\mathbf{r}}_i}{\lambda_i} \right) \Phi_i \Phi'(r_i) \right] \qquad (34)$$

Again using eq. (34) together with the identity

$$\frac{1}{|\mathbf{r}_i - \mathbf{r}_j|} = \frac{1}{2\pi^2} \int \frac{d^3k}{k^2} \exp\left[-i\mathbf{k}.(\mathbf{r}_i - \mathbf{r}_j)\right] \qquad (35)$$

in eqs. (32) and (33) we obtain

$$(\triangle E_B)_g^e = \frac{\alpha_c}{4\pi^2} \sum_{i,j} \left\langle \sum_{\alpha} \lambda_i^{\alpha} \lambda_j^{\alpha} \right\rangle N_i^2 N_j^2 \int \frac{d^3k}{k^2} F_i^e(k) F_j^e(k) \qquad (36)$$

and

$$(\triangle E_B)_g^m = \frac{-2\alpha_c}{\pi^2} \sum_{i<j} \left\langle \sum_{\alpha} \lambda_i^{\alpha} \lambda_j^{\alpha} \right\rangle N_i^2 N_j^2 \int \frac{d^3k}{k^2} \mathbf{F}_i^m(k).\mathbf{F}_j^m(k) \qquad (37)$$

where

$$F_i^e(k) = \frac{1}{2\lambda_i^2} \left[\{4E_i(E_i + m_i) - k^2)\} <<j_0(|\mathbf{k}|r_i)>> \right.$$
$$\left. - 4\lambda_i a <<\log(r_i/b) j_0(|\mathbf{k}|r_i)>> \right] \qquad (38)$$

and

$$\mathbf{F}_i^m(k) = \frac{i}{2} <<j_0(|\mathbf{k}|r_i)>> (\sigma_i \times \mathbf{k}) \qquad (39)$$

Table 1. The coefficients a_{ij} and b_{ij} used in calculation of the colour-electric and colour-magnetic energy contributions due to one-gluon exchange

Baryons	a_{uu}	a_{us}	a_{ss}	b_{uu}	b_{us}	b_{ss}
N	0	0	0	-3	0	0
Δ	0	0	0	+3	0	0
Λ	1	-2	1	-3	0	0
Σ	1	-2	1	1	-4	0
Ξ	1	-2	1	0	-4	1
Σ^*	1	-2	1	1	2	0
Ξ^*	1	-2	1	0	2	1
Ω^-	0	0	0	0	0	+3

Then eqs. (36) and (37) can be written as

$$(\triangle E_B)_g^e = \frac{\alpha_c}{\pi} \sum_{i,j} \left\langle \sum_\alpha \lambda_i^\alpha \lambda_j^\alpha \right\rangle N_i^2 N_j^2 I_{ij}^e \tag{40}$$

$$(\triangle E_B)_g^m = -\frac{4\alpha_c}{3\pi} \sum_{i<j} <\lambda_i^\alpha \lambda_j^\alpha (\vec{\sigma}_i \cdot \vec{\sigma}_j)> \frac{N_i^2 N_j^2}{\lambda_i \lambda_j} I_{ij}^m \tag{41}$$

where

$$I_{ij}^e = \int_0^\infty dk F_i^e(k) F_j^e(k) \tag{42}$$

$$I_{ij}^m = \int_0^\infty dk k^2 << j_o(|\mathbf{k}|r_i) >><< j_o(|\mathbf{k}|r_j) >> \tag{43}$$

Finally by taking into account the specific quark flavour and spin configuration in various ground state baryons and using the relations $<\sum_\alpha (\lambda_i^\alpha)^2> = \frac{16}{3}$ and $<\sum_\alpha \lambda_i^\alpha \lambda_j^\alpha>_{i \neq j} = -\frac{8}{3}$ for baryons, in general one can write the energy correction due to one gluon exchange as

$$(\triangle E_B)_g^e = \alpha_c(a_{uu} T_{uu}^e + a_{us} T_{us}^e + a_{ss} T_{SS}^e) \tag{44}$$

$$(\triangle E_B)_g^m = \alpha_c(b_{uu} T_{uu}^m + b_{us} T_{us}^m + b_{ss} T_{SS}^m) \tag{45}$$

where a_{ij} and b_{ij} are the numerical coefficients depending on each baryon listed in table 1 and the terms $T_{ij}^{e,m}$ are

$$T_{ij}^e = \frac{16(E_i + m_i)(E_j + m_j)}{3\pi (E_i + m_i + a)(E_j + m_j + a)} I_{ij}^e \tag{46}$$

$$T_{ij}^m = \frac{32}{9\pi (E_i + m_i + a)(E_j + m_j + a)} I_{ij}^m \tag{47}$$

From table 1 one can observe that the colour-electric contribution to the baryon masses vanishes when all the constituent quark masses in a baryon are equal, where as it is non zero otherwise. However, even in the case of strange baryons, it would subsequently be seen that the colour-electric contribution is quite-small. Therefore, the degeneracy among the baryons is essentially removed through the spin-spin interaction energy in the colour-magnetic part. It must be pointed out here that the central logarithmic confinement potential (eq. (1)) which shows coulombic behaviour at short distances includes at least some part of the colour-coulombic energy in the zeroth order calculations of quark binding energy. Therefore in the present model a part of the colour-coulombic energy comes from the central potential and rest part of it is contributed by the OGE interaction (eqs. (44) and (45)) which are computed by using the first order perturbation theory.

4. Results and Discussion

In the preceding sect. it has been shown that the zeroth order mass $M_B^0 = E_B$ of different states of baryon arising out of the binding energies of constituent quarks, confined independently by a phenomenological average potential $V_q(r)$ which presumably represents the dominant non-perturbative multigluon interactions, must be subjected to certain corrections due to the GBE and OGE interaction together with that due to spurious c.m. motion. All of these corrections can be treated independently as though they are of the same order of magnitude, so that the physical mass of a baryon can be obtained as

$$M_B = E_B + (\triangle E_B)_{cm} + (\delta M_B)_\chi + (\triangle E_B)_g^m + (\triangle E_B)_g^e \tag{48}$$

where $(\triangle E_B)_{cm}$ is the energy associated with spurious center of mass motion (eq. (20)), $(\delta M_B)_\chi$ with ($\chi = \pi, \eta$ and K) is the energy arising out of GBE interaction between quarks in a baryon (eq. (26)) and $[(\triangle E_B)_g^e + (\triangle E_B)_g^m]$ is the colour-electric and colour-magnetic interaction energies arising out of the residual one gluon exchange process (eqs. (44) and (45)).

The quantitative evaluation of these terms on right hand side of eq. (48) with in the framework of the model primarily involves the potential parameters (a,b) of the model, the quark masses m_q and corresponding binding energy E_{nl}^q along with other relevant model quantities. In the present phenomenological study, the potential parameter 'a' and the constituent quark masses m_q are suitably chosen and different values of the parameter 'b' are appropriately fixed for 1s, 2s and 1p states of baryons so as to obtain a reasonable fit to the mass spectra of the ground states as well as the excitation spectra of the baryons taking into account the energy corrections due to OGE and GBE interactions between constituent quarks together with that due to c.m. motion. Here we chose the quark masses m_q and potential parameter 'a' as

$$(m_u = m_d, m_s) = (0.275, 0.438) \; GeV \tag{49}$$

$$a = 0.094 \; GeV \tag{50}$$

and choose suitably the values of the parameter 'b' as

$$(b_{1s}, b_{2s}, b_{1p}) = (2.93, 3.48, 2.55) \; GeV^{-1} \tag{51}$$

for $1s, 2s$ and $1p$ states respectively.

Table 2. The calculated values of $T_{ij}^{e,m}$ and I_χ required, respectively for OGE and GBE contributions in MeV

States	T_{uu}^m	T_{us}^m	T_{ss}^m	T_{uu}^e	T_{us}^e	T_{ss}^e	I_π	I_K	I_η
1S	78.072	71.257	81.719	397.817	489.379	484.642	100.775	62.784	1.964
2S	2.167	-	-	213.689	231.652	251.913	11.233	0.337	0.245
1P	5.734	-	-	268.980	292.262	313.778	43.230	1.870	1.404

Table 3. GBE contributions $(\delta M_B)_\chi$ (where $\chi = \pi, \eta, K$) for the ground states and the excited states of baryons in MeV

LS Multiplet	$(\delta M_B)_\pi$	$(\delta M_B)_K$	$(\delta M_B)_\eta^{uu}$	$(\delta M_B)_\eta^{us}$	$(\delta M_B)_\eta^{ss}$	total	$(\delta M_B)_\chi$
N	-309.225	0	24.533	0	0	24.534	-284.692
Δ	-61.845	0	-24.533	0	0	-24.534	-86.378
Λ	-185.535	-58.553	24.534	0	0	24.534	-219.554
Σ	-20.615	-97.588	-8.178	-41.076	0	-49.254	-167.457
Σ*	-20.615	-39.035	-8.178	20.538	0	12.360	-47.290
Ξ	0	-97.588	0	-41.076	-12.895	-53.971	-151.559
Ξ*	0	-39.035	0	20.538	-12.895	7.643	-31.392
Ω^-	0	0	0	0	-38.685	-38.685	-38.685
$\frac{1}{2}^+$, N(1440)	-90.175	0	-20.188	0	0	-20.188	-110.363
$\frac{3}{2}^-$, N(1520); $\frac{1}{2}^-$, N(1535)	-163.181	0	60.403	0	0	60.403	-102.778
$\frac{3}{2}^+$, Δ(1600)	-18.035	0	20.188	0	0	20.188	2.153
$\frac{1}{2}^-$, Δ(1620); $\frac{3}{2}^-$, Δ(1700)	8.516	0	35.436	0	0	35.436	43.952
$\frac{1}{2}^-$, Λ(1405); $\frac{3}{2}^-$, Λ(1520)	-111.626	-85.202	28.457	-74.457	0	-46.000	-242.828
$\frac{1}{2}^+$, Λ(1600)	-54.105	1.058	-20.188	0	0	-20.188	-73.235
$\frac{1}{2}^-$, Λ(1670); $\frac{3}{2}^-$, Λ(1690)	-111.626	1.635	22.457	-34.138	0	-5.681	-115.672

Then with these parameters the solution of eqn.(7) yields the individual quark binding energies E_{nl}^q for $1s, 2s, 1p$ states respectively as

$$(E_{1s}^u = E_{1s}^d, E_{1s}^s) = (0.4558, 0.5870) \, GeV \tag{52}$$

$$(E_{2s}^u = E_{2s}^d, E_{2s}^s) = (0.5617, 0.6961) \, GeV \tag{53}$$

$$(E_{1p}^u = E_{1p}^d, E_{1p}^s) = (0.5798, 0.7146) \, GeV \tag{54}$$

We then evaluate the integral expressions for $I_{ij}^{e,m}$ in eqs. (42) and (43) using a standard numerical method and calculate the terms $T_{ij}^{e,m}$ from eqs. (46) and (47), which are presented in table 2 and are used for computing the corrections due to GBE and OGE interactions.

In eq.(26) we find that the calculation of mesonic self-energy $(\delta M_B)_\chi$ depends upon the baryon-meson coupling constant $f_{BB'\chi}$. Indeed, in the chiral limit there is only one coupling constant for all Goldstone bosons. Due to explicit chiral symmetry breaking the coupling constant for π, η and K may become different. However, in order to prevent a proliferation of the free parameters we try to keep the number of free parameters as small as possible and assume a single phenomenological pion-nucleon coupling constant $f_{NN\pi} = 0.283$ for all mesons (π, η, K) following the approach of Glozman et al. [20-22] and this value is used here to compute the contribution due to GBE interactions. The calculated results for the contribution from the GBE interaction to the ground states as well as the excited states of baryons are presented explicitly in table 3. It is evident from table 3 that, particularly, for the ground state baryons the contribution due to pion exchange is somewhat larger than those due to Kaon and eta exchanges in case of $N, \triangle$ and Λ baryons whereas the contributions due to pion, kaon and η-exchanges are comparable in case of the other baryons. This result of the model is due to the fact that the couplings of pions, kaons and etas are related to SU(3). The energy contributions and the results obtained for the ground states and excitation spectra of light-and-strange baryons are displayed in table 4. The calculated values of the ground state and excited state masses of light and strange baryons are found to agree reasonably well with the experiment. It is found that the OGE contribution in the present model requires a value of quark gluon coupling constant $\alpha_c = 0.2$ which is quite consistent with the idea of treating OGE contribution in low order perturbation theory. The calculated values of the color-energy contributions arising from OGE interaction at short distances are somewhat small in the present model due to the fact that the central logarithmic potential (eq.(1)) possessing Coulombic behaviour at short distances includes a larger part of the color Coulombic energy in zeroth order calculation of quark binding energy.

It must be mentioned here that the evaluation of the energy contribution from GBE interaction in the sect. 3.2 ignores to a large extent the short range part of the meson exchange interactions which are of crucial importance in baryonic mass splittings. However we include these energy contributions phenomenologically in the present model through the potential parameter 'b' which has been suitably fixed at different values for 1S; 2S and 1P states of baryons so as to obtain the baryonic mass spectra in reasonable agreement with the experiment.

In the present model we find that the SU(3), breaking effect due to the quark masses $m_u = m_d \neq m_s$ lifts the degeneracy in the ground state baryon masses through the center of mass corrected energy term E_B among the groups $(N, \triangle)$, $(\lambda, \Sigma, \Sigma^*)$, (Ξ, Ξ^*) and Ω^-. Then in the next step, the GBE corrections $(\delta M_B)_\chi$ in QCD remove the degeneracy partially between N and $\triangle$; λ, Σ and Σ^*, Ξ and Ξ^*. However, the energy contributions due to OGE, particularly the color magnetic interaction energy removes the mass degeneracy completely among these baryons. It should be pointed out here that the colour-electric interactions energy due to OGE being minimal in case of excited states are ignored in this model.

Table 4. Energy Contributions $(\triangle E_B)_{cm}$, $(\triangle E_B)_g$, $(\delta M_B)_\chi$ and physical masses (M_B) of ground states and excited states of baryons in MeV

LS multiplet	E_B^0	$(\triangle E_B)_{cm}$	$(\triangle E_B)_g^m$	$(\triangle E_B)_g^e$	Total	$(\delta M_B)_\chi$	M_B Cal.	M_B Expt.
N	1367.430	−95.895	−46.843	0	−46.843	−284.692	940	940
$\triangle$	1367.430	−95.895	46.843	0	46.843	−86.378	1232	1232
Λ	1498.629	−96.972	−46.843	−19.260	−66.103	−219.554	1116	1116
Σ	1498.629	−96.972	−41.390	−19.260	−60.650	−167.457	1173.55	1193
Σ^*	1498.629	−96.972	44.118	−19.260	24.858	−47.290	1379.23	1385
Ξ[1]	1629.829	−97.866	−40.661	−19.260	−59.921	−151.559	1320.48	1321
Ξ^*[1]	1629.829	−97.866	44.847	−19.260	25.587	−31.392	1526.16	1533
Ω^-	1761.028	−98.619	49.032	0	49.031	−38.685	1672.76	1672
$\frac{1}{2}^+$, N(1440)	1685.186	−86.364	−1.300	0	−1.300	−110.363	1487.16	1440
$\frac{3}{2}^-$, N(1520) ; $\frac{1}{2}^-$, N(1535)	1739.263	−85.107	−3.440	0	−3.440	−102.778	1547.94	1527
$\frac{3}{2}^+$, Δ(1600)	1685.186	−86.364	1.300	0	1.300	2.153	1602.28	1600
$\frac{1}{2}^-$, Δ(1620) ; $\frac{3}{2}^-$, Δ(1700)	1739.263	−85.107	3.440	0	3.440	43.952	1701.55	1660
$\frac{1}{2}^-$, Λ(1405) ; $\frac{3}{2}^-$, Λ(1520)	1874.156	−86.633	−3.440	−0.353	−3.793	−242.828	1540.90	1462
$\frac{1}{2}^+$, Λ(1600)	1819.597	−87.856	−1.300	0.459	−0.841	−73.235	1657.67	1600
$\frac{1}{2}^-$, Λ(1670) ; $\frac{3}{2}^-$, Λ(1690)	1874.156	−86.633	−3.440	−0.353	−3.793	−115.672	1668.06	1680

We thus find the logarithmic potential model which takes into account the GBE contributions together with the contributions from OGE and c.m. motion provides the ground states and excitation spectra of baryons in reasonable agreement with the experiment. This model also explains successfully the correct level ordering of positive-negative parity excitation in N, Δ and λ spectra. It may be pointed out that the mass spectra of baryons have been studied in non-coulombic power low potential model in an earlier work [56] using such an approach successfully. This model could reproduce successfully the ground state mass spectra of baryons but failed to explain the correct level ordering of excited states of $N, \triangle$ and λ spectra.Therefore we find that the present model which treat the meson degree of freedom perturbatively in the manner like it is done in CBM [26], can describe the ground states and excitation spectra of light and strange baryons in reasonable agreement with experiment.

Acknowledgments

The authors are grateful to Dr. N. Barik, Mayurbhanj Professor of physics, Utkal University, Bhubaneswar, India, for his valuable suggestions and useful discussions on this work.

References

[1] C. Quigg and J.L. Rosner, *Phys. Lett.* **B 71**, 153(1977).

[2] C. Quigg and J.L. Rosner, *Phys. Rep.* **56**, 167(1979).

[3] E. Magyari, *Phys. Lett.* **B95**, 295(1980).

[4] S.N. Jena and D.P. Rath, *Phys. Rev.* **D34**, 196(1986).

[5] S.N. Jena and D.P. Rath, Pramana- *J. Phys.* **27**, 773(1986).

[6] N. Barik, S.N. Jena and D.P. Rath, *Phys. Rev.* **D41**, 1568(1990).

[7] D.P. Rath and S.N. Jena, Pramana - *J. Phys.* **32**, 753 (1989).

[8] N. Barik, S.N. Jena and D.P. Rath, *Int. J. Mod. Phys.* **A7**, 6813(1992).

[9] N. Barik, S.N. Jena and D.P. Rath, *Int. J. Mod. Phys.* **A9**, 327(1994).

[10] S.N. Jena, K.P. Sahu and P. Panda, *Int. J. of Th. Phys.* **243**, 161(2002).

[11] S.N. Jena, K.P. Sahu and T.C. Tripathy, *Int. J. of Th. Phy.* **11**, 69 (2004).

[12] N. Isgur and G. Karl, *Phys. Rev.* **D 18**, 4187(1978).

[13] N. Isgur and G. Karl, *Phys. Rev.* **D20**, 1191(1979).

[14] Y. Nogami and N. Ohtsuka, *Phys. Rev.* **D26**, 261(1982).

[15] A. de. Rujula, H. Georgi and S.L. Glashow, *Phys.Rev.* **D12**, 147(1975).

[16] L.Ya. Glozman, Z. Papp, W. Plessas, K. Varga and R.F. Wagenbrunn, *Phys. Rev.* **C57**, 3406(1998).

[17] I.T. Obukhovsky and A.M. Kusainov,*Phys.Lett.* **B238**, 142(1990).

[18] A. Valcarce, P. Gonzalez, F. Fernandez and V.Vento, *Phys.Lett.* **B367**, 35(1996).

[19] Z. Dziembowski et al., *Phys. Rev.* **C53**, R2038(1996).

[20] L.Ya. Glozman, W. Plessas, K. Varga and R.F. Wagenbrunn, *Phy. Rev.* **D58**, 97030(1998).

[21] L.Ya. Glozman and D.O. Riska, *Phys. Rep.* **268**, 263(1996).

[22] L.Ya. Glozman, *Nucl. Phys.* **A 663**, 103(2000).

[23] A. Chodos, R.L. Jaffe, K. Johnson, C.B. Thorn and V.F. Weisskopf, *Phys. Rev.* **D9**, 1471(1974).

[24] A. Chodos, R.L. Jaffe, K. Johnson, C.B. Thorn and V.F. Weisskopf,*Phys. Rev.* **D10**, 2599(1974).

[25] T. De. Grand, R.L. Jaffe, K. Johnson and J. Kiskis, *Phys. Rev.* **D12**, 2060(1975).

[26] A.W. Thomas, Adv. *Nucl. Phys.* **13**, 1(1983).

[27] R. Tegen, R. Brockmann and W. Weise, *Z. Phys.* **A307**, 339(1982).

[28] R. Tegen and W. Weise, *Z. Phys.* **A 314**, 357(1983).

[29] R. Tegen, M. Schedle and W. Weise, *Phys. Lett.* **B125**, 9(1983).

[30] P. Leal. Ferreira and N. Zagury, *Lett. Nuovo Climento Soc. Ital. Fis.* **20**, 511(1977).

[31] P. Leal. Ferreira, J. A. Halayal and N. Zagury, *Lett. Nuovo Climento Soc. Ital. Fis.* **A 55**, 215(1980).

[32] N. Barik and B.K. Dash, *Phys. Rev.* **D 31**, 1652(1985).

[33] N. Barik and B.K. Dash, *Pramana - J. Phys.* **24**, 707(1985).

[34] N. Barik and B.K. Dash, *Phys Rev.* **D 33**, 1925(1986).

[35] P. Leal. Ferreira, *Lett. Nuovo Climento Soc. Ital. Fis.* **20**, 157(1977).

[36] S.N. Jena and S. Panda, Pramana - *J. Phys.* **35**, 21(1990).

[37] S.N. Jena and S. Panda, *Int. J. Mod. Phys.* **A7**, 2841(1992).

[38] S.N. Jena and S. Panda, *J. Phys.* **G18**, 273(1992).

[39] S.N. Jena, M.R. Behera and S. Panda, *Phys. Rev.* **D55**, 291(1997).

[40] S.N. Jena, M.R. Behera and S. Panda, *J. Phys.* **G 24**, 1089(1998).

[41] S.N. Jena and M.R. Behera, *Int. J. Mod. Phys.* **E 7**, 69(1998).

[42] S.N. Jena and M. R. Behera, *Int. J. Mod. Phys.* **E 7**, 425(1998).

[43] S.N. Jena and M.R. Behera, *Pramana- J. Phys.* **44**, 357(1995).

[44] S.N. Jena and M.R. Behera, *Pramana- J. Phys.* **47**, 233(1996).

[45] A. Martin, *Phys. Lett.* **B 93**, 338(1980).

[46] N. Barik and S.N. Jena, *Phys. Lett.* **B 97**, 261(1980).

[47] N. Barik and S.N. Jena, *Phys. Lett.* **B 97**, 265(1980).

[48] N. Barik and S.N. Jena, *Phys. Rev.* **D 28**, 612(1982).

[49] N. Barik and M. Das, *Phys. Lett.* **B 120**, 403(1983).

[50] N. Barik and M. Das, *Phys. Rev.* **D 28**, 2823(1983).

[51] N. Barik and M. Das, *Phys. Rev.* **D33**, 176(1986).

[52] N. Barik and M. Das, Pramana- *J. Phys.* **27**, 727(1986).

[53] S.N. Jena, P. Panda and K.P. Sahu, *Int.J. of Th. Phys.* **9**, 69 (2002).

[54] S.N. Jena, P. Panda and K.P. Sahu, *Int. J. of Th. Phys.* **8**, 351(2002).

[55] S.N. Jena, P. Panda and K.P. Sahu, *J.Phys. G. Nucl. Part. Phys.* **27**, 1519(2001).

[56] S.N. Jena, P. Panda and T.C. Tripathy, *Phys. Rev.* **D63**, 014011(2000).

[57] Y. Nambu and G. Jona- Lasino, *Phys. Rev.* **112**, 345(1961).

[58] Y. Nambu and G. Jona- Lasino, *Phy. Rev.* **124**, 246(1961).

[59] R. Brockmann, W. Weise and E. Werner, *Phys. Lett* **B 122**, 201(1983).

[60] S. P. Klevansky, *Rev. Mod. Phys.* **64**, 649(1992).

[61] S. Weinberg, *Physica* **A 96**, 327(1979).

[62] A. Manohar and H. Georgi, *Nucl. Phys.* **B 234**, 189(1984).

[63] D.I. Diakonov and V. Yu. Petrov, *Nucl. Phys.* **B 272**, 457(1986).

[64] M. Gell-Mann and M. Levy, *Nuovo Climento* **16**, 705(1960).

[65] J. Bartelski, A. Szymacha, L. Mankiewicz and S. Tatur, *Phys. Rev.* **D 29**, 1035(1984).

[66] E. Eich, D. Rein and R. Rodenberg, *Z. Phys.* **C 28**, 225(1985).

[67] S. Aoki et al., *Phys. Rev.Lett.* **82**, 4392(1999).

[68] N. Barik and M. Das, Pramana- *J. Phys.* **27**, 783(1986).

[69] N. Barik and M. Das, *J. Phys.* **G 13**, 567(1987).

[70] S.N. Jena, M.R. Behera and S. Panda, *J. Phys.* **G 24**, 1089(1998).

[71] S.N. Jena, M.R. Behera and S. Panda, *Phys. Rev.* **D 55**, 291(1997).

[72] L. Ya. Glozman, hep- ph / 0004229 (unpublished)(2000).

In: Theoretical Physics and Nonlinear Optics
Editors: Thomas F. George et al
ISBN 978-1-61122-939-4
© 2012 Nova Science Publishers, Inc.

Chapter 14

NON-LINEAR REFRACTIVE INDEX THEORY: (I) GENERALISED OPTICAL EXTINCTION THEOREM AND DISPERSION RELATION

S.S. Hassan[1,*], *R.K. Bullough*[2,†] *and R. Saunders*[3,‡]

[1]University of Bahrain, College of Science, Department of Mathematics, PO Box 32038, Kingdom of Bahrain.
[2]University of Manchester, Mathematics Department, Manchester M13 9PL, United Kingdom
[3]Manchester Metropolitan University, Faculty of Science and Engineering, Department of Computing and Mathematics, Manchester M1 5GD, United Kingdom.

Abstract

Starting from the Bloch-Maxwell equations for two-level atoms forming an extended system, taken as a parallel-sided slab for a Fabry-Perot cavity configuration, we generalize the optical extinction theorem to the non-linear regime. Generalized form of the Lorentz-Lorenz dispersion relation for the refractive index (m) is derived. Within the context of optical multistability phenomenon, the input-output field relationship is derived in terms of (m).

1. Introduction

In this two part series of papers we present a comprehensive report of the fundamental theoretical study of optical multistability (OM) in a non-linear Fabry-Perot (F-P) cavity in the normal vacuum state of the electromagnetic field. The starting point is the coupled Bloch-Maxwell equations for a system of 2-level atoms coupled to the electromagnetic field in an extended medium. The cavity is defined by the region (a parallel sided slab) occupied by the atoms and there are no additional mirrors. A *self-consistent* scheme is developed

---

[*]E-mail address: Shoukryhassan@hotmail.com (Corresponding Author)
[†]Ceased on 30th August 2008
[‡]E-mail address: R.Saunders@mmu.ac.uk

which allows subsequent numerical analysis. Very brief reports of some of the analysis presented here have been made in the past [1]-[4] but a complete report of our theory with all the necessary details has never been made.

In our theory the F-P cavity is defined by the parallel-sided slab occupied by atoms. No macroscopic Maxwell boundary conditions are imposed at the surface of this slab. Instead, the more fundamental boundary conditions of outgoing free fields at infinity are imposed. The theory yields a non-linear refractive index m, together with the Fresnel reflection and transmission coefficients expressed in terms if m. In the many-body sense, the theory is therefore a microscopic theory rather than a macroscopic theory. A comparable theory of the F-P cavity but in the linear optical regime has been reported at some length [5]-[8] in an all order many-body theory involving all orders of the correlation between the atoms. In this extension, to incident fields intense enough to create non-linear optical multi-bistability, we restrict the many-body side of the analysis to a one-body theory, except for the occasional remarks connecting the results [5]-[8] of the linear theory. Thus the starting point is the Bloch-Maxwell system defined smoothly at points $\underline{x}$ inside the slab of the F-P cavity [9]. The presented non-linear theory of this paper is therefore a coarse-grained theory where no inter-atomic correlations (2-body, three body etc.) appear. Within the one-body approximations the non-linear theory developed in the paper is essentially exact for two –level atoms.

The analytical results of the theory, presented in part (I) of this series, and confirmed by the subsequence numerical work in part (II), show that the cavity detuning depends on the input intensity. The non-linear behavior of the F-P cavity arises from the intensity dependency of the non-linear refractive index, m, that emerges from the theory through an application of the fundamental "optical extinction theorem" originally due to Ewald-Oseen (cf. references [10]-[14] and references [5]-[8] for the equivalent linear theory). The non-linear action of the F-P system at the boundaries of the slab-like cavity are derived from the fundamental Bloch-Maxwell system of equations with the outgoing boundary conditions. Also, we show how the derived non-linear refractive index m consistently describes the F-P cavity action in a non-linear way, leading to multistability, and we see certain corrections to this explicitly in solving atomic inversions. This non-linear action of the F-P cavity contrasts with the non-linear action of a ring cavity where the non-linearity is provided by optical feedback.

Part (I) of this series presents the theoretical generalization of the optical extinction theory to the non-linear regime in the normal vacuum, while part (II) concentrates on a self –consistent numerical scheme to investigate multistable behavior, utilizing the analytical formulas derived in part (I).

The organization of this paper, part (I), is as follows. In section 2 we write down the Bloch-Maxwell system of equations in the one-body form explained above. These are coupled partial differential equations in which Maxwell's equations are linear and can be conveniently put into integral form. In section 3 we make the necessary generalization of the optical extinction theory to the non-linear regime, together with the generalized form of the Lorentz-Lorenz dispersion relation for the refractive index m. In section 4 we consider a low atomic density approximation to the analytical expressions derived in section 3. This way we draw a corresponding between the present analysis and previous work [15], [16] on OM for a Fabry-Perot cavity which involve spatial mean-field [17]-[19] or other approximations:

2. Model Equations

Our starting point is the coupled Bloch-Maxwell system of equations for an extended system of many 2-level atoms coupled to the electromagnetic field and taken in the "one-body" approximation in which elements $r_\pm(\underline{x}, t)$, $r_3(\underline{x}, t)$ of the atomic Bloch vectors are densities with a smooth dependence on the position $\underline{x}$ of the atoms. In the first instance these are taken without any slowly varying envelope approximation. Later we find precisely what envelope description should be used-one which is essentially exact.

In a frame rotating at the angular frequency ω (namely the frequency of a harmonic incident laser field) the c-number Bloch equations for the atoms, in the rotating wave approximation and in the electric dipole approximation are [9] (also see later derivations in the general case of squeezed vacuum state of the electromagnetic field [21], [22]).

$$\frac{\partial r_+(\underline{x},t)}{\partial t} + \gamma(1+i\delta)r_+(\underline{x},t) = 2i\left[\mathrm{E}^*(\underline{x},t) - \Omega e^{-ik_0 z}\right] r_3(\underline{x},t) \tag{2.1a}$$

$$\frac{\partial r_3(\underline{x},t)}{\partial t} + \gamma(1+2r_3(\underline{x},t)) = -i\left[\mathrm{E}^*(\underline{x},t) - \Omega e^{-ik_0 z}\right] r_-(\underline{x},t) + c.c. \tag{2.1b}$$

with $r_- = (r_+)^*$

The symbols are: γ is the A-coefficient, $\delta = \frac{(\omega-\omega_0)}{\gamma}$ is the normalized atomic detuning, ω_0 is the two-level atomic frequency, $\Omega \equiv p\hbar^{-1}E_o$ is the amplitude of the incident laser field expressed as a Rabi frequency, p is the magnitude of the atomic dipole matrix element. In the laboratory frame, the dipole atomic density is $n_0 p\hat{\underline{u}}\left(r_+ e^{i\omega t} + r_- e^{-i\omega t}\right)$ where $\hat{\underline{u}}$ is a unit vector along p, n_0 is the atomic number density: $r_3(\underline{x}, t)$ is the density of the atomic inversion. In this laboratory frame, Maxwell's equation is

$$\operatorname{curl}\operatorname{curl}\underline{\mathrm{E}}(\underline{x},t) + \frac{1}{c^2}\frac{\partial^2}{\partial t^2}\underline{\mathrm{E}}(\underline{x},t) = -4\pi n_0 \frac{\partial^2}{\partial t^2}\underline{\mathrm{P}}(\underline{x},t) \tag{2.2}$$

The Maxwell equation (2.2) is linear and has a well known Green's function [5]-[9],[13], [14]. As long as we are concerned throughout with time dependence $e^{\pm i\omega t}$ we can work with the Green's functions $\underline{\underline{F}}$ and $\underline{\underline{F}}^*$ which are the second rank tensors

$$\underline{\underline{F}}(\underline{x}, \underline{x}'; \omega) = (\nabla\nabla + k_0^2 \underline{\underline{U}}) \frac{e^{ik_0 r}}{r} \tag{2.3}$$

in which $r = |\underline{x} - \underline{x}'|$, $\underline{\underline{U}}$ is the unit tensor, and $k_0 = \omega c^{-1}$. The form of the harmonic laser field assumed is evidently $\mathcal{E}_0\left[e^{i(\omega t - k_0 z)} + e^{-i(\omega t - k_0 z)}\right]$ with $\mathcal{E}_0$ a real amplitude. Thus the total field driving the Bloch equations (1) expressed as a Rabi frequency has negative frequency part $\left(\underline{\mathrm{E}}^*(\underline{x},\omega) - \Omega e^{-ik_0 z}\right)$ in which [9]

$$\underline{\mathrm{E}}^*(\mathrm{x},\mathrm{t}) = n_0 p^2 \hbar^{-1} \int_{V-v} \underline{\underline{F}}^*(\underline{x}, \underline{x}'; \omega) : \hat{\underline{u}}\hat{\underline{u}}\, r_+(\underline{x}', \omega)\, d\underline{x}' \tag{2.4}$$

For simplicity we assume that all fields have the polarization direction $\hat{u}$ of the atomic dipole matrix elements. In principle it is possible to average over the assumed isotropic distribution of atomic polarization directions but this complication, in the one-body theory, is easily introduce later in the expressions we shall reach in section 3 for the non-linear refractive index m . .Notice that in the expression (2.4) for E^*, the complex conjugate of the atomic polarizing field, a small sphere ν is extracted from the region of integration (V is the parallel-sided slab) at the point $\underline{x}$ where an atom is supposed to be. This sphere acts like the famous Lorentz sphere introduced by Lorentz in linear dielectric theory and is the source of the Lorentz polarizing field (different from the Maxwell field inside the medium [5]-[8]) appearing in the present theory. However, it is mathematically necessary to extract ν from V in (2.4) since the integral is only conditionally convergent [5]-[8]: $\underline{\underline{F}}$ behaves like r^{-3} as $r \to 0$. Thus ν is a sphere of vanishingly small radius.

Although this situation surrounding the polarization field appears to be non-physical this is not the case. The small sphere acts in the theory precisely as does the Lorentz's physically based sphere and there is a correction involving the local correlation between the atoms (as in the linear theory [5]-[8]). This correction which is believed small even in the non-linear theory is beyond the one-body theory and is neglected in the present non-linear analysis. All the many-body theoretical details, involving correlation that make physical sense of the sphere ν, in the linear case are carried out in references [5]-[8].

In the following section 3 we outline the steps of the solution of the Bloch equations (1) driven by the polarizing field of Eq. (2.4). Since Eq. (2.4) depends under the integral sign on $r_+(\underline{x},\omega)$ this system of equations is a non-linear integro-differential systems. As we carry out these steps in section III we make further references to the comparable linear theory [5]-[8] whenever this seems to be helpful for the understanding of the non-linear theory.

Notice that there are radiative damping terms already contained in the Bloch equations (1) These describe optical scattering from non-linear medium. The simplest of the many-body theoretical correlation corrections to (2.4) adds a further integral like that in (2.4) in which however the pair correlation function for pairs of atoms $g(\underline{x},\underline{x}')$ appears in the form $g_2(\underline{x},\underline{x}') - 1$ under the integral sign (compare the linear theory in [5]-[8] and [9]). This has the effect of changing the damping of the Bloch system to include pair-wise optical scattering from pairs of atoms in the dielectric. In the non-linear context corrections like this become still more complicated. Thus we neglect this at the one-body level in the non-linear theory.

We are now almost in a position to develop the non-linear theory at one-body level analytically, that is we formally solve the system of equations (1) with Eq. (2.4).

We move the differential operator under the integral sign in (2.4) outside the integral so that [5]-[8],[12]

$$\frac{\underline{E}^*(\underline{x},\omega)}{n_0\, p^2\, \hbar^{-1}} = \frac{4\pi}{3}\underline{\underline{U}}\, r_+(\underline{x},\omega) + \hat{u}\,\hat{u} : (\underline{\nabla}\,\underline{\nabla} + k_o^2\underline{\underline{U}}) \int_V \frac{e^{ik_o r}}{r} r_+(\underline{x}',\omega)\, d\underline{x}'$$

(2.5)

The first term is entirely due to the small sphere and is the source of the Lorentz-Lorenz form of the expression we derive for the non-linear refractive index m. The resulting integral is convergent and no small sphere need be removed from the region of integration V.

We now develop the optical extinction theorem and derive the expression for a non-linear refractive index m.

3. Generalized Optical Extinction Theorem

In the present non-linear theory the atomic inversion density $r_3(\underline{x}) > -\frac{1}{2}$, its ground state value in the steady state, and depends on $\underline{x}$, the position in the cavity. It is this complication which makes the theory non-linear, for as a consequence of the radiation damping in the Bloch equations (1), all the field intensities must damp inside the cavity and since $r_3(\underline{x},t)$ depends on these intensities $r_3(\underline{x},t)$ necessarily depends on $\underline{x}$. Consistency then requires that the polarization fields $r_\pm(\underline{x},t)$ are no longer simply of harmonic form in space. In the linear theory $r_\pm(\underline{x},t)$ depend on waves $e^{\pm imk_0 z}$ where m proves to be the linear refractive index. In the present non-linear theory we find we need to take as ansatz

$$r_+(\underline{x},t) = P_f(z) e^{-imk_0 z} + \Lambda P_b(z) e^{imk_0 z} \tag{3.1}$$

in response to the incident laser field $e^{\pm ik_0 z}$. The axis z is the axis of the F-P cavity, taken to be the slab bounded by the two surfaces $z = 0$ and $z = L > 0$, and the laser field is, for simplicity, normally incident upon the slab. This slab is the region of integration V (see equation (4)) and is infinite in value. However, one should really use a parallel-sided cylinder, for example, as the cavity with a large radius R orthogonal to z. Corrections due to the additional diffraction effects from the surface of the cylinder are not considered in our theory within the context of OM. In the linear theory [5],[6] these corrections are very complicated but are of small order macroscopically.

Notice now that Eq. (3.1) means that a non-linear refractive index m has entered the theory. It is the purpose of the analysis to follow to find an expression for m and so find the input-output relation for the cavity expressed as far as possible in terms of this refractive index m.

Notice also that (3.1) has envelope functions $P_f(z)$ and $P_b(z)$ depending on z. $P_f(z)$ and $P_b(z)$ are envelopes assumed to be 'slowly varying' on the scale of a wavelength $\lambda = 2\pi k_0^{-1}$. In practice we approximate the analysis by truncating the second and higher derivatives in z of both $P_f(z)$ and $P_b(z)$ (see below). The quantity Λ in (3.1) is a reflection coefficient which is to be determined.

We now turn to the essence of the extinction theorem. The essential physical content of this theorem is that light in the absence of matter travels as a free field [5]-[8]. If it is a harmonic wave along z it has z–dependences $e^{\pm ik_0 z}$. If such waves enter a region occupied by matter, such as the region V of the F-P cavity, a new wave number $mk_0 \neq k_0$ arises in both linear and non-linear theory inside V. Then scattering from the surface of V exactly cancels the free fields which in the absence of the matter would be inside V and replaces these free fields with fields with new wave numbers. In mathematical terms this exact cancellation of free fields provides the connection between the amplitude of the original free field and the amplitudes of the field induced, with a new wave number, inside the cavity V. At the same time the 'refractive index' m has to satisfy a second condition which means in practice that there is a formula (of Lorentz-Lorenz type) which fixes the values of m.

The mathematical analysis implementing this physics is carried out in various ways in the linear case in [5]-[8] and their references. How is it that the theory can be extended to the non-linear case? The point is that the Maxwell equations are linear and the field E^* of (2.5) depends on $\underline{x}$ through the Green's function $\frac{e^{-ik_0 r}}{r}$ under the integral (taken over $\underline{x}'$: $r = |\underline{x} - \underline{x}'|$). By using the result that

$$\int_V \frac{e^{-ik_0 r}}{r} e^{\pm imk_0 z'} d\underline{x}' = \frac{1}{k_0^2(m^2-1)} \int_V \left[e^{\pm imk_0 z'} \nabla_{\underline{x}'}^2 \frac{e^{-ik_0 r}}{r} - \frac{e^{-ik_0 r}}{r} \nabla_{\underline{x}'}^2 \frac{e^{\pm imk_0 z'}}{r} \right] d\underline{x}'$$

and applying Green's Theorem one gets

$$\int_V \frac{e^{-ik_0 r}}{r} e^{\pm imk_0 z'} d\underline{x}' = \frac{4\pi}{k_0^2(m^2-1)} e^{\pm imk_0 z}$$
$$+ \frac{1}{k_0^2(m^2-1)} \int_\Sigma \left[e^{\pm imk_0 z'} \nabla_{\underline{x}'} \left(\frac{e^{-ik_0 r}}{r} \right) . d\underline{s} - \frac{e^{-ik_0 r}}{r} d\underline{s} . \nabla_{\underline{x}'} \left(e^{\pm imk_0 z'} \right) \right] \quad (3.2)$$

where $\int_\Sigma$ is a surface integral, taken over the surface of V, which depends on the field point $\underline{x}$ through the Green's function $\frac{e^{-ik_0 r}}{r}$. This surface integral satisfies the free-field wave equation $(\nabla^2 + k_0^2) \int_\Sigma = 0$. On the other hand, with $m \neq 1$, the remaining term in (3.1) plainly satisfies a differential wave equation $(\nabla^2 + k_0^2) e^{\pm imk_0 z} = 0$. It follows that the terms in the theory split into two parts : one satisfying the free field wave equation and the other satisfying the different wave equation with $m \neq 1$. In the linear theory the first free field condition implements the connection between scattering from dipole sources over the surface of V and the incident free field. The second condition fixes the parameter m. All of the analysis goes through in the non-linear case also except that we need to take careful account of the envelope functions $P_f(z)$ and $P_b(z)$ modulating the waves $e^{\pm imk_0 z}$.

In the appendix A we show how by inserting the ansatz (3.1) into the expression (2.4) for the inter-atomic 'polarizing field' $\underline{E}^*(\underline{x},\omega)$ we find the following expression (up to first derivatives):

$$\frac{\underline{E}^*(\underline{x},t)}{n_0 \, p^2 \, \hbar^{-1}} = \frac{4\pi}{(m^2-1)} \left[\left(\frac{m^2+2}{3}\right) P_f(z) - \frac{2im}{k_0(m^2-1)} P_f'(z) \right] e^{-imk_0 z}$$
$$+ \frac{1}{(m^2-1)} \left[I(P_f(z)) - \frac{2im}{k_0(m^2-1)} I(P_f'(z)) \right]$$
$$+ \Lambda(\text{same terms with } P_f \text{ replaced with} P_b \text{ and m replaced by -m})(3.3)$$

where

$$P_f'(z) = \frac{dP_f(z)}{dz}, P_b'(z) = \frac{dP_b(z)}{dz}$$

and

$$I(h(x)) = \int_{\Sigma} \left[h(\underline{z}') e^{-imk_0 z'} \left(\nabla_{\underline{z}'} \left(\frac{e^{-ik_0 r}}{r} \right) . d\underline{s} \right) - \frac{e^{-ik_0 r}}{r} (d\underline{s} . \nabla_{\underline{z}'}) h(\underline{z}') e^{-imk_0 z'} \right] \quad (3.4)$$

Note that in (3.3) all of the surface integrals as defined from (3.4), depend only on $e^{\pm i k_0 z}$ and satisfy the free field wave equation $(\nabla^2 + k_0^2) e^{\pm i k_0 z} = 0$, with wave number k_0 as we show below. On the other hand the remaining terms are envelope functions multiplying $e^{\pm i m k_0 z}$ and these cannot satisfy a wave equation with $m \neq 1$. When (3.3) is inserted in the first of the steady state Bloch equations in (1a) with $\dot{r}_{\pm,3} = 0$, we have to equate separately the terms involving $e^{\pm i m k_0 z}$ and those involving $e^{\pm i k_0 z}$. In consequence we get, for the terms involving $e^{\pm i m k_0 z}$, the first order differential equations,

$$\gamma(1+i\delta) P_f(z) = \frac{8\pi i n_0 p^2}{\hbar(m^2-1)} \left[\left(\frac{m^2+2}{3} \right) P_f(z) - \frac{2im}{k_0(m^2-1)} P_f'(z) \right] r_3(z) \quad (3.5a)$$

$$\gamma(1+i\delta) P_b(z) = \frac{8\pi i n_0 p^2}{\hbar(m^2-1)} \left[\left(\frac{m^2+2}{3} \right) P_b(z) + \frac{2im}{k_0(m^2-1)} P_b'(z) \right] r_3(z) \quad (3.5b)$$

In the linear case where $P_f'(z) = P_b'(z) = 0$ and $r_3(z) \simeq -\frac{1}{2}$, Eq. (3.5) then yields exactly the usual Lorentz-Lorenz dispersion relation

$$\frac{m^2-1}{m^2+1} = \frac{4\pi n_0 p^2}{3\hbar\gamma(i-\delta)}. \quad (3.6)$$

But in the present non-linear case, Eqs. (3.5) couples to the spatially dependent $r_3(z)$, and m is no longer determined. Since m is to be simply a number in the ansatz (6) and cannot depend on z, we expect that m can nevertheless be related to some z-independent inversion R_3 by a relation like (3.6). The idea here is to introduce some 'average' inversion which best describes $r_3(z)$ at all points z inside the slab. To do this we re-write Eq. (3.5a), for example, in the form

$$\gamma(1+i\delta) P_f(z) = \frac{8\pi i n_0 p^2}{\hbar(m^2-1)} \left[\left(\frac{m^2+2}{3} \right) P_f(z)(R_3 - R_3 + r_3(z)) - \frac{2im}{k_0(m^2-1)} P_f'(z) r_3(z) \right] \quad (3.7)$$

and chose m to satisfy a generalized form of (11), namely,

$$\frac{m^2-1}{m^2+1} = \frac{-8\pi n_0 p^2}{3\hbar\gamma(i-\delta)} R_3 \quad (3.8)$$

Then,

$$P_f'(z) = f(z) P_f(z) \quad (3.9a)$$

A similar argument for the Eq. (10b) gives

$$P_b'(z) = -f(z) P_b(z) \quad (3.9b)$$

where $f(z)$ is given by

$$f(z) = \frac{k_0 \left(m^2 - 1\right) \left(m^2 + 2\right)}{6mi} \left(1 - \frac{R_3}{r_3(z)}\right) \qquad (3.10)$$

Equations (3.9) are formally integrated

$$P_f(z) = P_f(0) \exp\left[\int_0^z f(z')\, dz'\right] \qquad (3.11a)$$

$$P_b(z) = P_b(0) \exp\left[-\int_0^z f(z')\, dz'\right] \qquad (3.11b)$$

To proceed further we treat $m(\omega)$ as an adjustable parameter, so we choose it so that the F-P action survives in the non-linear case. It is sufficient that we take

$$P_{f,b}(0) = P_{f,b}(L) = P_o(\omega) \qquad (3.12)$$

This way we find that the parameter R_3 is exactly given in the form

$$R_3 = \frac{L}{\int_0^L r_3^{-1}(z')\, dz'}, \qquad (3.13)$$

that is $\frac{1}{R_3}$ is an average of the inverse of $r_3(z)$, but it is not the mean field average [18] (see also [9] and references therein).

Note that in the linear theory $r_3(z) = -\frac{1}{2} = R_3$. Also, in the non-linear theory, (3.8) with (3.13) is still an exact result as we are not approximating R_3 as an approximate average.

Now from Eq. (2.1b) and within the same argument applied to (1a) and the use of Eqs. (3.9), and (3.10) we get

$$2r_3(z)\left(r_3(z) + \tfrac{1}{2}\right) = -\left[|P_f(z)|^2 e^{-i(m-m^*)k_0 z} + |\Lambda|^2 |P_b(z)|^2 e^{i(m-m^*)k_0 z}\right]$$
$$+ \text{terms in } e^{\pm i(m+m^*)k_0 z}$$
$$(3.14)$$

The cross terms in $e^{\pm i(m+m^*)k_0 z}$ are due to the fact that $r_3(z)$ depends on the intensity in the cavity which is proportional to $\left|P_f(z) e^{-imk_0 z} + \Lambda P_b(z) e^{imk_0 z}\right|^2$, and are indications of the standing wave effects. Since this causes considerable complications we simply ignore the standing waves in this paper. The other terms in (3.14), $e^{\pm i(m-m^*)k_0 z}$, are damped and growing (damped backward going) wave profiles. From (3.8) the scale on which the damping takes place is roughly

$$-i(m - m^*) \simeq \alpha_0 \frac{4R_3}{1 + \delta^2}; \qquad \alpha_0 = \frac{4\pi n_0 p^2}{\hbar \gamma} \qquad (3.15)$$

which is negative since R_3 is negative. For Na vapour with $n_0 \sim 10^{12} cc^{-1}$, $\alpha_o \sim 10^{-4}$, so $-i(m - m^*)$ leads to damping by $e^{-2\pi}$ in a length $L \sim 10^4$ cm and wave length ~ 0.5 cm. Of course, the envelopes $P_{f,b}$ are to describe this damping, but we expect that their precise form do not change this picture too much.

Thus, within the neglect of the standing waves, Eq. (3.14) with the use of (3.11) can be put in the form,

$$2r_3(z)\left(r_3(z) + \tfrac{1}{2}\right) = -|P_o|^2\left[e^{h_1(z)} + |\Lambda|^2 e^{-h_1(z)}\right] \tag{3.16a}$$

where

$$h_1(z) = -i(m - m^*)k_0 z + 2\mathrm{Re}\left[\frac{(m^2+2)(m^2-1)}{6mi}\right]k_0 z$$
$$+ 2\mathrm{Re}\left[\frac{(m^2-1)^2(1+i\delta)}{m}\right]\left(\frac{\gamma k_0}{16\pi n_0 p^2 \hbar^{-1}}\right)\int_0^z r_3^{-1}(z')\,dz'. \tag{3.16b}$$

Now we make the second step in the extinction theorem and equate the terms in $e^{\pm ik_0 z}$ in Eq. (1a). First we recall that the surface integrals in (3.3) are evaluated for the parallel-sided slab geometry $((0 \le z \le L)$. The results are already available [5].

At the surface $z' = 0$,

$$I(P_{f,b}(0)) = -2\pi P_{f,b}(0)(1 \pm m)e^{-ik_0 z} \tag{3.17a}$$

and similarly

$$I(P'_{f,b}(0)) = -2\pi P'_{f,b}(0)(1 \pm m)e^{-ik_0 z}. \tag{3.17b}$$

At the surface $z' = L$

$$I(P_{f,b}(L)) = -2\pi P_{f,b}(L)(1 \pm m)e^{\pm imk_0 z}e^{-ik_0(L-z)} \tag{3.18a}$$
$$I(P'_{f,b}(L)) = -2\pi P'_{f,b}(L)(1 \pm m)e^{\pm imk_0 z}e^{-ik_0(L-z)} \tag{3.18b}$$

After substituting (22), (23) into (3.3) and then inserting the result in (1a) and using Eqs. (3.9), (3.12) and hence comparing the coefficients of $e^{\pm ik_0 z}$ we find respectively that

$$\Lambda = \left(\frac{m-1}{m+1}\right)e^{-2imk_0 L} \tag{3.19}$$

$$p_0 = \frac{\frac{-\Omega}{2\pi n_0 p^2 \hbar^{-1}}\left(\frac{m^2-1}{m^2+1}\right)}{\left[1 - \left(\frac{m^2+2}{3}\right)\left(1 - \frac{R_3}{r_3(0)}\right)\right]\left[1 - \left(\frac{m-1}{m+1}\right)\Lambda\right]} \tag{3.20}$$

Equation (3.19) represents the Fresnel reflection coefficient, times the phase shift, while (3.20) means that the usual F-P cavity action takes place. For normal incidence on the F-P cavity $(0 \le z \le L)$ we use the results (22), (23), (3.20), into (3.3) to obtain the input-output relation for $z > L$

$$E_{out} = E_{in}\frac{\frac{4\pi}{(1+m)^2}e^{-i(m-1)k_0 L}}{\left[1 - \left(\frac{m-1}{m+1}\right)^2 e^{-2imk_0 L}\right]}\left\{\frac{\left[1 - \left(\frac{m^2+2}{3}\right)\left(1 - \frac{R_3}{r_3(L)}\right)\right]}{\left[1 - \left(\frac{m^2+2}{3}\right)\left(1 - \frac{R_3}{r_3(0)}\right)\right]}\right\} \tag{3.21}$$

where $E_{in} = \frac{\hbar\Omega}{n_0 p^2}$. E_{out} is scaled by $\frac{n_0 p^2}{\hbar}$. This is together with a free field contribution from the surface at $z' = 0$ is just sufficient to annihilate the incident field everywhere.

The factors in the inner square brackets that contain $r_3(0)$ and $r_3(L)$ reflects the non-linear nature of the relations (3.21): indeed the non-linearity is in m(ω) since it depends on the number R_3. The F-P action is represented by the term in $\left[1 - \left(\frac{m-1}{m+1}\right)^2 e^{-2imk_0L}\right]$.

The analytic formulae derived in this section namely (3.8), (3.11), (3.13), (21) and (24)-(26) are the main results of this paper, part (I). In the following section we show how the conventional optical bistability (OB) behavior lies within the obtained formulae of the present theory.

4. Optical Bistability in the Small Density Limit

We now derive some transparent form for the input-output relationship of Eq. (3.21), when the atomic density is small. First, we may write the expression

$$\left|1 - \left(\frac{m-1}{m+1}\right)^2 e^{-2imk_0L}\right|^2 = \left(1 - |R|^2 e^{-B}\right)^2 + 4|R|^2 e^{-B} \sin^2\left(\frac{\theta - A}{2}\right) \quad (4.1)$$

where $R \equiv \left(\frac{m-1}{m+1}\right) = |R|e^{i\lambda}$, $\theta = 2(\lambda + \nu\pi - k_0L)$
with ν an integer and we have put

$$2(m-1)k_0L = A - iB \quad (4.2)$$

For small atomic density n_0, such as in metal vapour, where m $\simeq 1$, we assume that $m^2 - 1 \simeq 2(m-1)$ and $m^2 + 2 \simeq 3$. Hence from the generalized dispersion relation (3.8) one finds that

$$m - 1 \simeq \left(\frac{4\pi n_0 p^2}{\hbar\gamma}\right)\left(\frac{\delta + i}{1 + \delta^2}\right) R_3 \quad (4.3)$$

From (4.2) and (4.3) we then have $A = \frac{C\delta R_3}{1+\delta^2}$ and $B = \frac{-CR_3}{1+\delta^2} > 0$ where the 'cooperation' number [17], [18] $C = \frac{2}{\tau_R \gamma}$ and $\tau_R \equiv \frac{\hbar}{4\pi n_0 p^2 k_0 L}$ is the well-known cooperative super-fluorescence characteristic time [25]. For the second term in (4.1), θ is a scaled cavity detuning and A depends on both the atomic detuning δ and R_3 and hence, via (3.13), (21), (3.20) and (3.21), on the input field intensity $|E_{in}|^2$. Consequently this second term is the source of dispersive OB. Note that C can be small for small n_0 or small k_0L or both. Further, if we consider the envelopes $P_{f,b}(z)$ and hence $r_3(z)$ to be interpreted as mean average values [18], so in this case $R_3 \simeq r_3$ and from (3.20) we approximately have

$$p_0 \simeq \frac{-\hbar\Omega(m-1)}{2\pi n_0 p^2 \left[1 - \left(\frac{m-1}{m+1}\right)\Lambda\right]} \quad (4.4a)$$

Using (3.21) and (4.3) into (4.4a) we get

$$|p_0|^2 \simeq \left|\frac{E_{out}}{\delta}\right|^2 \frac{r_3^2 e^B}{\left|\frac{2im}{(m+1)^2}\right|^2 (1+\delta^2)} \tag{4.4b}$$

With (4.4a) substituting into (3.14) we get for small enough n_0 that,

$$r_3 = -\frac{1}{2}\frac{1+\delta^2}{1+\delta^2+|X|^2} \tag{4.5}$$

where $|X|^2 \simeq \left|\frac{E_{out}}{\delta}\right|^2 \left(\frac{1+|\Lambda|^2}{2\left|\frac{2im}{(m+1)^2}\right|^2}\right)$ is a scaled output intensity from the cavity.

Further, for small n_o we put $4\sin^2\left(\frac{\theta-A}{2}\right) \simeq (\theta-A)^2$ and $e^{-B} \simeq 1$ (small absorption) and the use of (4.5) and insertion of (4.1) into (26) yields then the usual cubic expression for the dispersive OB [19], [23], [24]

$$|Y|^2 = |m|^2|X|^2\left[1+\left\{\frac{|R|^2}{\left(1-|R|^2\right)^2}\left(\theta+\frac{C\delta}{1+\delta^2+|X|^2}\right)\right\}^2\right] \tag{4.6}$$

where $|Y|^2 = \left|\frac{E_{in}}{\delta}\right|^2 \left(\frac{1+|\Lambda|^2}{2(1-|R|^2)^2}\right)$ is the scaled input intensity. If the sine term in is retained in (4.1) then the relation (4.6) becomes a multiply-branched function of $|X|^2$ with the possibility of optical multistability. Also, if e^{-B} is retained as such in (4.1), (4.6) is then corrected by additional absorptive bistability.

5. Summary

The central theme of the present work is to present formal analytical results for a non-linear refractive index theory of optical multistability(OM). Within the formulation of the theory the optical extinction theorem has been generalized to the non-linear regime. Apart from our preliminary results [1]-[4] and an earlier work by Bloembergen and Pershan [26] (but not within the context of OM) we know of no other application of this extinction theorem to non-linear optics. Specifically, our derived formulae are:

1. Formula (3.8) for the generalized Lorentz-Lorenz dispersion relation of the non-linear refractive index (m) which depends on the intensity of radiation entering the F-P cavity. This dependence arises through the parameter R_3 where R_3^{-1} is an average of the inverse of the atomic inversion $r_3(z)$ arising in the steady state at different points inside the cavity when a sufficient intense field enters the cavity. The actual atomic inversion depends on the cavity action and hence on the non-linear refractive index (m). Thus, a *self-consistent scheme* can be created which converges to essentially exact results for the input-output intensity relation of the F-P cavity,

2. Formula (16) for the spatially dependent forward and backward polarization envelopes $P_{f,b}(z)$,

3. Formula (3.13) for the parameter R_3,

4. Formulae (3.16) determines $r_3(z)$,

5. Formula (3.19) is the usual form for the Fresnel reflection coefficient,

6. Formula (3.20) for the dipole envelopes at the boundaries, and

7. Formula (3.21) for the input-output field relation outside the F-P cavity $(z > L)$.

In low atomic density limit our analytical results yield the conventional bistable behavior. The self-consistent analytical results derived in this part (I) are investigated numerically in part (II) of the series of papers.

Finally, we mention that the self-consistent analytical results presented here have been extended to the case of optical rotation via magnetic dipole interactions [27],[28]-but these require a more tedious computational work than that presented in part II.

Appendix A

Here we derive the analytical expression for the atomic polarization field $\underline{E}(\underline{x}, \omega)$, Eq. (8).

From (4)

$$\underline{E}^*(\underline{x}, t) = n_0 p^2 \hbar^{-1} \int_{V-v} \underline{\underline{F}}^*(\underline{x}, \underline{x}'; \omega) : \hat{\underline{u}}\hat{\underline{u}}\, r_+(\underline{x}', \omega)\, d\underline{x}' \tag{A.1}$$

where

$$\underline{\underline{F}}(\underline{x}, \underline{x}'; \omega) = (\nabla\nabla + k_0^2 \underline{\underline{U}}) \frac{e^{ik_r r}}{r}; \quad r = |\underline{x} - \underline{x}'| \tag{A.2}$$

If we consider, in the first instance, a one wave ansatz, that is we put

$$r_+(\underline{x}, \omega) = P_f(z) e^{-imk_0 z} \tag{A.3}$$

then (A.1) becomes

$$\begin{aligned}\underline{E}^*(\underline{x}, t) &= n_0 p^2 \hbar^{-1} \int_{V-v} P_f(z') e^{-imk_0 z'} \underline{\underline{F}}^*(\underline{x}, \underline{x}'; \omega) : \hat{\underline{u}}\hat{\underline{u}}\, r_+(\underline{x}', \omega)\, d\underline{x}' \\ &= n_0 p^2 \hbar^{-1} \left(\int_{V-v} P_f(z') e^{-imk_0 z'} \hat{\underline{u}}\hat{\underline{u}} : \nabla\nabla \left(\frac{e^{-ik_0 r}}{r} \right) d\underline{x}' \right. \\ &\quad \left. + k_0^2 \int_{V-v} P_f(z') e^{-imk_0 z'} \hat{\underline{u}}\hat{\underline{u}} : \underline{\underline{U}} \left(\frac{e^{-ik_0 r}}{r} \right) d\underline{x}' \right) \end{aligned} \tag{A.4}$$

Using Gauss' Theorem to the first term on the right hand side of Eq. (A.4) we get

$$\int_{V-v} P_f(z') e^{-imk_0 z'} \hat{\underline{u}}\,\hat{\underline{u}} : \underline{\nabla}\,\underline{\nabla} \left(\frac{e^{-ik_0 r}}{r}\right) d\underline{x}' =$$
$$\int_{\Sigma} P_f(z') e^{-imk_0 z'} \hat{\underline{u}}.d\underline{s}\, \nabla_{\underline{x}'} \frac{e^{-ik_0 r}}{r} - \int_{\sigma} P_f(z') e^{-imk_0 z'} \hat{\underline{u}}.d\underline{s}\, \nabla_{\underline{x}'} \frac{e^{-ik_0 r}}{r}$$
(A.5)

where $\sum$ is the surface of the volume V in which the medium is contained and σ represents a small sphere at $\underline{x}'$ inside V. The second term in (A.5) is the Lorentz local field [4],[5],[12],

$$-\int_{\sigma} P_f(z') e^{-imk_0 z'} \hat{\underline{u}}.d\underline{s}\, \nabla_{\underline{x}'} \frac{e^{-ik_0 r}}{r} = \tfrac{4\pi}{3}\underline{\underline{U}} P_f(z) e^{-imk_0 z}$$
(A.6)

For the condition $\hat{\underline{u}}.d\underline{s} = 0$, the first term in (A.5) then vanishes [5] and hence (A.5) gives

$$\int_{V-v} P_f(z') e^{-imk_0 z'} \hat{\underline{u}}\hat{\underline{u}} : \underline{\nabla}\,\underline{\nabla} \left(\frac{e^{-ik_0 r}}{r}\right) d\underline{x}' = \tfrac{4\pi}{3}\underline{\underline{U}} P_f(z) e^{-imk_0 z}$$
(A.7)

Now for the second term in (A.4) we have

$$\int_{V-v} P_f(z') e^{-imk_0 z'} \hat{\underline{u}}\hat{\underline{u}} : \nabla^2_{\underline{x}'} \left(\frac{e^{-ik_0 r}}{r}\right) d\underline{x}' = -k_0^2 \int_{V-v} P_f(z') e^{-imk_0 z'} \left(\frac{e^{-ik_0 r}}{r}\right) d\underline{x}'$$
(A.8)

Also

$$-\int_{V-v} \frac{e^{-ik_0 r}}{r} \nabla^2_{\underline{x}'} \left(P_f(z') e^{-imk_0 z'}\right) d\underline{x}'$$
$$= -\int_{V-v} \frac{e^{-ik_0 r}}{r} \frac{\partial^2}{\partial z'^2} \left(P_f(z') e^{-imk_0 z'}\right) d\underline{x}'$$
$$= -\int_{V-v} \frac{e^{-ik_0 r}}{r} \left(m^2 k_o^2 P_f(z') + 2im\, k_0 P_f'(z') - P_f''(z') e^{-imk_0 z'}\right) d\underline{x}'$$
(A.9)

From (A.8) and (A.9) we have

$$\int_{V-v} P_f(z') e^{-imk_0 z'} \nabla^2_{\underline{x}'}\left(\frac{e^{-ik_0 r}}{r}\right) - \frac{e^{-ik_0 r}}{r}\nabla^2_{\underline{x}'}\left(P_f(z') e^{-imk_0 z'}\right) d\underline{x}'$$

$$= (m^2 - 1) k_0^2 \int_{V-v} P_f(z') e^{-imk_0 z'} \frac{e^{-ik_0 r}}{r} d\underline{x}'$$
$$+ 2imk_0 \int_{V-v} P_f'(z') e^{-imk_0 z'} \frac{e^{-ik_0 r}}{r} d\underline{x}'$$
$$- \int_{V-v} P_f''(z') e^{-imk_0 z'} \frac{e^{-ik_0 r}}{r} d\underline{x}'.$$

So

$$k_0^2 \int_{V-v} P_f(z') e^{-imk_0 z'} \frac{e^{-ik_0 r}}{r} d\underline{x}' =$$

$$\frac{1}{(m^2-1)} \int_{V-v} \left(P_f(z') e^{-imk_0 z'} \nabla_{\underline{x}'}^2 \left(\frac{e^{-ik_0 r}}{r} \right) \right.$$
$$\left. - \frac{e^{-ik_0 r}}{r} \nabla_{\underline{x}'}^2 \left(P_f(z) e^{-imk_0 z'} \right) \right) d\underline{x}'$$
$$- \frac{2imk_0}{(m^2-1)} \int_{V-v} P_f'(z') e^{-imk_0 z'} \frac{e^{-ik_0 r}}{r} d\underline{x}'$$
$$+ \frac{1}{(m^2-1)} \int_{V-v} P_f''(z') e^{-imk_0 z'} \frac{e^{-ik_0 r}}{r} d\underline{x}'.$$

(A.10)

Applying Green's theorem for the first integral on the right side of (A.10) gives,

$$k_0^2 \int_{V-v} P_f(z') e^{-imk_0 z'} \frac{e^{-ik_0 r}}{r} d\underline{x}' = \frac{4\pi}{(m^2-1)} P_f(z) e^{-imk_0 z}$$
$$+ \frac{1}{(m^2-1)} \int_\Sigma \left(\begin{array}{c} P_f(z) e^{-imk_0 z} \nabla_{\underline{x}'} \left(\frac{e^{-ik_0 r}}{r} \right) .d\underline{s} \\ - \frac{e^{-ik_0 r}}{r} d\underline{s}.\nabla_{\underline{x}'} \left(P_f(z') e^{-imk_0 z'} \right) \end{array} \right)$$
$$- \frac{2imk_0}{(m^2-1)} \int_{V-v} P_f'(z') e^{-imk_0 z'} \frac{e^{-ik_0 r}}{r} d\underline{x}'$$
$$- \frac{1}{(m^2-1)} \int_{V-v} P_f''(z') e^{-imk_0 z'} \frac{e^{-ik_0 r}}{r} d\underline{x}' .$$

(A.11)

Hence the use of (A.7) and (A.11) into (A.4) gives

$$\frac{\underline{E}^*(\underline{x},\omega)}{n_0 p^2 \hbar^{-1}} = \frac{4\pi}{3} \left(\frac{m^2+2}{m^2-1} \right) P_f(z) e^{-imk_0 z}$$
$$+ \frac{1}{(m^2-1)} \int_\Sigma \left(P_f(z') e^{-imk_0 z'} \nabla_{\underline{x}'} \left(\frac{e^{-ik_0 r}}{r} \right) .d\underline{s} \right.$$
$$\left. - \frac{e^{-ik_0 r}}{r} d\underline{s}.\nabla_{\underline{x}'} \left(P_f(z') e^{-imk_0 z'} \right) \right)$$
$$- \frac{2imk_0}{(m^2-1)} \int_{V-v} P_f'(z') e^{-imk_0 z'} \frac{e^{-ik_0 r}}{r} d\underline{x}'$$
$$- \frac{1}{(m^2-1)} \int_{V-v} P_f''(z') e^{-imk_0 z'} \frac{e^{-ik_0 r}}{r} d\underline{x}'$$

(A.12)

In a similar way we get for the second integral on the right –hand side of (A.12)

$$k_0^2 \int_{V-v} P_f'(z') e^{-imk_0 z'} \frac{e^{-ik_0 r}}{r} d\underline{x}' = \frac{4\pi}{(m^2-1)} P_f'(z) e^{-imk_0 z}$$
$$+ \frac{1}{(m^2-1)} \int_\Sigma \left(P_f'(z') e^{-imk_0 z'} \nabla_{\underline{x}'} \left(\frac{e^{-ik_0 r}}{r} \right) . d\underline{s} - \frac{e^{-ik_0 r}}{r} d\underline{s} . \nabla_{\underline{x}'} \left(P_f'(z') e^{-imk_0 z'} \right) \right)$$
$$- \frac{2imk_0}{(m^2-1)} \int_{V-v} P_f''(z') e^{-imk_0 z'} \frac{e^{-ik_0 r}}{r} d\underline{x}' - \frac{1}{(m^2-1)} \int_{V-v} P_f'''(z') e^{-imk_0 z'} \frac{e^{-ik_0 r}}{r} d\underline{x}'$$
(A.13)

The use of (A.13) into (A.12) gives, up to ignoring the terms in $P_f''(z)$ and higher derivatives,

$$\frac{\text{E}^*(\underline{x},t)}{n_0 \text{p}^2 \hbar^{-1}} = \frac{4\pi}{(\text{m}^2-1)} \left[\left(\frac{\text{m}^2+2}{3} \right) P_f(z) - \frac{2im}{k_0(\text{m}^2-1)} P_f'(z) \right] e^{-k_0 z}$$
$$+ \frac{1}{(\text{m}^2-1)} \left[I(P_f(z)) - \frac{2im}{k_0(\text{m}^2-1)} I(P_f'(z)) \right]$$
(A.14)

with $I(h(z))$ given by (3.4). Clearly, the ansatz (3.1)

$$r_+(\underline{x},t) = P_f(z) e^{-imk_0 z} + \Lambda P_b(z) e^{imk_0 z}$$

will give along the same line that lead to (A.14) the inter-atomic field, of (3.3), . Hence we have the required result.

References

[1] R. K. Bullough and S. S. Hassan, *Optical Extinction Theorem in the Non-Linear Theory of Optical Multistability in The Max Born Centenary Conference Proceedings* (September, 1982). eds. M. J. Colles and D. W. Swift, 369, (SPIE, Billingham, W.A., 1982), pp. 257-363.

[2] R. K. Bullough, S. S. Hassan and S. P. Tewari, *Refractive Index Theory of Optical Bistability in Quantum Electronics and Electro Optics*, ed. P. L. Knight, (Wiley, N.Y., 1982), pp. 229-332.

[3] R. K. Bullough, S. S. Hassan, G. P. Hildred and R. R. Puri, *Mirrored and Mirrorless Optical Bistability: Exact C-Number Theory of Atoms forming a Fabry-Perot Cavity in Optical Bistability* **II**, eds C.M. Bowden, H. M. Gibbs and S. L. McCall, (Plenum,1984), pp. 445-62.

[4] R. K. Bullough and F. Hynne in *P. P. Ewald and his Dynamical theory of X-Ray Diffraction*, eds. D.Q.J.Cruickshank, M.J. Juretschke and N. Kato, (OUP, Oxford, 1992), pp. 98-100

[5] R. K. Bullough, *Phil. Trans. Roy. Soc. (London), Series A*, **254** (1962) 397

[6] F. Hynne and R. K. Bullough, *Phil. Trans. Roy. Soc. (London), Series A*, **321** (1987) 251

[7] F. Hynne and R. K. Bullough, *Phil. Trans. Roy. Soc. (London), Series A*, **321** (1987) 305

[8] F. Hynne and R. K. Bullough, *Phil. Trans. Roy. Soc. (London), Series A*, **330** (1990) 253

[9] S S Hassan and R K Bullough in *Optical Bistability,* eds. C. M. Bowden, M. Ciftan and H. Robl, (Plenum, N.Y., 1981), pp. 367-404

[10] O.W.Oseen, *Ann. Phys. (Leipzig)*, **48**, (1915) 1.

[11] P. P. Ewald, *Ann. Phys. (Leipzig)*, **49**, (1916) 1, 117.

[12] C. G. Darwin, *Trans. Camb. Phil. Soc.* **23** (1924) 137.

[13] L. Rosenfeld, *Theory of the Electrons* (North Holland, Amsterdam, 1951); Also reprinted with new prefaces by Dover, N.Y., 1965.

[14] M. Born and E.Wolf, *Principles of Optics*, 6th Edition (Pergamon, Oxford,1980).

[15] J. H. Marburger and F. S. Felber, *Phys. Rev. A* **17** (1978) 335.

[16] D. A. B. Miller, *IEEE Journal of Quantum Electronics* **QE-17**, (1981), 306.

[17] R. Bonifacio and L.A. Lugiato, *Opt. Commun.* **19** (1976) 172.

[18] R. Bonifacio and L. A. Lugiato, *Phys. Rev. A* **18** (1978) 1729

[19] S. S. Hassan, P. D. Drummond and D. F. Walls, *Opt. Commun.* **27** (1978). 480

[20] M. Orenstein, S. Speiser and J. Katriel, *Opt. Commun.* **48**, (1984) 367; M. Orenstein, S. Speiser and J. Katriel, *IEEE J. Quant. Electron.* **21** (1985). 1513.

[21] S. S. Hassan , H. A. Batarfi, R. Saunders and R. K. Bullough , *Eur. Phys. J. D*, **8**. (2000) 417.

[22] S. S. Hassan, R. K. Bullough and H. A. Batarfi, *in Studies in Classical and Quantum Nonlinear Optics*, ed.O. Keller (Nova Sci. Pub., NY, 1995) pp 609-623.

[23] R. Bonifacio and L.A. Lugiato, *Lett. Nuovo. Cim.*, **21** (1978) 517

[24] E. Abraham,.S. S.Hassan, R. K. Bullough,.*Opt. Commun.*, **33** (1980) 93.

[25] R. Saunders , R. K. Bullough and C. Feuillade C, *The Theory of Far Infrared Superfluorescence in Coherence and Quantum Optics* **IV**, ed. L. Mandel and E. Wolf (Plenum, New York, 1978), pp. 263-286.R. Saunders R and R. K. Bullough R K, *Theory of FIR Superfluorescence and Radiation in Coherence and Quantum Optics* **III**, ed. C. M. Bowdon, D. Howgate, H. .P Roble (Plenum New York, 1977), pp. 209-256.

[26] N. Bloembergen and P. S. Pershan *Phys. Rev.* **128** (1962) 606.

[27] A-S. F. Obada, S. S. Hassan and M. H. Mahran, .A-S F Obada, *.Physica A,* **142** (1987) 601.

[28] S. S. Hassan,.M. H. Mahran and A-S. F. Obada, *Physica A,* **142** (1987) 619

In: Theoretical Physics and Nonlinear Optics
Editors: Thomas F. George et al

ISBN 978-1-61122-939-4
© 2012 Nova Science Publishers, Inc.

Chapter 15

NON-LINEAR REFRACTIVE INDEX THEORY: II - NUMERICAL STUDIES

S.S. Hassan[1,*], *R.K. Bullough*[2,†] *M.N. Ibrahim*[3], *R. Saunders*[4,‡], *T. Jarad*[5,§] *and N. Nayak*[6,¶]

[1]University of Bahrain, College of Science, Department of Mathematics,
PO Box 32038, Kingdom of Bahrain
[2]University of Manchester, Mathematics Department, Manchester
M13 9PL, United Kingdom
[3]Ex-Postgraduate student at UMIST, Department of
Mathematics, Manchester M60 1QD, UK
[4]Manchester Metropolitan University, Faculty of Science and Engineering,
Department of Computing and Mathematics,
Manchester M1 5GD, United Kingdom
[5]University of Bahrain, College of Applied Sciences,
P.O. Box 32038, Kingdom of Bahrain
[6]Satyendra Nath Bose National Centre for Basic Sciences,
DB-17, Sector-1, Salt Lake City, Calcutta 700064, India

Abstract

The analytical formulae derived in the preceding paper [1] of this series, for the input-output field relation that is nonlinearly dependent on the refractive index is treated numerically through a self-consistent procedure scheme. The multi-stable behavior is exhibited for system data suitable for dense medium.

1. Introduction

In part (I) of this series of papers [1] we developed a non-linear refractive theory for an extended system of two level atoms, confined to a Fabry-Perot (F-P.) cavity with the ge-

---

[*]E-mail address: shoukryhassan@hotmail.com (Corresponding author).
[†]Ceased on 30th August, 2008.
[‡]E-mail address: R.Saunders@mmu.ac.uk
[§]E-mail address: tariqjarad@hotmail.com
[¶]E-mail address: nayak@bose.res.in

ometry of a parallel-sided slab. The derived analytical formulae for the input-output field relation shows the F-P action and its non-linear nature dependent on the refractive index (m). This (m) is governed by a generalized type of Lorentz-Lorenz dispersion relation that is indirectly dependent on the intensity of the field entering the F.-P. cavity through a parameter R_3 reflecting the fact that the atomic inversion is field-dependent inside the F-P. cavity. The analytical results of paper [1] have been derived in analogy to the linear regime by generalizing the optical extinction theorem en route.

The purpose of the present paper, part (II) of this series, is to present a self-consistent numerical scheme based on the analytical formulation derived in [1]. In section 2 we cast these analytical formulae in normalized forms, followed by the detail of the computational steps involved in implementing that scheme numerically in section 3. A discussion of the numerical results is given in section 4. In the Appendix (A) we outline the numerical treatment for the average of the inverse of the atomic inversion.

2. Normalized Analytical Formulas

The relevant non-linear system of formulae derived in section 3 of [1] are mainly equations (13), (16), (18), (21) and (24)-(26). These can be put in the following normalized forms:

(A) The generalized Lorentz-Lorenz dispersion relation, Eq. (13) of [1]

$$\frac{m^2 - 1}{m^2 + 1} = \frac{\alpha(-2R_3)}{(i - \delta)} \tag{1}$$

where $\alpha = \frac{4\pi n_0 p^2}{3\hbar\gamma}$, $\delta = \frac{\omega - \omega_0}{\gamma}$

Also, Eq. (1) may be put into the form

$$\frac{m^2 - 1}{m^2 + 1} = \alpha p(\omega_0) \tag{2a}$$

where $p(\omega_0)$ is the *non-linear* polarizability,

$$p(\omega_0) = \frac{(-2R_3)}{(i - \delta)} \tag{2b}$$

and

$$R_3 = \frac{1}{\int_0^1 r_3^{-1}(\bar{z}) \, d\bar{z}} \tag{3}$$

$(\bar{z} = z/L)$.

(B) The reflection coefficient, Eq. (24) of [1],

$$\Lambda = \left(\frac{m - 1}{m + 1}\right) e^{-2im\bar{k}} \tag{4}$$

where $\bar{k} = k_0 L$, $k_0 = 2\pi \lambda_0^{-1}$ ($\lambda_0 \sim 10^{-5}$ cm in the optical range).

(C) The absolute value of the dipole envelopes at the boundaries (see Eq. (17) of (I) in terms of the input or output intensity, Eq. (25) of [1]).

$$|p_0|^2 = \frac{I_i}{(2\pi)^2} \left| \frac{\frac{m^2-1}{m+1}}{\left(1 - \left(\frac{m^2+2}{3}\right)\left(1 - \frac{R_3}{r_3(0)}\right)\right)\left(1 - \left(\frac{m-1}{m+1}\right)\Lambda\right)} \right|^2 \quad (5)$$

$$= \frac{I_0}{(2\pi)^2} \left| \frac{\left(\frac{m^2-1}{m+1}\right)\left(\frac{(m+1)^2}{4m}\right) e^{i(m-1)\bar{k}}}{\left(1 - \left(\frac{m^2+2}{3}\right)\left(1 - \frac{R_3}{r_3(1)}\right)\right)} \right|^2 \quad (6)$$

where the input and output normalized intensities are defined as $I_i = \left|\frac{E_{in}}{\gamma}\right|^2$ and $I_0 = \left|\frac{E_{out}}{\gamma}\right|^2$. with $r_3(0)$, $r_3(1)$ the values of $r_3(\bar{z})$ at the boundaries $\bar{z} = 0, 1$ respectively.

(D) The input-output intensity relation, Eq. (26) of [1]

$$I_0 = I_i \left| \frac{\left(\frac{4m}{(m+1)^2}\right) e^{-i(m-1)\bar{k}}}{\left(1 - \left(\frac{m-1}{m+1}\right)\Lambda\right)} \left(\frac{1 - (\frac{m^2+2}{3})(1 - \frac{R_3}{r_3(1)})}{1 - (\frac{m^2+2}{3})(1 - \frac{R_3}{r_3(0)})} \right) \right|^2 \quad (7)$$

(E) The atomic inversion density $r_3(z)$, Eq. (21) of [1],

$$2r_3(\bar{z})\left(r_3(\bar{z}) + \tfrac{1}{2}\right) = -|p_0|^2 \left[e^{h_1(\bar{z})} + |\Lambda|^2 e^{-h_1(\bar{z})}\right] \quad (8)$$

where

$$h_1(\bar{z}) = -i(m - m^*)\bar{k}\bar{z} + Re\left[\frac{(m^2+2)(m^2-1)}{3mi}\right]\bar{k}\bar{z}$$
$$+ \bar{k} Re\left[\frac{(m^2-1)^2(1+i\delta)}{6\alpha m}\right] \int_0^{\bar{z}} r_3^{-1}(z')\, dz'. \quad (9)$$

The special cases of Eq. (9) are:
(i) At the input end of the cavity ($\bar{z} = 0$)

$$2r_3(0)\left(r_3(0) + \tfrac{1}{2}\right) = -|p_0|^2 \left[1 + |\Lambda|^2\right] \quad (10)$$

(ii) At the output end of the cavity ($\bar{z} = 1$)

$$2r_3(1)\left(r_3(1) + \tfrac{1}{2}\right) = -|p_0|^2 \left[e^{2\,\text{Im}(m)\bar{k}} + |\Lambda|^2 e^{-2\,\text{Im}(m)\bar{k}}\right] \quad (11)$$

The above set of equations in (A)-(F) are solved numerically for given values of the scaled parameters α, δ, $\bar{k}$ by a self-consistent scheme as explained in the next section.

3. Self-consistent Numerical Scheme

The intention is to compute the input intensity I_i for given values of the output intensity I_0 once the values of R_3, m, Λ and p_0 are determined. For R_3 $\left(-\frac{1}{2} \leq R_3 < 0\right)$ a choice is made and then the complex refractive index m (ω) is determined from Eq. (1) and hence the reflection coefficient is computed from Eq. (4). Then from equations (6) and (11) we get $|p_0|$ and $r_3(1)$. Eq. (10) can then give $r_3(0)$. Once Eq. (8) is solved, as explained below, for $r_3(\bar{z})$; $0 \leq \bar{z} < 1$ for the given value of R_3, we can use Eq. (3) to calculate numerically R_3, say $(R_3)_1$. By taking the value $(R_3)_1$ as the new choice in Eq.(1) and repeating the same steps of computation n times until we obtain the values $(R_3)_{n+1}$. If the value $(R_3)_{n+1}$ is close to $(R_3)_n$ within an accuracy of $\sim 10^{-6}$ (say) then $(R_3)_{n+1}$ is the *best* value for the parameter R_3 that is corresponding to the given output intensity. The best value is used to calculate $r_3(0)$, $r_3(1)$ and hence the input intensity I_i is calculated from Eq. (7). By changing the values of the output intensity I_0 and carrying out this consistent procedure we get the corresponding values of I_i. The Fig.(1) shows a flow chart to calculate the parameter R_3 self consistently.

We notice the following points about the numerical scheme in Fig.1:

1. For a guessed value of R_3 the Lorentz-Lorenz relation of Eq (1) can be solved numerically to give two complex roots for m. Our investigations show that both roots give the same results in the numerical calculation.

2. To find $r_3(1)$, the atomic inversion at $\bar{z} = 1$, we see from equations (6) and (11) that $r_3(1)$ is given through a cubic equation. Analysis of this cubic, supported by the numerical results, show that there are one real root and two complex roots. We choose only the physical real root with $-\frac{1}{2} \leq r_3(1) \leq 0$.

3. To find $r_3(0)$, the atomic inversion at $\bar{z} = 0$, the quadratic equation (10) gives two solutions, namely

$$r_3^{\pm}(0) = -\frac{1}{4} \pm \frac{1}{4}\sqrt{1 - 8|p_0|^2\left(1 + |\Lambda|^2\right)} \qquad (12)$$

   We chose only the solution where $-\frac{1}{2} \leq r_3(0) \leq 0$. We comment further on this point below.

4. To calculate $r_3(\bar{z})$, the atomic inversion at any point throughout the F-P cavity, from Eq. (8) we have formally two solutions,

$$r_3^{\pm}(\bar{z}) = -\frac{1}{4} \pm \frac{1}{4}\sqrt{1 - 8|p_0|^2\left(e^{h_1(\bar{z})} + |\Lambda|^2 e^{-h_1(\bar{z})}\right)} \qquad (13)$$

   where $h_1(\bar{z})$ is given by Eq. (9).

Now the procedure to calculate numerically the integral in Eq. (9) is to divide the cavity length into N equal segments, each with lengths $q = \frac{1}{N}$ so $\bar{z} = nq$: $n = 0, 1, ...N$. Hence at $\bar{z} = nq$ Eq. (13) can be written as

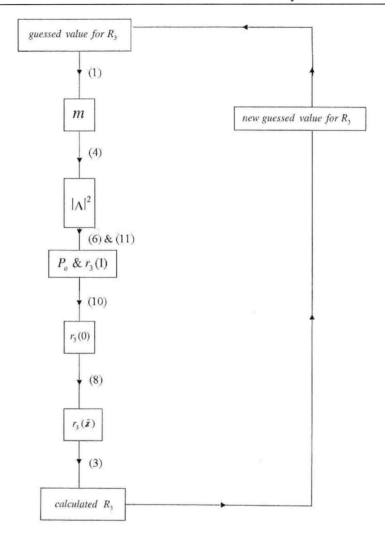

Figure 1. A flow chart to evaluate the parameter R_3 self-consistently.

$$r_3^\pm (nq) = -\frac{1}{4} \pm \frac{1}{4}\sqrt{1 - 8|p_0|^2 \left(e^{\,h_1(nq)} + |\Lambda|^2 e^{-h_1(nq)}\right)} \qquad (14)$$

The integral $\int_0^{nq} r_3^{-1}(z')\,dz'$ in Eq.(9) can be calculated numerically with $r_3(0) \equiv r_3^-(0)$ to an error $O(10^{-6})$, say, for example by using Euler's method: see Appendix (A). Having calculated $h_1(nq)$ the two possible roots $r^\pm(nq)$ in Eq.(14) are then obtained. Note that, in calculating $r_3(\bar{z}) \equiv r_3(nq)$ with the initial value $r_3(0) \equiv r_3^-(0)$, the value of $r_3(\bar{z})$ at $\bar{z} = nq$, that is $r_3(1)$, was consistent with that obtained through the cubic equation as noted in (2) above, up to an error $O(10^{-6})$. The other possible initial value for $r_3(0)$, namely $r_3^+(0)$ gives a value of $r_3(1)$ which is not consistent with that obtained by the cubic equation This confirms our choice for $r_3(0) = r_3^-(0)$ as noted in (3) above.

Now, the rule to choose either $r_3^-(\bar{z})$ or $r_3^+(\bar{z})$ in Eq.(13) or Eq.(14) is based on the calculation of the gradient function for both $r_3^\pm(\bar{z})$ from the two different expressions and then comparison is made consistently, and an error $O(10^{-6})$ if needed. From Eq.(8) we find the gradient $G(\bar{z}) = \frac{dr_3(\bar{z})}{d\bar{z}}$ as

$$G(\bar{z}) = \frac{-|p_0|^2 \frac{dh_1(\bar{z})}{d\bar{z}} \left[e^{h_1(\bar{z})} - |\Lambda|^2 e^{-h_1(\bar{z})} \right]}{(4r_3(\bar{z}) + 1)} \quad (15)$$

The gradient functions $O^{\pm}((n-1)q)$ for $r_3^{\pm}(nq)$ are defined by

$$O^{\pm}((n-1)q) = r_3^{\pm}(nq) - r_3^{\pm}((n-1)q) \quad (16)$$

where $r_3^{\pm}(nq)$ are evaluated from Eq.(14). These gradient functions are used to decide which of the values of $r_3^{\pm}(nq)$ should be chosen by comparing the sign of the gradient $G(\bar{z} = nq)$ with that of $O^{\pm}((n-1)q)$.

If

$$sign\, O^{+}((n-1)q) \neq sign\, O^{-}((n-1)q) \quad (17)$$

and

$$sign\, O^{+}((n-1)q) \neq G(nq) \quad (18)$$

then $r_3(nq) = r_3^{+}(nq)$ or else it is $r_3^{-}(nq)$.

However if

$$sign\, O^{+}((n-1)q) = sign\, O^{-}((n-1)q) \quad (19)$$

and

$$\left| O^{+}((n-1)q) - G(nq) \right| \leq \left| O^{-}((n-1)q) - G(nq) \right| \quad (20)$$

where

$$\left| O^{+}((n-1)q) - G(nq) \right| \lesssim O(10^{-2}) \quad (21)$$

then $r_3(nq) = r_3^{+}(nq)$ or else it is $r_3^{-}(nq)$. In practice the numerical work shows that r_3^{-} is the chosen solution according to the above.

4. Discussion of the Numerical Results

Implementing the self-consistent numerical scheme shows the following results:

1. At exact atomic resonance, $\delta = 0$, and for the parameter values $\alpha = 10^{-4}$ and $k_0 L = 10^4$, relevant for Na vapor, the input-output intensity relation shows essentially a linear behavior, even for high input intensities. Consistent with this the real part of the refractive index is nearly one and the effective parameter $R_3 \simeq -\frac{1}{2}$. There is no change in the phase shift $e^{-2imk_0 l}$; that is no sufficient positive feedback to produce saturation of the medium. For values of $\alpha = 10^{-3}$ and $k_0 L = 10^5$ in the off-resonance case with $|\delta| = 100$ the results still show no intensity-dependent refractive index and hence no non-linearity behavior of the input-output relationship.

2. For a range of parameters suitable to dense materials ($L \sim 1\mu$) we have found an intensity dependent refractive index. The non-linearity then is large enough to account for multi-stable behavior in a highly dispersive case. The dependence of the real part of the refractive index, Re(m), on the input intensity is shown in Fig. 2 for the parameter values $\alpha = 330$, $k_0 L = 10^3$ and $\delta = -1460$. The imaginary part Im(m) is small ($\sim 10^{-4}$) and it shows similar non-linear behavior.

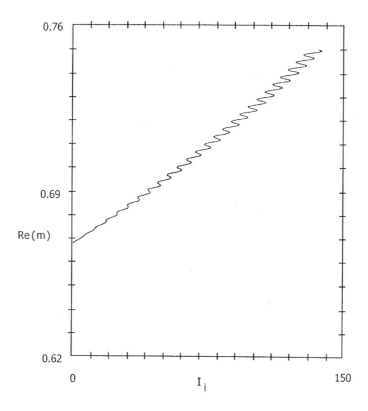

Figure 2. The real part of the refractive index, Re (m) versus the input intensity I_i, for $\alpha = 330, \delta = -1460, k_0 L = 10^3$.

3. The variation of the atomic inversion $r_3(z)$ along the cavity length for different values of the input intensity I_i is shown in Fig.(3a). For low values of I_i, $r_3 \simeq -\frac{1}{2}$, the linear regime, and all atoms are essentially in the ground state. For increasing I_i, $r_3(z)$ increases and becomes intensity dependent where it tends to zero, showing saturation, with further increase of I_i. Further, for fixed values I_i, δ the increase of the absorption coefficient α produces increase in $r_3(z)$ (Fig.3b). Also, it is found that the effect of changing the atomic detuning δ on $r_3(z)$ for fixed I_i, $r_3(z)$ decreases in value as might be expected. In this case, the radiation damping decreases due to a decrease in absorption. Also, the decrease in the value of δ has the dispersive effect of increasing both the polarization and refractive index.

4. As for the input-output relationship (Fig.4) the non-linearity in (m) detunes the cavity through the periodic phase shift, $e^{-2imk_0 L}$, and this leads to a multi-branched OB

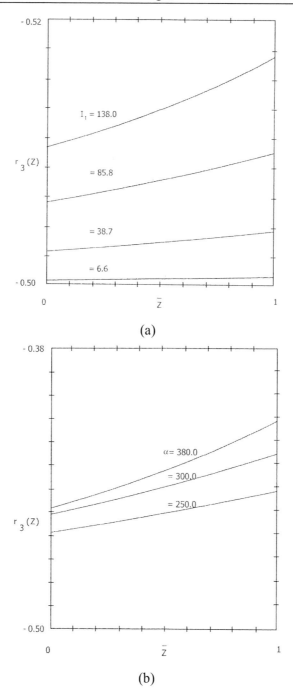

Figure 3. (a) The spatially dependent atomic inversion $r_3(z)$ versus the normalized cavity length $\bar{z}$ for fixed values of $\alpha = 330$, $\delta = -1460$, $k_0 L = 10^3$ and for different values of the input intensity I_i. (b) as (a) but for fixed $\delta = -1460$, $k_0 L = 10^3$, $I_i = 80$ and for different values of α.

curve (dispersive multi-stability.). The effect of changing the absorption coefficient α on the multi-stable behavior is shown also in Fig.(4): increasing α deceases (m) and the variation of the phase shift gets smaller. This in turn yields smaller hysteresis loops. The increase of the parameter $k_0 L$ has a similar effect. Note that by changing the sign of δ no multi-stable behavior is found, as we explain in (5) below, and hence positive feedback alone is insufficient to explain the occurrence of OM.

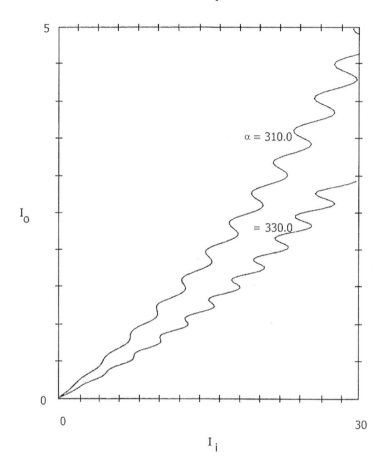

Figure 4. The input-output intensity for fixed values of $\delta = -800, k_0 L = 10^3$ and for $\alpha = 310, 330$.

5. The asymmetric behavior, as a function of δ, of the input-output relation (7) can be explained as follows. From (1) and for $\delta^2 \gg 1$ we have

$$\frac{m^2 - 1}{m^2 + 1} \simeq \frac{2\alpha R_3}{\delta} + i\frac{2\alpha R_3}{\delta^2} \qquad (22)$$

The ratio Re(m)/Im(m)$\sim o(\delta)$ —which is also confirmed numerically-so, we may take m real in this case. If we denote $m_\pm$ for the cases $\delta \gtrless 0$ we then have

$$m_\pm^2 \simeq \frac{|\delta| \pm 4\alpha R_3}{|\delta| \pm 2\alpha R_3} \qquad (23)$$

i.e. $m_+ m_- \simeq 1$ for $\delta^2 > (\alpha R_3)^2$.

As well, we have in this case $\left|e^{-i(m-1)k_0 L}\right|^2 \simeq 1$ for $\delta \gtrless 0$ and the factor $\left|\frac{(m+1)^2}{4m}\right|^2$ is a symmetric function of δ. But the factor

$$\left|1-\left(\frac{m-1}{m+1}\right)^2 e^{-2imk_0 L}\right|^2 = \begin{cases} \left|1-\left(\frac{m_--1}{m_-+1}\right)^2 e^{-2im_-k_0 L}\right|^2, & \delta < 0 \\ \left|1-\left(\frac{m_--1}{m_-+1}\right)^2 e^{\frac{-2ik_0 L}{m_-}}\right|^2, & \delta > 0 \end{cases} \quad (24)$$

and hence it is an asymmetric function of δ. (Note that in this case $r_3(0) \simeq r_3(1)$)

5. Summary

In this paper, part (II), the analytic formulae derived in paper (I) of this series of papers on an extended system of 2-level atoms placed inside a slab-like geometry (F-P cavity) for the generalized Lorentz-Lorenz dispersion relation and input-output field relationship in the non-linear regime have been treated computationally using a self-consistent numerical scheme investigated in detail. The chosen values for the physical system parameters which show the non-linear behavior of optical multi-stability (OM) are suited for a dense medium and for a highly dispersive situation.

In a following work we intend to examine the dynamical stability of OB/OM within our theory.

Appendix A

The numerical calculation of the integral $\int_0^{nq} r_3^{-1}(z') \, dz'$, $(n = 1, 2, ...N; \; q = \frac{1}{N})$ is outlined as follows. According to Euler's method,

$$\int_a^b f(x)\,dx \simeq (b-a)f(a) \qquad (A.1)$$

so we can approximate

$$\int_0^{nq} r_3^{-1}(z')\,dz' = \sum_{i=1}^{i=n} \int_{(i-1)q}^{iq} r_3^{-1}(z')\,dz'$$

$$\simeq \sum_{i=1}^{i=n} q r_3^{-1}((i-1)q) \qquad (A.2)$$

where $r_3((i-1)q)$ is evaluated from Eq.(14).

References

[1] S. S. Hassan, R. K. Bullough and R Saunders, *Int. J. Theoret. Phys., Group Theory and Nonlinear Optics,* Vol. 13, No 3/4 (The preceding paper, part (I)).

INDEX

A

Abraham, 226
access, 41
accounting, 73, 105, 170, 195
algorithm, 41
amplitude, 45, 46, 50, 213, 215
angular momentum, vii, 3, 4, 12, 14, 21, 23, 24, 25, 26, 31
annihilation, 13, 17, 55
atoms, ix, 56, 57, 211, 212, 213, 214, 229, 235, 238

B

Bahrain, 211, 229
baryon, vii, ix, 70, 71, 72, 73, 74, 76, 77, 78, 79, 80, 81, 99, 100, 101, 102, 103, 104, 105, 106, 107, 109, 110, 112, 165, 166, 167, 168, 169, 170, 171, 172, 175, 176, 177, 178, 181, 182, 184, 189, 190, 191, 192, 193, 194, 195, 197, 198, 201, 202, 204
baryons, viii, ix, 69, 70, 71, 72, 73, 76, 77, 78, 80, 81, 82, 99, 100, 101, 102, 103, 104, 106, 107, 109, 110, 111, 112, 165, 167, 169, 171, 175, 176, 177, 181, 182, 183, 184, 189, 190, 191, 192, 193, 194, 196, 197, 201, 202, 203, 204, 205, 206
base, 14
binding energies, 104, 110, 181, 182, 193, 202
binding energy, 73, 80, 104, 105, 109, 169, 182, 194, 202, 204
black hole, vii, 21, 22, 23
Bloch-Maxwell equations, ix, 211
boson, viii, ix, 69, 70, 71, 99, 102, 106, 131, 132, 134, 136, 139, 145, 147, 149, 167, 173, 190, 191, 192
bosons, vii, 70, 71, 77, 80, 100, 102, 108, 109, 110, 118, 136, 137, 142, 145, 146, 147, 148, 151, 191, 192, 196, 204
boundary value problem, 44
bounds, 137, 143, 145, 150, 156
branching, 143, 146, 147, 148, 149, 150, 151, 152, 155

breakdown, 30
Bulgaria, 1, 18, 19

C

calculus, 39
causality, 37, 85, 91, 96
center-of-mass motion., viii, 69, 167
CERN, 132
charm, 70, 101
chiral symmetry, viii, ix, 69, 70, 71, 80, 99, 100, 101, 102, 106, 108, 109, 110, 165, 166, 167, 168, 171, 184, 191, 192, 195, 196, 204
chiral transformation, 105, 170
chirality, 117
City, 165, 189, 229
classes, vii, 39
closed string, 26
closure, 57
color, 204
common sense, 137
compatibility, ix, 17, 123
complications, 7, 31, 218
computation, 110, 182, 192, 232
computing, 11, 80, 183, 203
condensation, 21
configuration, ix, 32, 42, 43, 44, 72, 181, 201, 211
confinement, 70, 71, 82, 100, 101, 103, 112, 166, 167, 168, 190, 191, 192, 193, 198, 202
conformity, 101, 166, 183, 192
conjugation, 2
conservation, 2, 4, 16, 105, 108, 133, 147, 170, 196
constituents, 123
construction, 2, 88
convention, 3, 5
cooperation, 220
correlation, 86, 212, 214
correlation function, 214
correlations, 212
cost, 65
coupling constants, 175
CPT, 37
criticism, 85

D

damping, 214, 215, 218, 235
D-branes, 21, 22, 26, 27, 30
decay, ix, 50, 70, 78, 101, 106, 107, 115, 119, 121, 136, 137, 144, 145, 146, 147, 148, 149, 150, 152, 154, 165, 170, 183, 197
defects, vii, viii, 39, 40, 41, 42, 43, 44, 46, 48, 49, 50, 52, 53
degenerate, 55, 59, 61, 62, 63, 64, 65, 66, 78, 107, 121, 175, 176, 184, 197
derivatives, 3, 6, 8, 10, 11, 13, 17, 215, 216, 225
destruction, 172
detection, 132, 144
differential equations, 217
diffraction, 215
Dirac equation, 71, 72, 101, 103, 124, 166, 167, 168, 184, 192, 193
dislocation, viii, 39, 40, 42, 43, 45, 46, 47, 48, 49, 50, 52, 53
dispersion, x, 211, 212, 217, 220, 221, 230, 238
displacement, 44, 46, 49, 52
distribution, 45, 88, 214
divergence, 29, 106, 170
duality, 21, 26, 27, 30, 36, 37

E

editors, 19
effective field theory, 21
electric charge, 126, 153
electric field, 75, 200
electromagnetic, 17, 70, 101, 167, 190, 211, 213
electron, 86, 139
energy, viii, ix, 24, 25, 40, 49, 55, 56, 59, 60, 61, 62, 63, 64, 65, 66, 70, 71, 73, 74, 75, 76, 77, 78, 79, 80, 81, 100, 102, 103, 105, 107, 108, 109, 110, 111, 112, 119, 132, 139, 141, 143, 152, 166, 167, 168, 169, 170, 171, 175, 176, 178, 179, 180, 181, 182, 183, 184, 189, 190, 191, 192, 194, 195, 196, 197, 199, 200, 201, 202, 204
engineering, 53
enlargement, 119
EPR, 85, 86
equality, 6, 8, 34
equilibrium, vii, 39, 40, 42, 43, 45, 46, 49, 50, 52, 53
Euler-Lagrange equations, vii, 1, 2, 3, 10, 11, 15
evolution, 8, 9, 15, 17, 27, 126
excitation, viii, ix, 70, 99, 100, 102, 112, 189, 190, 191, 192, 202, 204, 206
exclusion, 70, 75, 100, 180, 200
extinction, x, 211, 212, 215, 219, 221, 230
extraction, 91, 92

F

Fabry-Perot cavity configuration, ix, 211
fermions, ix, 30, 116, 120, 121, 131, 135, 136, 137, 139, 140, 144, 145, 147, 148, 149, 151
Feynman diagrams, 136
field, vii, viii, ix, x, 1, 2, 3, 4, 5, 6, 7, 8, 9, 10, 11, 12, 13, 14, 15, 17, 18, 19, 21, 23, 27, 29, 30, 31, 32, 33, 35, 36, 37, 39, 49, 50, 53, 71, 75, 101, 106, 118, 123, 131, 134, 137, 166, 170, 171, 172, 173, 175, 180, 191, 200, 211, 213, 214, 215, 216, 217, 218, 220, 221, 222, 223, 225, 229, 230, 238
field theory, vii, 1, 2, 9, 13, 17, 18, 19, 21, 27, 29
films, 55
flavor, 71, 132, 133, 190, 196
flavour, 70, 72, 76, 100, 101, 102, 106, 108, 170, 171, 173, 176, 181, 193, 201
fluid, 27, 38
Fock space, 57
force, 49, 50, 74, 135, 178, 198
formation, 119
formula, 7, 11, 57, 58, 59, 124, 154, 156, 216
foundations, 39
four dimensional black hole, vii, 21
free fields, 2, 5, 9, 212, 215
freedom, vii, 29, 31, 32, 33, 34, 35, 36, 70, 71, 81, 82, 100, 102, 111, 112, 123, 133, 191, 192, 197, 206
fusion, 139, 144

G

gauge group, vii, 39
gauge invariant, 33, 35
gauge theory, viii, 30, 32, 33, 36, 42, 43, 49, 116, 121
geometry, 41, 43, 219, 238
gluon self-couplings, ix, 72, 74, 189, 192
gluons, 74, 124, 178, 199
graph, 43
grouping, 117
guidelines, 102, 192

H

hadrons, 123
Hamiltonian, 8, 12, 14, 30, 34, 37, 55, 56, 57, 70, 77, 100, 106, 171, 172, 173, 191, 196
Heisenberg picture, 2, 3, 7, 8, 10, 11, 12, 14, 15, 16
HHS, 118
Higgs boson, ix, 131, 132, 135, 137, 139, 143, 144, 146, 148, 149, 150, 151, 154, 156, 163
Higgs field, 118, 134
Higgs particle, ix, 122, 131, 152
Higgsinos, 119, 120
higher dimensional model spaces, vii

I

Hilbert space, 2, 3, 5, 14
hybrid, 100, 102, 190, 191
hyperfine interaction, 99, 190
hysteresis, 237

I

ideal, 56
identification, 17, 33, 34, 35, 36, 42, 43
identity, 6, 40, 42, 43, 76, 180, 200
images, 13
immersion, 41, 43
impurities, 55
incidence, 219
independent variable, 95
India, 29, 69, 82, 99, 112, 115, 165, 184, 189, 206, 229
inequality, viii, 60, 61, 65, 85, 91, 93, 94, 95, 97
insertion, 221
integration, viii, 50, 85, 86, 89, 91, 92, 93, 95, 96, 214, 215
invariants, 132
inversion, 16, 213, 215, 217, 221, 230, 231, 232, 235, 236
Iran, 21, 123
isospin, 78, 107, 153, 197

J

Japan, 67
justification, 101, 166, 192

K

kinematic equations of defects, vii, 39, 42, 43, 46, 48, 52, 53

L

Lagrangian density, 72, 74, 101, 103, 105, 106, 166, 167, 168, 170, 173, 178, 193, 199
Lagrangian formalism, 2, 4, 5, 9, 11, 17, 182
Lagrangian quantum field theory, 1, 5, 18
Large Hadron Collider, 131, 132
lattices, 55
laws, 2, 4
lead, vii, 30, 34, 39, 49, 73, 105, 133, 170, 195, 225
Lie algebra, 39, 42
light, viii, 2, 70, 71, 72, 79, 80, 99, 100, 101, 102, 103, 106, 109, 110, 112, 115, 116, 119, 120, 121, 122, 137, 144, 145, 150, 165, 171, 190, 191, 196, 198, 204, 206, 215
low-lying baryons, viii, 69, 167
lying, 196

M

magnet, 55, 86
magnetic field, 32
magnetic moment, 70, 101, 126, 167, 190
magnetic properties, 56
magnets, 56, 86, 90
magnitude, 80, 109, 182, 202, 213
manifolds, 41
manipulation, 24
many-body theory, 212
mapping, 6, 30, 31, 35
mass, vii, viii, ix, 13, 31, 34, 69, 70, 71, 73, 74, 78, 79, 80, 81, 82, 99, 100, 101, 102, 103, 104, 105, 106, 107, 108, 109, 110, 112, 115, 119, 120, 121, 122, 125, 126, 131, 132, 133, 134, 135, 137, 139, 140, 141, 142, 143, 147, 149, 150, 151, 152, 165, 166, 167, 168, 169, 170, 171, 173, 175, 176, 177, 178, 179, 181, 182, 184, 189, 190, 191, 192, 193, 194, 195, 196, 197, 199, 202, 204, 206
materials, 43, 44, 53, 235
matrix, 26, 40, 41, 42, 44, 51, 52, 53, 63, 121, 177, 213, 214
matter, 26, 27, 41, 45, 118, 120, 215
Maxwell equations, 216
measurement, viii, 85, 91, 92, 93
measurements, 85, 86, 90, 91
memory, 121
mesons, 70, 80, 101, 108, 110, 196, 204
Minimal Supersymmetric Model (MSSM), 131
Minkowski spacetime, 2, 3
mixing, 120
models, vii, ix, 23, 37, 48, 70, 71, 99, 100, 101, 102, 110, 115, 131, 132, 135, 136, 137, 142, 165, 166, 171, 190, 191, 192
momentum, vii, 1, 2, 3, 4, 5, 6, 7, 8, 9, 10, 11, 12, 13, 14, 15, 16, 17, 18, 21, 23, 24, 25, 26, 31, 73, 74, 105, 108, 126, 127, 170, 173, 195, 196
momentum picture of motion, vii, 1, 2
Moon, 130
Moscow, 18, 19
motivation, 32, 115

N

neglect, 171, 214, 219
Netherlands, 85
neutral, 116, 121, 122, 133, 134, 136, 150
neutrinos, 121
next generation, ix, 131
noncommutativity, 30
nonlinear optics, vii, 221
normalization constant, 105, 168, 193, 194
nucleons, 70, 101, 167, 190
null, 43, 49
numerical analysis, 124, 212

Index

O

one-gluon exchange, viii, 69, 181, 190, 201
open string, 21, 23, 27, 30, 33
orbit, 25

P

parallel, 44, 211, 219
parameter vectors, 86
parity, viii, 115, 118, 119, 120, 122, 206
partial differential equations, 15, 212
particle mass, 134
particle physics, 30
PGE, 127, 128
phenomenology, ix, 30, 70, 100, 118, 119, 121, 131, 191
physics, vii, 19, 37, 55, 95, 115, 130, 133, 206, 216
pions, 70, 100, 106, 167, 171, 173, 204
pleasure, 36
Poincare group, 33
polar, 24
polarizability, 230
polarization, 213, 214, 215, 222, 235
polynomial functions, 11
Portugal, 131
positive feedback, 234, 237
positron, 86
prejudice, 101, 166, 192
preparation, 162
principles, 19, 32, 166, 168, 191
probability, 85, 86, 88, 89, 90, 96
probability density function, 85
probe, vii, 21, 22, 23
project, 91, 118
proliferation, 80, 110, 204
propagation, 48
proposition, 57, 119

Q

QCD, ix, 70, 71, 77, 81, 100, 102, 106, 108, 109, 110, 124, 141, 165, 183, 191, 192, 195, 196, 204
quantization, 17, 31, 34
quantum field theory, vii, 1, 2, 3, 5, 6, 7, 9, 12, 14, 17, 18, 19, 137
quantum fields, 2, 3, 19
quantum mechanics, 1, 2, 7, 8, 12, 14, 17, 18, 19, 29, 85, 97
quantum theory, 34, 85
quarks, viii, ix, 69, 70, 71, 72, 73, 74, 77, 78, 79, 80, 99, 100, 101, 102, 103, 104, 105, 106, 108, 109, 110, 123, 124, 125, 133, 135, 139, 165, 166, 167, 168, 169, 170, 171, 172, 173, 174, 175, 176, 177, 178, 179, 181, 189, 190, 191, 192, 193, 194, 195, 196, 198, 199, 202

R

Rabi frequency, 213
radial motion of the system, vii, 21, 26
radiation, 86, 215, 221, 226, 235
radius, viii, 51, 82, 111, 123, 124, 126, 127, 128, 129, 198, 214, 215
real time, 21
reality, 97
reasoning, 40
recall, 9, 40, 91, 219
refractive index, vii, x, 211, 212, 214, 215, 221, 229, 230, 232, 234, 235
relevance, 47, 48, 121
renormalization, 30, 74, 179, 199, 200
response, 46, 215
restoration, 136
restrictions, 2, 7, 9, 12, 17, 43
root, 63, 71, 73, 101, 104, 123, 127, 166, 169, 191, 194, 232
root-mean-square, 127
roots, 73, 104, 232, 233
rotations, 41, 44, 46
rules, 3, 136, 158

S

saturation, 234, 235
scalar field, 10, 22
scalar particles, 134, 151
scalar sector, ix, 131
scaling, 25
scattering, 123, 124, 136, 214, 215, 216
Schrödinger picture, vii
screw dislocations, 47, 53
Seiberg-Witten map, vii, 29, 36, 37
showing, 116, 235
signs, 91
solution, 6, 9, 10, 15, 16, 17, 25, 32, 42, 43, 44, 45, 49, 50, 59, 63, 66, 70, 73, 100, 104, 169, 171, 191, 194, 203, 214, 232, 234
space-time, vii, 8, 15, 17, 22, 29, 30, 31, 32, 33, 34, 36, 37
spectroscopy, 99, 190
spin, vii, viii, 3, 4, 29, 31, 32, 33, 34, 36, 55, 56, 57, 71, 76, 78, 85, 86, 102, 103, 107, 124, 181, 193, 197, 201
spinning particle, vii, 29, 31, 32, 34, 36
spinor fields, 18
stability, 238
standard deviation, 86, 89, 95
Standard Model, ix, 131, 132
state, vii, viii, ix, 1, 2, 3, 5, 8, 9, 13, 14, 15, 17, 55, 57, 59, 69, 70, 71, 73, 75, 77, 79, 80, 81, 82, 101, 102, 103, 104, 105, 107, 108, 109, 110, 111, 112, 139, 147, 165, 167, 168, 169, 170, 171, 175, 177,

Index

180, 181, 182, 184, 190, 193, 194, 195, 196, 200, 201, 204, 206, 211, 213, 215, 217, 221, 235
states, viii, ix, 2, 3, 14, 41, 53, 55, 56, 57, 70, 71, 72, 74, 76, 78, 79, 80, 81, 86, 99, 100, 101, 102, 103, 104, 105, 107, 109, 110, 111, 112, 167, 168, 170, 171, 172, 176, 177, 178, 183, 184, 189, 192, 193, 194, 195, 197, 198, 202, 203, 204, 205, 206
statistics, viii, 85
stress, 24, 40, 49, 53
string theory, 22, 26, 27
strong interaction, 173
structure, 12, 29, 41, 42, 71, 101, 118, 123, 133, 166, 190, 191, 192
subgroups, 116
substitution, 16, 73, 104
substitutions, 44
supersymmetry, 27, 116
suppression, 121
SUSY, viii, 115, 116, 117, 118, 121
symmetry, viii, ix, 24, 30, 37, 69, 70, 71, 80, 91, 99, 100, 101, 102, 105, 106, 108, 109, 110, 115, 116, 117, 118, 119, 120, 122, 132, 133, 165, 166, 167, 168, 170, 171, 173, 184, 190, 191, 192, 195, 196, 204

T

target, 35
techniques, 30, 48
technology, 55
tension, 22, 25
top quark, 147
torus, viii, 58, 115, 116
total energy, 23
transformation, vii, 5, 29, 30, 31, 32, 33, 36, 41, 42, 43, 51, 58, 120, 133
transformations, 2, 4, 5, 34, 38, 41, 43, 116, 133
translation, 18, 19, 23, 33, 40, 44, 46, 51, 53
transmission, 212
treatment, 157, 230
Two-Higgs Doublet Models (2HDM), ix, 131

U

uniform, vii, 21, 26, 89, 92, 96
United, 211, 229
United Kingdom, 211, 229
universe, 133
Uzbekistan, 55

V

vacuum, 2, 57, 108, 123, 132, 133, 195, 211, 212, 213
valence, 108, 125, 196
vapor, 234
variables, viii, 2, 3, 12, 13, 14, 17, 30, 31, 32, 33, 34, 35, 41, 42, 50, 51, 85, 86, 87, 88, 89, 92, 94, 95, 96, 97
vector, viii, ix, 2, 5, 8, 9, 13, 17, 18, 33, 45, 52, 57, 58, 69, 71, 78, 101, 105, 107, 117, 118, 124, 131, 137, 139, 145, 146, 147, 148, 149, 165, 166, 167, 170, 172, 173, 190, 191, 192, 197, 213
velocity, 2, 37
Volterra solutions, vii, 39

W

wave number, 215, 217
workers, 170, 184

Y

Yang-Mills, 37, 116
yield, 24, 110, 120, 222